WILD FLOWERS

OF BRITAIN AND NORTHERN EUROPE

Text by

Richard Fitter · Alastair Fitter

Illustrated by

Marjorie Blamey

Collins, St James's Place, London

William Collins Sons & Co Ltd
London · Glasgow · Sydney · Auckland
Toronto · Johannesburg

First published 1974
© 1974 Marjorie Blamey, Richard Fitter, Alastair Fitter
ISBN 0 00 211278 7
Filmset by Jolly & Barber Ltd, Rugby
Reproduced and printed by F. E. Bording, Copenhagen

Contents

Introduction 7

Main Key 13

Key to Trees and Tall Shrubs 23

Conifer Families 24
Willow Family: Willows, Sallows, Poplars 26
Other catkin-bearing Families 30
Other tree Families 34

Birthwort Family to **Nettle** Family 38
Dock Family: Bistorts, Docks, Sorrels etc 40-44
Purslane Family 42
Goosefoot Family: Goosefoots, Oraches etc 46

Key to Chickweeds and Allies 50

Pink Family: Chickweed-like plants 52
Campions, Pinks etc 58

Water-lily Family 66
Buttercup Family: Hellebores etc 66
Buttercups 68
Water Crowfoots 70
Meadow-rues, Clematises 72
Columbine, Pheasant's-eyes etc 74
Anemones 76
Fumitory Family: Fumitories, Corydalises 78
Poppy Family 80

Key to Crucifers 82

Cabbage Family: Yellow Crucifers 84
White and Mauve Crucifers 90

Mignonette Family 100
Sundew Family 100
Stonecrop Family 100, 102
Saxifrage Family 100, 104
Rose Family: Meadowsweet to Lady's Mantles 106
Roses, Brambles 108
Strawberries, Cinquefoils 110
Cherries, Pears etc 114

Key to Peaflowers 116

Pea Family: Milk-vetches 118
    Shrubs 120
    Vetches 122
    Peas, Vetchlings 124
    Clovers 130

Wood-sorrel Family 132
Flax Family 132
Geranium Family: Cranesbills 132
Spurge Family 136
Balsam Family 138
Milkwort Family 138
Daphne Family 138
Currant and other shrub Families 140
Mallow Family 142
St John's Wort Family 144
Violet Family: Violets, Pansies 146
Rock-rose Family 148
Willowherb Family: Willowherbs, Evening Primroses 148
Dogwood Family 154

Key to Umbellifers 152

Carrot Family: Sanicle, Sea Holly 154
    White Umbellifers 156
    Yellow Umbellifers 166

Wintergreen Family 168
Heath Family 170
Primrose Family 174
Sea-lavender Family 178
Gentian Family: Centauries, Gentians 180
Bindweed Family: Bindweeds, Dodders 184
Bedstraw Family: Bedstraws, Woodruffs 184
Borage Family: Comfreys, Gromwells, Buglosses 188
    Forgetmenots 190

Key to Two-lipped Flowers (Labiate and Figwort Families) 194

Labiate Family: Skullcaps, Germanders 196
    Self-heals, Catmints 198
    Dead-nettles 200
    Claries, Woundworts 202
    Mints, Thymes 204
Nightshade Family 206

Figwort Family: Mulleins 208
Figworts, Toadflaxes 210
Foxgloves, Bartsias 212
Speedwells 214
Louseworts, Cow-wheats 216

Broomrape Family 218
Plantain Family 220
Valerian Family 222
Honeysuckle Family 222
Teasel Family: Teasels, Scabiouses 224
Bellflower Family: Bellflowers, Campions 226

Key to Composites (Daisy Family) 228

Daisy Family: Hemp Agrimony, Golden-rod etc 230
Daisy-like Flowers, white and mauve 232
Cudweeds 234
Daisy-like Flowers, yellow 236
Butterbur, Yarrow, Mugwort 238
Ox-eye Daisy, Marigolds, Coltsfoot 240
Ragworts, Groundsel 242
Thistles, Knapweeds 244
Dandelion-like Flowers 250

Water-plantain Family 258
Lily Family: Lilies, Asphodels 260
Fritillary, Gageas 262
Leeks, Garlics 264
Squills, Grape Hyacinths 266
Solomon's Seals, Stars of Bethlehem 268
Daffodil Family: Daffodils, Snowflakes 270
Iris Family: Flags, Irises, Crocuses 272
Orchid Family 274

Key to Waterweeds 288

Waterweed Families 290

Appendix 1 Additional species 298
Appendix 2 Introduced shrubs 307
Appendix 3 Aggregates 308
Ecology 309
Glossary 314
Plant Photography 318
Societies to Join 318
Further Reading 318
Index of English Names 319
Index of Scientific Names 327

S

B

G

F

	region not covered by this book
	boundary between distribution areas
	major outcrops of chalk and limestone
	scattered outcrops of chalk and limestone
	land over 1000 m. (3250 ft.)

The map does not show limestone in the Alps.

Introduction

The Text

All the native flowering plants of the north-western quadrant of Europe are described in the text, with the exception only of grasses, sedges and rushes. A small number of plants penetrating westwards only as far as Finland, and of alpine species descending down the river shingle to the neighbourhood of Munich are also omitted. Ferns, horsetails and their allies are not flowering plants and are therefore omitted, although seven species which might be confused with flowering plants are briefly mentioned. All widely established alien herbaceous plants and trees are also described, and some less widely established alien shrubs most of them familiar in gardens, are listed in Appendix 2, p.307. It has proved impossible to include more than a few of the very large number of alien trees, shrubs and herbaceous plants that are locally established in a small part of the area covered by the book.

The **area chosen** lies roughly between Arctic Norway in the north and the river Loire in western and central France and the river Danube in Germany in the south. Its boundary runs along the Loire from its mouth on the Atlantic coast of France eastwards to Dijon and Basle and thence along the Danube and the foothills of the Alps to Munich, so as to include the Vosges and the Black Forest, but excluding the Jura and the Alps. At Munich it turns northwards across Germany through Regensburg, Bayreuth, Erfurt and Braunschweig, to reach the Baltic at Lubeck, and thence up the Gulf of Finland and the eastern frontier of Finland to the Arctic Ocean. Iceland is included. These boundaries are shown on the map opposite.

To indicate distribution in the text, this north-western segment of Europe has been divided into four regions. (1) Great Britain, Ireland and the Isle of Man; (2) France, Belgium, Luxemburg and the Channel Isles; (3) Germany, the Netherlands, and Denmark; and (4) Norway, Sweden, the Faeroes, and Iceland. See page 9 for the symbols used to indicate these regions.

Description. Since the illustration is immediately opposite the text relating to each plant, a detailed description is not given, and descriptive details are largely confined to points such as height, which cannot easily be brought out in the illustration, to factual information such as habitat and distribution, which, of its nature, cannot be illustrated, and to stressing certain points which are crucial for identification.

Each text description includes certain standardised information, such as the height of the plant, flowering period and the habitat. The major diagnostic features distinguishing it from similar plants with which it might be confused, are shown in italics. A number of assumptions are made in the descriptions: unless otherwise stated plants are erect and herbaceous, leaves are stalked, and flowers are open.

The descriptions are compiled as follows:

Group descriptions of families and genera (see p.10) are given wherever possible, so as to avoid constant repetition of common characters, for example the shape of a peaflower or the nature of a composite flower. They should therefore always be read carefully and in conjunction with the species descriptions. The group descriptions refer only to species mentioned in the text.

English names follow, in most cases and where possible, the list approved by the Botanical Society of the British Isles. Where no English names exist, we have not invented any. Nor have we always included English names which are mere translations of the Latin ones.

Latin scientific names in general follow *Flora Europaea* (published up to Umbelliferae at time of writing), otherwise as in Polunin's *A Field Guide to the Flowers of Europe* or Dandy's *List of British Vascular Plants* or appropriate regional Floras.

British status: plants native to, or commonly naturalised in the British Isles are preceded by an asterisk.

Number: each plant has a number corresponding with its number on the plate opposite, and elsewhere may be referred to, where necessary, by this number preceded by the page number, so that the Primrose becomes 174/1. Where several plants are very similar, only one is illustrated and the others are described in the same paragraph. They are given the number of illustrated species and a letter. The descriptions of all such **subsidiary species,** however many are listed, refer back to the main species unless otherwise stated, and only the principal characters that differ are mentioned. On a few pages the descriptions necessary to identify the plants overrun the page. Rather than cut these to a point which would endanger identification, some of the subsidiary species on these pages have been transferred to Appendix I, p.298-306.

Hairiness or hairlessness is indicated in most cases, often an important clue to identification

Height is indicated as follows:

Tall
over 60 cm (2 ft)

Medium
30-60 cm
(1-2 ft)

Short
10-30 cm
(4-12 in)

Low
0-10 cm (0-4 in)

Plant sizes can vary greatly according to altitude, climate and soil. Plants are more likely to be found smaller than is indicated here than larger.

Status: annual, biennial or perennial status is shown in each case. Perennials are usually stouter than annuals, and of course are much more likely to be seen above ground in winter.

Leaf-shapes: see Glossary (p.314) for definition of terms such as *pinnate*.

Flower shapes and arrangements: see Glossary (p.314). Flower sizes are diameters unless otherwise stated. Flower colour refers to petals or to sepals when there are no petals. It is more variable in some species than others. Most pink, mauve, purple and blue flowers produce white forms from time to time, and many white flowers can be tinged pink. Some, such as milkworts (p.138), are exceptionally variable. But a normally coloured flower is usually to be found nearby.

Flowering-time refers to the central part of the area, and may be earlier in the south, near the sea and in forward seasons, and later in the north, on mountains and in backward seasons.

Fruits are usually only described where important for identification.

Habitat: for definition of terms such as fen and bog, here used strictly, see Glossary (p.314). The major mountain regions and areas with lime (chalk or limestone) in the soil are shown in the map on p.6.

Distribution: the following symbols show whether the plant occurs, either commonly or uncommonly, within each region. If it has only a very few localities, "rare" is added, and if it is mainly found, e.g. in the south of the region, the word "southern" is also added.

    T – Throughout the area covered by the book (see p.7).
    B – Great Britain, Ireland, Isle of Man.
    F – France, Belgium, Luxemburg, Channel Isles.
    G – Germany, the Netherlands, Denmark.
    S – Norway, Sweden, Iceland, the Faeroes.

If parentheses enclose the symbol – for example, (G) – the plant is introduced, not native. The great majority of introduced plants occur on waste or disturbed ground, roadsides, or other habitats much affected by human activity. Only rarely, as with New Zealand Willowherb (p.150), do they succeed in invading natural or semi-natural habitats, such as heaths, moors and calcareous grassland.

The Illustrations

Over 1000 species are illustrated in colour. Many illustrations are life-size, but, where they are not, a single flower of the plant in question is also shown life-size. Only the *Iris* plate (page 273) is painted to a smaller scale. Sometimes the distinctive character of a subsidiary species is also illustrated. The symbol □ is used after the scientific name of all subsidiary species with an illustration either on the plate opposite or on the text page.

Since the illustrations are arranged in the scientific order, it will often be necessary to identify a plant via the keys. A plant should never be identified simply from the picture, especially since several similar species may be described under the one illustrated. Always remember, too, that a flower may sometimes have a different colour from the one illustrated, and that some important identification features, such as smell or hairiness, may not be of the type that can be shown in an illustration.

Classification and Scientific Nomenclature

The seed plants or Spermatophyta are in two main groups, a small number of gymnosperms or conifers and a much larger number of angiosperms or flowering plants. The European conifers (p. 24) are all trees or shrubs, evergreen (except the larches) with narrow leaves ("needles"), male and female flowers always separate but usually on the same plant, and the seeds, unlike those of the flowering plants, not enclosed in an ovary. The flowering plants are themselves divided into two groups, dicotyledons and monocotyledons. Most of the plants in the book are dicotyledons (pp. 26-256), which have two seed-leaves (cotyledons), mature leaves which are usually broad, often stalked, and nearly always net-veined, and flower-parts in fours or fives. The monocotyledons (pp. 258-286) are represented mainly by lilies, orchids, and allied plants. They have only one seed-leaf, their later leaves are usually narrow and unstalked, often parallel-sided and nearly always parallel-veined, their flower-parts are almost always in multiples of three. Except for Butcher's Broom (p. 268), none of them are woody.

Within these major divisions flowering plants, like all other plants, are arranged in families, genera, and species, groups of increasingly close affinity, in what, according to current scientific opinion, corresponds to the way in which they have evolved. Family names usually end in . . . aceae. In the book the plants are mainly arranged in the currently accepted order of families. The chief exceptions to this rule are the grouping together, for the reader's convenience, of trees and shrubs, mainly at the beginning, and of waterweeds at the end.

Individual scientific plant names are binomial; the generic name comes first, always with a capital initial, and the specific name second, always with a small initial, and since it is an adjective, agreeing in gender with the generic name. This system, devised by the great 18th-century Swedish naturalist, Linnaeus, represents the international scientific nomenclature, and using Latin or Latinised Greek, enables naturalists in any country to know exactly what plant is being referred to.

How to Identify Wild Flowers

A great deal of pleasure may be had in identifying wild flowers without any equipment at all, and as far as possible in this book we have relied on characters that can clearly be seen by the naked eye. Nevertheless a small pocket lens, magnifying 5 or 10 times, is an invaluable aid, and is no trouble to carry, unlike the binoculars used by birdwatchers – though these can be useful to the botanist in helping to identify inaccessible plants. With the aid of a lens such details as hairiness or the number of styles and stamens, sometimes critical for distinguishing otherwise similar plants, can more readily be determined.

Wild flowers, being static, have many advantages over animals for those wishing to identify them. It is much easier to bring a knowledgeable friend back to see an unusual plant, than to hope a rare bird or butterfly will still be there next day. We would urge our readers to do this wherever possible, and *not to pick specimens*. Be careful not to wait too long, as some plants wither after a short time in flower. If a discovery is really rare, the plant's survival may be endangered by picking a piece to show somebody for identification, so the slogan is:

> "Take the book to the plant, not the plant to the book."

If you are not likely to be able to return to the plant, perhaps because you are on a day trip into the country, make a sketch on the spot and take detailed notes on the following points:

- Height and shape of the plant; whether annual or perennial.
- Shape of leaves, arrangement on stem, length of stalk.
- Shape and colour of flowers; number of petals, stamens and styles.
- Arrangement of flowers, e.g. solitary or in clusters or spikes.
- Shape and colour of fruits.
- Degree of hairiness or downiness on all parts of the plant.
- Habitat and abundance.

For this reason always carry a small field note-book with you. A good colour photograph may also be valuable (see p. 318).

Wherever possible we have confined ourselves to easily ascertainable characters, as above, which do not require the plant to be picked to see if the stem is hollow or its rootstock tuberous, for example. In a few instances, however, where these are the best known or most reliable characters, we have given such details.

Abbreviations used in the Text

Agg.	Aggregate		
Fl, fls	Flower, flowers	Sp.	Species (singular)
Flg.	Flowering	Spp.	Species (plural)
Flhds	Flowerheads	Ssp.	Subspecies
Lf, Lvs	Leaf, leaves	Var.	Variety
×	Indicates a hybrid, e.g. 1 × 2 is a hybrid between species 1 and species 2.		
*	Native or commonly naturalised in Britain.		
□	This symbol after the name of a subsidiary species (see p. 8) indicates that it is illustrated.		
B, F, G, S, and T	Found in Britain, France, Germany, Scandinavia, and Throughout respectively. For precise definition of these areas see introduction (pp. 7, 9).		
( )	Indicates the plant is introduced, not native.		

The Keys

The Main Key (opposite) is based on flower-shape and colour. It includes all plants in the book except: trees and tall shrubs with catkins or small flowers; water weeds with inconspicuous flowers; a few, such as those on p. 296, which have been included for comparative purposes only; and the introduced shrubs in Appendix 2, p. 307. Subsidiary species are covered by the reference to the main species, even when they have been removed to Appendix 1, p. 298.

INDIVIDUAL FLOWERS LARGE OR CONSPICUOUS   13

Open, star-like or saucer-shaped flowers   13

Two petals   13

Three petals   13

Four petals   13

Five petals   14

Six or more petals   15

 Bell- and cup-shaped flowers   16

One-lipped flowers   17

Two-lipped flowers   17

Large flowers of miscellaneous shapes   18

INDIVIDUAL FLOWERS SMALL, IN HEADS OR SPIKES   19

Globular heads, tightly packed   19

Brush-like flhds   19

Dandelion-like flowers   19

Daisy-like flowers   19

Flat heads of small flowers   19

 Flowers in lfless spikes   20

Flowers in one-sided spikes   20

Flowers in leafy whorls   21

 Small spikes at base of leaves   21

INDIVIDUAL FLOWERS SMALL, IN CLUSTERS OR SOLITARY   21

 Flowers in loose, stalked clusters   21

Flowers in small clusters at base of leaves   21

Flowers solitary, or in pairs at base of leaves   21

In addition, eight special keys are scattered through the text, appearing with the families or groups concerned. They are cross-referenced in the main key.

Trees and tall shrubs   23

Chickweeds and allies   50-1

Crucifers (Cabbage Family)   82-3

Peaflowers   116-17

Umbellifers (Carrot Family   152-3

Two-lipped flowers of the Mint and Figwort Families   194-5

Composites (Daisy Family)   228-9

Waterweeds with small flowers   288

The Main Key

Individual flowers large or conspicuous   see below
Individual flowers small   see page 19
(n.b. Composites, actually tight heads of small flowers, also have a separate key, pp. 228-9)

INDIVIDUAL FLOWERS LARGE OR CONSPICUOUS

**Open, star-like or
Saucer-shaped flowers**

Two petals

 Enchanter's Nightshades **148**

 Goldilocks Buttercup **70**

Three petals

 Mossy Stonecrop **102**, Water Plantains, Star-fruit, Arrow head, Frogbit, Water Soldier **258**

 Ranunculus hyperboreus **68**, Goldilocks Buttercup **70**, Rannoch Rush **220**

 Water Plantains **258**

 Water Plantains **258**

Four petals

 Bastard toadflaxes **38**, Mossy Sandwort **52**, Water Tillaea **102**, Dwarf cornel **154**, Squinancywort, Woodruff **184**, Bedstraws, Cleavers **186**

 Cranberry **172**, Squinancywort **184**, Slender Marsh Bedstraw **186**

 Wall Bedstraw **186**

 Crosswort, Lady's Bedstraw **186**

 Spurge Laurel **138**, False Cleavers, Wall Bedstraw **186**, Herb Paris **268**

 Blue Woodruff **184**

 Mezereon **138**, Dwarf Cornel **154**, Field Madder **184**

 Pearlworts **56**, Crucifers **82**, Allseed **132**, Water Chestnut **258**, May Lily **260**

 Poppies, Red-horned Poppy **80**, *Capsella rubella* **96**, Garden Cress **98**

 Red-horned Poppy **80**, Wallflower **86**

 Poppies, Greater Celandine **80**, Crucifers **82**, Tormentil **112**, Evening Primroses **148**, Yellow Centauries **180**

 Mistletoe **38**, Pearlworts **56**, Narrow-leaved Pepperwort **98**, Golden Saxifrages **100**

 Opium Poppy **80**, Crucifers **82**

 Meadow-rues, Baneberry, *Clematis recta* **72**, Shoreweed **127**

 Lesser Meadow-rue **72**, Roseroot **100**

 Traveller's Joy **72**

 Meadow-rues, Alpine Clematis **72**

 Crucifers **82**, Willowherbs **150**

Willowherbs **150**

13

Willowherbs **150**

 Speedwells **214**

 Pink Water Speedwell **214**, *Veronica urticaefolia* **303**

 Speedwells **214**

 Heath Speedwell **214**

Five petals

 Spring Beauty **42**, Chickweeds and allies **50**, Catchflies **58** and **62**, White Campion **60**, Tunic Flower **62**, Roses **108**, Rock Cinquefoil **110**, Hollyhock **142**

 Pink Purslane **42**, Catchflies **58-62**, Red Campion **60**, Cow Basil, Proliferous Pink, Tunic Flower **62**, Roses **108**, Pink Barren Strawberry **110**, Cranesbills **134**, Mallows **142**, Birdseye Primrose **174**

 Sticky catchfly **60**, Small-flowered Catchfly **62**

 Cowslip **174**

 Portulaca oleracea **42**, Cinquefoils **112**, Primrose, Oxlip, Cowslip **174**

 Mallows **142**, Primroses **174**

 Chickweeds and allies **50**, Bladder Campion **58**, Catchflies **58-62**, White Campion **60**

 Catchflies **58-62**, Red Campion **60**, Cranesbills **134**

 Spanish Catchfly **58**

Berry Catchfly **62**

Northern Catchfly **58**

 Wild Pink **64**

 Ragged Robin **60**, Pinks **64**, Mallows **142**

 Pinks **64**

 Bastard Toadflax **38**, Orpine, Stonecrops **102**, Purging Flax **132**, Burning Bush **138**, Chickweed Wintergreen **176**, Bogbean, Vincetoxicum **178**, Nightshades **206**

 Orpine, Pink Stonecrop **102**, Burning Bush **138**, Bogbean **178**

 Bastard Toadflax **38**, Mousetail **74**, Orpine, Stonecrops **102**, Vincetoxicum **178**, Wild Madder **186**, Tomato **206**

 Hop **38**, Saltwort **48**, Lesser Chickweed **54**, Lady's Mantles **106**, Gooseberry **140**, Ivy **154**, False Cleavers, Wall Bedstraw **186**

 Lomatogonium **182**

 Fastigiate Gypsophila **62**, Marsh Cinquefoil **110**, Marsh Felwort **182**, Bittersweet **206**

 Knotgrass **40**, Gypsophila **62**, Buttercups, Water Crowfoots **70**, Sundews, Grass of Parnassus **100**, Saxifrages **104**, Meadowsweet, Goatsbeard Spiraea **106**, Strawberries **110**, Brambles **108**, White Rock-rose **148**, Common Wintergreen **168**, Labrador Tea **172**, Brookweed **176**, Heliotrope **188**, White Mullein **208**, *Linum tenuifolium*, White Flax **301**.

 Knotgrass **40**, Soapwort **60**, Annual Gypsophila **62**, Bramble **108**, Cotoneasters **114**, Pink Oxalis **132**, Cranesbills **134**, Sea-heath **148**, Umbellate Wintergreen **168**, Sea Milkwort **176**, Nonea **188**, Lungworts **190**, Viper's Bugloss, Oyster Plant **192**, *Linum viscosum*, *L. tenuifolium* **301**

Houndstongues **188**, Hawkweed Saxifrage **300**

 Nonea **188**

 Amsinckia **188**, Orange Mullein **208**

 Buttercups, Spearworts **68-70**, Marsh Saxifrage **104**, Agrimonies **106**, Cinquefoils, Silverweed **112**, Yellow Oxalis, Bermuda Buttercup, Yellow Flax **132**, Rockroses, Common Fumana **148**, Yellow Wintergreen, Diapensia **168**, Fringed Water-lily **178**, Amsinckia, Yellow Alkanet **188**, Mulleins **208**, *Tribulus terrestris* **301**

 Hemp **38**, Polycnemum **46**, Hairy Seablite **48**, Tufted Saxifrage **104**, White Bryony **148**, Hawkweed Saxifrage **300**

 Flaxes **132**, Meadow Cranesbill **134**, Pimpernels **176**, Jacob's Ladder **184**, Lungworts, Blue-eyed Mary, Forgetmenots, Madwort **190**, Buglosses, Alkanets, Borage, Oyster Plant **192**, Large Speedwell **214**

 Corncockle **60**, Purple Saxifrage **104**, Dusky Cranesbill **134**, Sea-heath **148**, Water Violet **176**, Jacob's Ladder **184**, Houndstongues, Heliotrope **188**, Purple Viper's Bugloss, Abraham Isaac and Jacob, Alkanet, Vervain **192**, Purple Mullein **208**.

 Chickweeds and allies **50**, Snowdrop Windflower **76**, Saxifrages **104**, Brambles, Raspberry, Cloudberry **108**, Storksbills **132**, One-flowered Wintergreen **168**, Androsaces **176**, Gromwells **188**

 Sea Spurrey **56**, Storksbills **132**, Wild Azalea **172**, Northern Androsace **176**, Centauries **180**

 Arctic Bramble **108**, Scarlet Pimpernel **176**

 Cyphel **52**, Pearlworts **56**, Marsh Marigold **68**, Yellow Saxifrage **104**, Herb Bennet

110, Sibbaldia **112**, St John's Worts, Tutsan, Rose of Sharon **144**, Loosestrifes, Yellow Pimpernel **174**, Changing Forgetmenot **190**

 Sea Sandwort **52**, Knawels **56**, Green Hellebore **66**, Sea Storksbill **132**

 Love-in-a-mist **66**, Periwinkles **178**, Gentians **182**, Purple Gromwell **188**, Forgetmenots **190**

 Cranesbills **134**, Purple Gromwell **188**, Venus's Looking Glasses **226**

 Violets **146**

 Yellow Wood Violet, Pansies **146**

 Violets **146**

 Violets, Pansies **146**

Six or more petals

 White Water-lily **66**, Wood Anemone, Pale Pasque Flower **76**, Mountain Avens **110**, Chickweed Wintergreen **176**, Snowdon Lily, St. Bernard's Lily, False Helleborine **260**, Drooping and Common Stars of Bethlehem **268**

 Winter Aconite **66**, Lesser Celandine **68**, Yellow Anemone **76**, Alpine Avens **110**, Great Yellow Gentian **180**

 False Helleborine, Asphodels **260**

 Blue Anemone **76**, Blue-eyed Grass **272**

 Pasque Flowers, Hepatica **76**, Gladiolus **272**

 Corn Mignonette **100**, Kerry Lily **260**, Spiked Star of Bethlehem, Butcher's Broom **268**

 Montbretia **272**

Mousetail **74**, Wild Migno-nette **100**, Black Bryony **266**, Spiked Star of Bethlehem, Butcher's Broom **268**

 Purple Loosestrife **150**, Kerry Lily **260**

 Meadowsweet **106**, Mountain Avens **110**

 Grass Poly **150**, Crowberry **172**, Flowering Rush **258**

 Pheasant's-eyes **74**

 Yellow Water-lily **66**, Pheas-ant's eyes **74**

 St. Bernard's Lily, Kerry Lily **260**, Ramsons **264**, Spiked Star of Bethlehem **268**

 Reflexed Stonecrop, Hen-and-chickens Houseleek **102**, Yellowwort **180**, Bog Asphodel **262**

 Gageas **262**

 Squills **266**, Blue-eyed Grass **272**

 Pasque Flowers **76**, Kerry Lily **260**

Bell- and cup-shaped flowers

 Wood Sorrel **132**, Cassiope **172**, Alpine Snowbell **174**, Snowdon Lily **260**

 Pink Oxalis **132**, Bog Pimper-nel **176**, Twinflower **222**

 Yellow Oxalis, Bermuda But-tercup **132**, Creeping Jenny **174**

 Stinking Hellebore **66**

 Alpine Snowbell **174**, *Polemonium acutiflorum* **184**, Harebell, Bellflowers **226**

 Heather **170**

 Round-leaved and Inter-mediate Wintergreens **168**

 Spurges **136**, Yellow Winter-green **168**

 Navelwort **100**, Common and Toothed Wintergreens **168**, Cowberry, Black Bearberry, Cassiope **172**, Comfreys **188**, Lily of the Valley **260**, Wild Leek **264**, Solomon's Seals, Wild Asparagus **268**, Snow-flakes **270**

 Common Wintergreen **168**, Bog Rosemary, Heaths, Dod-ders **184**, Fringe Cups **300**

 Pick-a-back Plant **300**

 Hen-and-chickens Houseleek **102**, Yellow Birdsnest **168**, Wild Asparagus **268**

 Navelwort **100**, Bilberry **172**, Deadly Nightshade **206**, Solo-mon's Seals **268**, Fringe Cups **300**

 Comfreys **188**, Harebell **226**, Bluebells, Grape Hyacinths **266**

 Asarabacca **38**, Arctic Rhododendron, Heather, Heaths **170**, Comfreys **188**, Deadly Nightshade **206**, Wild Leek **264**

 Alpine Snowbell **174**, Spring Crocus **272**

 Foxglove **212**, Meadow Saf-fron **266**

 Spotted Gentian **182**

 Foxgloves **212**

 Deadly Nightshade **206**

 Alpine Snowbell **174**, Gen-tians **182**, Viper's Bugloss **192**, Harebell, Bellflowers **226**

 Asarabacca **38**, Gentians **182**, Bellflowers **226**, Meadow Saf-fron **266**, Crocuses **272**

 Wild and 3-cornered Leeks, Welsh Onion **264**, Drooping Star of Bethlehem, Wild Asparagus **268**, Snowflakes, Snowdrop **270**

Small Pasque-flower **76**, Water Avens **110**, Fritillary **262**, German Garlic **264**

Creeping Jenny **174**, Wild Tulip **262**, Welsh Onion **264**, Wild Asparagus **268**

 Water Avens **110**, Fritillary **262**, Wild Leek, German Garlic **264**

 Bindweeds **184**, Thorn-apple **206**

 Dodders, Bindweeds **184**

 Flax Dodder **184**, Henbane, Small Tobacco Plant, **206**

 Gentians **182**, Apple of Peru **206**, Harebell, Bellflowers **226**

 Apple of Peru **206**, Bellflowers, Venus's Looking Glasses **226**

Lipped flowers
One-lipped flowers

 Germanders **196**

 Mountain Germander, Ground Pine, Wood Sage **196**

 Bugles **196**

 Birthwort **38**

Two-lipped flowers (see also Peaflowers 116)

 Labiates and Figworts **194**, Globularia **218**

 Labiates and Figworts **194**, Toothwort **218**, Black-berried Honeysuckle **222**

 Figworts **208**, Broomrapes **218**, *Lonicera alpigena* **222**

Labiates and Figworts **194**, Yellow Figwort **208**, Broomrapes **218**, Honeysuckles **222**, Bladderworts **127**

Labiates and Figworts **194**, Globularia **218**

 Vervain **192**, Labiates and Figworts **194**, Broomrapes **218**

 Corydalises, Fumitories **78**

 Bulbous Corydalis, Fumitories **78**

 Yellow Corydalis **78**, Skullcaps **196**, Cow-wheats **216**

 Skullcaps **196**

  Milkworts **138**

 Alpine Butterwort **308**

 Himalayan Balsam **138**, Pale Butterwort **218**

 Orange Balsam **138**

 Balsams **138**

 Butterworts **218**

 Toadflaxes **210**

 Toadflaxes **210**

 Dense-flowered Orchid **280**, Butterfly Orchids, Creeping Lady's Tresses **282**, Violet Helleborine **284**, Coralroot **286**

 Bee and Spider Orchids **274**, Lady, Loose-flowered, Early Purple and Green-winged Orchids **276**, Marsh and Spotted Orchids **278**, Black Vanilla Orchid, Calypso **286**

 Dark Red and Broad-leaved Helleborines **284**

Lady's Slipper Orchid **274**, Pale-flowered Orchid **276**, Early and Broad-leaved Marsh Orchids **278**, Bog Orchids **286**

 Dense-flowered Orchid **280**, Bog Orchids **286**

 Lady's Slipper, Spider, Fly and Black Vanilla Orchids **274**, Violet Bird's-nest Orchid **286**

 Military, Burnt-tip, Monkey and Bug Orchids **276**, Pyramidal and Fragrant Orchids **278**

 Bug Orchid **276**, Lesser Twayblade **280**

 Man and Lizard Orchids **280**, Bird's-nest Orchid **286**

 Man and Lizard Orchids, Twayblades **280**

 Ghost Orchid **286**

 Dense-flowered Orchid **280**, White and Narrow-leaved Helleborines, Creeping Lady's Tresses **282**

 False Musk Orchid **280**, Red Helleborine **282**

 Frog Orchid **280**

 False Musk, Musk and Frog Orchids **280**

 White Frog Orchid **282**, Coralroot Orchid **286**

 Coralroot, Fen, Bog and One-leaved Bog Orchids **286**

 Coralroot, Fen, Bog and One-leaved Bog Orchids **286**

 Heath Lobelia **226**

Water Lobelia **290**

Large flowers of miscellaneous shapes

 Sowbread **176**

 Cyclamen **176**

 Columbine **74**

 Daffodils **270**

  Foxgloves **212**

 Bellflowers **226**

 Martagon Lily **262**

 Pyrenean Lily **262**

 Calypso **274**

 Lady's Slipper Orchid **274**

 Globe Flower **68**

 Aconitum variegatum **74**

 Wolfsbane **74**

Monkshood, Northern Wolfsbane, Larkspurs **74**

Gladiolus **272**

 Irises **272**

 Lords and Ladies **270**

INDIVIDUAL FLOWERS SMALL, IN HEADS OR SPIKES

Globular Heads tightly packed

 Sticky Mouse-ear **54**, Common Cudweed **234**, Leeks **264**

 Proliferous Pink **62**, Thrift **74**, Garlics, Chives **264**

 Common Cudweed **234**

 Cockleburs **230**, Crow Garlic **264**, Onion **305**

 Globularia **218**, Cornsalads **222**, Grape Hyacinths **266**

 Chives, Garlics, Leeks **264**, Small Grape Hyacinth **266**

 Pirri-pirri Bur **106**, Clovers **130**, Small Teasel **224**

 Clovers **130**

 Great Burnet **106**, Crimson Clover **130**

 Branched Plantain **220**

 Dwarf Willow **28**, Medicks, Trefoils **128**, Mountain Clover **130**, Pineapple Mayweed **236**

 Hop **38**, Salad Burnet **106**

 Scabiouses, Rampions **224**

 Red Clover **130**, Devilsbit Scabious **224**

 Sanicle, Astrantia, Field Eryngo **154**

 Sanicle, Astrantia **154**

 Tufted Loosestrife **174**

 Sanicle, Field Eryngo **154**

 Sea Holly **154**, Globularia **218**, Scabiouses, Rampions **224**

 Devilsbit Scabious **224**

 Marsh Pennywort **154**

 Marsh Pennywort **154**, Moschatel **220**

 Dodders **184**

Brush-like heads

 Astrantia, Sanicle **154**, Composites **228**, Pipewort **260**

 Pipewort **260**

 Astrantia, Sanicle **154**, Composites **228**

 Cudweeds **234**, *Artemisia austriaca* **238**

 Safflower **248**

 Slender Hare's-ear **164**, Yellow Scabious **224**, Composites **228**

 Ragweed, Cockleburs **230**

 Composites **228**

 Cornsalads, Scabiouses, Teasels **224**, Composites **228**

Dandelion-like flowers

Composites **228, 250-56**

Daisy-like flowers

  Composites 228

Flat heads of small flowers

 Hoary Cress **98**, Orpine, White Stonecrop **102**, Umbellifers **152**, Dwarf Elder **222**, Yarrow, Sneezewort **238**

 Orpine **102**, Umbellifers **152**, Dwarf Elder **222**, Hemp Agrimony **230**

 Roseroot **100**, Reflexed Stonecrop, Orpine **102**, Spurges **136**, Umbellifers **152**, Goldilocks, Ploughman's Spikenard **230**, Yarrow **238**

 Petty Spurge **136**, Umbellifers **152**

 Valerianella coronata 304

 Ploughman's Spikenard **230** Cornsalads **222**

Flowers in leafless spikes

 Redshank, Waterpepper, Buckwheat **40**, Copse Bindweed, Knotweeds **42**

 Redshank, Waterpepper, Buckwheat **40**, Black Bindweed, Himalayan Knotweed **42**

 Pale Persicaria **40**, Docks, Sorrels **44**

 Willows **28**, Bog Myrtle, Dwarf Birch **30-32**, Rannoch Rush **220**, Mugworts, Sea Wormwood **238**

 Bulrushes **296**

 Willows **26-28**, Pale Persicaria, Waterpepper **40**, Black Bindweed **42**, Docks, Sorrels **44**, Goosefoots, Oraches, Sea Beet **46**, Sea Purslane **48**, Dog's Mercury **136**, Currants **140**, Arrowgrasses **220**, Sweet Flag **296**

 Sea-lavenders **178**

 Alpine Bistort **40**, Goatsbeard Spiraea **106**, Haresfoot Clover **130**, Hoary Plantain **220**, Spiked Rampion **224**, White Frog Orchid **282**

 Bistorts **40**, Haresfoot and Crimson Clovers **130**

 Green Amaranth **48**, Crimson Clover **130**

 Ribwort and Buckshorn Plantains **220**

 Mousetail **74**, Greater and Buckshorn Plantains **220**, Asphodels **260**, Musk Orchid **280**, Bog Orchids **286**

 Hop, Hemp, Nettles **38**, Goosefoots, Oraches **46**, Glassworts, Amaranths **48**, Sea Plantain, Arrow-grasses **220**

 Sea-lavenders **178**, Spearmint **204**, Black Vanilla Orchid **274**

 Fumitories **78**, Mignonettes **100**, White Melilot **126**, Thyme-leaved Speedwell **214**, May Lily **260**

 Fumitories **78**, Pink Water Speedwell **214**

 Spanish Catchfly **58**, Alpine Meadow-rue **72**, Barberry **78**, Weld, Wild Mignonette **100**, Melilots **126**, Goldenrods **230**

 Hop, Hemp **38**

 Speedwells **214**, Grape Hyacinth **266**

 Alpine Meadow-rue **72**, Purple Fumitory **78**, Heath Speedwell **214**, Small Grape Hyacinth **266**, Black Vanilla Orchid **274**

20

Flowers in one-sided spikes

 Nottingham Catchfly **58**, Vetches **122**, Navelwort **100**, Toothed Wintergreen **168**, Three-cornered Leek **264**, Lily of the Valley, Drooping Star of Bethlehem **268**, Lady's Tresses **282**

 Vetches **122**, Heaths **170**, Toothwort **218**

 Montbretia **272**

Yellow Bird's-nest **168**, Canadian Goldenrod **230**

Navelwort **100**

 Scutellaria columnae **196**, *Campanula bononiensis* **226**, Bluebell **266**

Vetches **122**, Heaths **170**, Sea-lavenders **178**

Flowers in leafy whorls

 Bedstraws **186**, Labiates **194**

 Labiates **194**

 Docks **44**

 Crosswort **186**

 Docks **44**

 Labiates **194**

 Milk-vetches **120**, Water Violet **176**, Labiates **194**

Flowers in small spikes at base of leaves

 Hemp, Nettles **38**, Goosefoots, Oraches **46**, Amaranths **48**

 Amaranths **48**

INDIVIDUAL FLOWERS SMALL, IN CLUSTERS OR SOLITARY

Flowers in loose stalked clusters

 Wall Bedstraw **186**, Red Valerian **222**

 Meadow-rues **72**, Meadowsweet **106**, Bedstraws **186**, Red Valerian **222**, Canadian Fleabane **230**

 Lesser Meadow-rue **72**, Lady's Bedstraw, Wild Madder **186**, Black Bryony **266**

Slender Marsh Bedstraw **186**, Valerians **222**, Hemp Agrimony **230**

 Maple-leaved Goosefoot **46**, 4-lvd Allseed **56**, Lady's Mantles **106**, Spurges **136**, False Cleavers, Wall Bedstraw **186**

Lesser Meadow-rue 72

Flowers in small clusters at base of leaves

 Strapwort, Coral Necklace 56, Swinecresses 98, Mossy Stonecrop 102, Cudweeds 234

 Cudweeds 234

 Heath Cudweed 234

 Narrow Hare's-ear 166, Cudweeds 234, Joint Pine 298

 Pellitory 38, Iceland Purslane 42, Knawels, Ruptureworts 56, Golden Saxifrages 100, Parsley Piert 106, Annual Thymelaea 138

 Field Madder 186

Flowers solitary or in pairs at base of leaves

 Bastard Toadflax 38, Knotgrass 40, Blinks 42, Strapwort 56, Mossy Stonecrop 102, Guernsey Centaury 180, Cleavers 186, Mudworts 290

 Mind Your Own Business 38, Knotgrass 40, Black Bindweed 42, Grass Poly 150, Crowberry 172, Chaffweed, Sea Milkwort 176, Guernsey Centaury 180, Cornish Moneywort 216, Water Purslane, Mudworts 290

 Juniper, Joint Pine 24, Guernsey Centaury 180, Cornish Moneywort 216

 Juniper 24, *Polygonum patulum* 40, Iceland Purslane 42, Polycnemum 46, Glassworts, Seablites, Saltworts 48, Annual Mercury 136, Annual Thymelaea 138, Gooseberry 140, Hampshire Purslane 290.

Key to Trees and Tall Shrubs

(Pages 26-36)

This is a key, based on leaf shape, to all trees and tall shrubs (shrubs more than 1 m high) in the book.

A tree is a tall woody plant with a single, usually stout, stem. A shrub is a woody plant with several fairly stout stems from a single rootstock, usually not so tall as a small tree. When trees are coppiced, however, as beech and hornbeam sometimes are, they may have several stems like a shrub. A pollarded tree, on the other hand, is one which has been beheaded 6-7 ft from the ground, so as to produce a crop of thin branches. An undershrub is a low woody plant, such as heather or rock-rose.

Leaves oval or pointed oval

Bay Willow, Goat Willow, Sallows 26-28; Hornbeam, Beech, Evergreen Oak 32; Elms, Holly 34-36; *Apples, Pears, Whitebeams, Blackthorn, Cherries* 114; Spindle-tree, Buckthorns 140; *Dogwood,* Cornelian Cherry 154; *Strawberry Tree* 172; *Wayfaring Tree* 222; Shrubby Orache, Escallonia, Pittosporum, Spiraea, *Quince, Pyracantha, Cherries, Aronia,* Fuchsia, Lilac, *Shallon,* Prickly Heath 307.

N.B. Those in italics have white flowers.

Leaves ace-of-spades

Black Poplar, Birches 30; Limes 36; Flowering Nutmeg 307

Leaves elliptical

Sweet Chestnut 32; Privets 178; Medlar, Rhododendrons, Butterfly Bush 307

Leaves rounded

Aspen, Dwarf Birch, Alders, Hazel 30-32; Barberry 78; Cotoneasters, Amelanchiers 114 and 307; Box 140; Physocarpus, Snowberry 307

Leaves jagged

Oaks 32; Holly 36; Wild Service Tree 300; Hawthorns 114

Leaves palmately lobed

Grey Poplar 30; London Plane, Sycamore, Maples 34; Currants 140 and 307; Guelder Rose 222; Fig, Vine 307

Leaves long and narrow

Yew 24; Willows 26-28; Bog Myrtle 30; Sea Buckthorn 140

Leaves needles

Firs, Spruces, Larches, Pines, Cypress, Juniper, Yew 24; Gorses 120; Tamarisk 140; Western Red Cedar 298

Leaves pinnate or trefoil

Walnut 30; Ash, Tree of Heaven 36; Roses, Brambles 108 and 307; Rowan 114; Broom, Laburnum, Bladder Senna, False Acacia 120; Bladder-nut 140; Elders 222; Oregon Grape, Sorbaria, Box Elder 307

Leaves pinnate

Horse Chestnut 36; Tree Lupin 120; Virginia Creeper 307

CONIFERS   *Coniferae* (see p. 23).

Pine Family   Pinaceae

Fr a woody cone. Widely planted.

1* SILVER FIR *Abies alba.* Pyramidal tree to 50 m, branches regularly whorled; bark scaly, greyish. Lvs single, whitish beneath, in two rows, grooved, leaving an *oval scar.* Fl Apr-May. Cones erect, 10-20 m, with bracts between the scales. Mainly in mountains. (B),F,G,(S). *A. procera* and *A. grandis* from N America are widely planted. **1a* Douglas Fir** *Pseudotsuga menziesii* □ is taller, to 100 m, with branches irregularly whorled, bark ridged and red-brown, lvs fragrant when bruised and shorter, hanging cones. (T) from N America.

2* NORWAY SPRUCE *Picea abies.* Pyramidal tree to 60 m, branches regularly whorled; bark scaly, red-brown. Lvs single, in two rows, 4-sided, grass-green, falling to leave a *peg-like projection.* Fl May-June. Cones hanging, 10-18 cm, with no bracts protruding. Forests, especially in mountains. T, but (B). **2a* Sitka Spruce** *P. sitchensis* has sharply pointed blue-green lvs, and shorter cones with toothed scales. (T) from N America. **2b* Western Hemlock** *Tsuga hetero-pylla* □ has branches irregularly whorled, ir shoots drooping, flat lvs on cushion-like projections, and much shorter oval cones. (T) from N America.

3* EUROPEAN LARCH *Larix decidua.* Pyramidal tree to 35 m; bark rough, grey-brown. Lvs single and tufted, pale green, *deciduous.* Male fls yellow, female pink; Mar-Apr. Cones short, egG-shaped. Mountain forests. (B),F,G,(S). **3a***, p. 298.

4* SCOTS PINE *Pinus sylvestris.* Dome-shaped tree to 40 m, but pyramidal when young; bark reddish, flaking. Lvs in *pairs,* greyish, twisted, 30-70 mm; buds sticky, resinous. Male fls orange-yellow, female pinkish-green; May-June. Cones conical, hanging, 30-60 mm. Forests, moors, heaths.T. **4a* Austrian Pine** *P. nigra* is more pyramidal, with grey-black bark, and longer (to 190 mm) straighter greener lvs. (T) from S Europe. **4b* Maritime Pine** *P. pinaster* □ has more ridged bark, lvs much longer (to 250 mm), curved and greener, buds not sticky and larger, broader cones. (B,F) from S Europe. *P. contorta, P. radiata,* and *P. strobus* from N America, and *P. mugo* from the Alps are also widely planted.

Cypress Family   Cupressaceae

5* LAWSON'S CYPRESS *Chamaecyparis lawsoniana.* Evergreen columnar tree to 65 m; bark red-brown. Lvs *scale-like,* opposite, closely pressed to flattened stems, often greyish, parsley-scented, leading shoots drooping. Male fls pink, female yellow-brown; Apr. Cones globular, 8 mm. Widely planted, sometimes naturalised. (T) from N America. **5a***, p. 298.

6* JUNIPER *Juniperus communis.* Evergreen shrub or small tree to 6 m, bushy, columnar or (ssp. *nana)* prostrate. Lvs greyish, in whorls of 3, *spine-tipped.* Fls yellow, male and female on separate plants; May-June. Fr berry-like, green at first, blue-black in its second year. Coniferous woods, moors, heaths, scrub; ssp. *nana* on mountains and by sea. T.

Yew Family   Taxaceae

7* YEW *Taxus baccata.* Evergreen tree or shrub to 20 m; bark red-brown, flaking. Lvs dark green, in two rows. Fls green; male, with many yellow stamens, and female on separate trees; Feb-Apr. Fr in a fleshy *reddish-pink* cup. Woods, scrub, screes, often on lime; widely planted.T. **8 (Joint Pine),** p. 298.

1

1a

2

2b

4

3

4b

5

7

6

Willow Family   Salicaceae

Deciduous trees or shrubs. Lvs usually alternate and finely toothed. Fls small, petalless but with a scale, in catkins, often appearing before the lvs; male and female on different plants, male usually yellow, female usually green. Fr woolly with long hairs. Willows and Sallows *Salix* have buds with only one outer scale, catkins usually on lfy shoots, scales untoothed and usually 2 stamens. Hybrids are very frequent.

1* BAY WILLOW *Salix pentandra*. Shrub or small tree to 7 m, *hairless;* bark grey, rugged; twigs shiny. Lvs usually *broad* elliptical, dark glossy green, sticky and fragrant when young. Catkins slender, stamens 5 or more, stigmas purple; May-June, with lvs. By fresh water. T.

2* CRACK WILLOW *Salix fragilis*. Spreading tree to 25 m, often pollarded; bark grey, rugged; twigs hairless, easily breaking. Lvs *narrow* elliptical, hairless, paler beneath; stipules soon falling. Catkins slender; Apr-May, with lvs. By fresh water. T. **2a* White Willow** *S. alba* □ is less spreading and looks silvery-grey from silky hairs on young lvs and beneath mature lvs; twigs not fragile; May. Distinct subspecies, often planted, have blue-green lvs (ssp. *coerulea,* the Cricket-bat Willow) and bright orange-yellow twigs (ssp.*vitellina* □, the Golden Osier). **2** × **2a*** is common. **2b* Weeping Willow** *S. babylonica* with drooping branches and narrower lvs is also widely planted. **2c* Almond Willow** *S. triandra* is a shrub or small tree to 10 m, with smooth flaking cinnamon-brown bark, darker green lvs with persistent stipules and fatter catkins with 3 stamens.

3* GOAT WILLOW or GREAT SALLOW *Salix caprea* agg. Small tree or shrub to 10 m; twigs downy at first, later (except in the N) hairless. Lvs *oval with pointed tip,* usually rounded at base, hairless above (except in the N), softly grey downy beneath. Catkins stout, unstalked, the scales black-brown with white hairs, which show silvery white in bud, when known as 'pussy willow' or 'palm'; Mar-Apr, before lvs. Woods, scrub, by fresh water. T. **3a* Grey Willow or Grey Sallow** *S. cinerea* agg. □ is a shrub to 6 m, with twigs ridged under the bark and often continuing downy, and lvs narrower, tapering to base and either greyish above or with only a few rust-coloured hairs beneath. Usually in damp places. **3b* Eared Willow** or **Eared Sallow** *S. aurita* □ is a much smaller shrub, to 2m, with twigs widely angled and ridged under the bark, more rounded, markedly wrinkled lvs, and green catkin scales. **3c** *S. appendiculata* has faint ridges under bark, and narrower wrinkled lvs. G, southern.

4* PURPLE WILLOW *Salix purpurea*. Shrub to 5 m, with straight slender twigs, often purplish at first. Lvs *bluish-green,* lanceolate, often opposite, hairless, blackening when dried. Male catkins with dark, purple-tipped scales and reddish or purplish anthers, female often red-purple; Mar-Apr, before lvs. Fens, by fresh water. T, but (S). **4a* Violet Willow** *S. daphnoides* □ is taller, to 10 m, and may be a tree, with bluish waxy bloom on twigs, narrow, more sharply toothed lvs, not blackening, and yellow anthers. (B, F, G), S.

5* OSIER *Salix viminalis*. Shrub or small tree to 5 m; twigs long, straight, flexible, used in basket-making as withies. Lvs very *long and narrow,* to 25 cm, silky white beneath, untoothed, the margins down-turned. Catkins longer than 3 scales, scales brown with white hairs; Mar-Apr, before lvs. By fresh water. T, but (S). Many hybrids are cultivated and become naturalised. **5a** *S. elaeagnos* has twigs reddish or yellowish and lvs finely toothed. G, southern.

Golden Osier

Willow Family *(contd.)*

1* DARK-LEAVED WILLOW *Salix nigricans* agg. Shrub or small tree to 4 m; twigs usually downy. Lvs *pointed oval,* mostly greyish beneath, almost hairless, usually blackening when dried. Catkins short, scales green and black-brown; Apr-May, usually before lvs. By fresh water; in south of range only in the hills. T. **1a* Tea-leaved Willow** *S. phylicifolia* agg. is always a shrub, with twigs almost hairless and lvs shiny above, all greyish beneath and not blackening. B, G, S. **1b** *S. starkeana* agg. may have rounder lvs, often reddish when young, sometimes downy, not blackening. G, S.

2* DOWNY WILLOW *Salix lapponum.* Shrub to 4 m, with *grey down.* Lvs pointed oval, to broad lanceolate, *untoothed,* grey on both sides. Catkin scales brown with white hairs, anthers reddish at first; May-June, with lvs. Tundra, mountains. B, S, northern. **2a Bluish Willow** *S. glauca* agg. is shorter, with longer sparser hairs, narrower blunter lvs bluish beneath, and yellow catkins. S.

3* WOOLLY WILLOW *Salix lanata* agg. Shrub to 3 m, twigs *thickly downy.* Lvs broad oval, *untoothed,* yellow-hairy at first, then grey-hairy, finally almost hairless above. Catkins with long yellow hairs; May-July, with lvs. Tundra, mountains. B, S, northern.

4* MOUNTAIN WILLOW *Salix arbuscula.* Shrub to 2 m; twigs hairless, ridged under the bark. Lvs *pointed oval,* shiny above, greyish and often hairless beneath. Catkin scales with rusty hairs, anthers reddish at first, stigmas purple-brown; May-June, with lvs. Mountains. B, S. **4a** *S. hastata* □ has mature lvs larger, sometimes roundish, not shiny or hairy, and stipules much larger and more persistent. S. **4b* Whortle-leaved Willow** *S. myrsinites* □ is prostrate with twigs downy at first, not ridged, lvs shiny on both sides, hairy only on margins and sometimes untoothed, and anthers purple.

5* CREEPING WILLOW *Salix repens* agg. Shrub with *creeping* stems up to 2 m high; twigs silky hairy at first. Lvs narrow to broad oval, white usually with *silky hairs* beneath and sometimes above, almost untoothed. Catkins short, scales dark green tipped purple-brown, female green to purple-brown; Apr-May, before lvs. Swamps, bogs, fens, dune slacks. T. **5a** *S. myrtilloides* is more prostrate, with twigs brown-hairy at first, and lvs often rounder and not silky hairy, the margins down-turned. G, S.

6* DWARF WILLOW *Salix herbacea.* Prostrate *patch-forming* undershrub; twigs 2-3 cm long. Lvs *rounded,* toothed, hairless, shiny above. Catkins short, scales yellowish, sometimes tinged red; June, after lvs. Mountains, tundra. B, G, S. **6a** *S. polaris* has narrower, longer-stalked, often untoothed lvs, purplish scales and purple male catkins. S. **6b* Net-leaved Willow** *S. reticulata* □ has larger, darker green, wrinkled, longer-stalked, untoothed lvs, whitish and conspicuously net-veined beneath, and brownish scales with grey hairs; anthers and stigmas purple.

All species on this and the next page (32) are deciduous (except Evergreen Oak, p.32) and have alternate lvs and separate male and female fls – Poplars and Bog Myrtle on separate plants, the others on the same plant.

Willow Family (contd.)

POPLARS Populus have buds with several unequal outer scales, and catkins, appearing before the broad lvs, with toothed scales, the male reddish and with 4 or more stamens, the female greenish-yellow.

1* GREY POPLAR Populus canescens. Spreading tree to 30 m, with black gashes on smooth grey bark, suckering freely; young twigs and buds white downy. Spring lvs pointed oval, broadest above the middle, with large teeth, almost hairless in summer; summer and sucker lvs more sharply lobed and maple-like, always downy. Catkin scales deeply cut; March. Damp woods; widely planted. B,F,G. **1a* White Poplar** P. alba □ has young twigs and buds more thickly downy, spring lvs rounder, broadest below middle, always whitely downy beneath, summer and sucker lvs more deeply palmately lobed□, and catkin scales less deeply cut. (B,F),G.

2* ASPEN Populus tremula. Spreading tree to 20 m, suckering freely; bark smooth, grey-brown, twigs and slightly sticky buds almost hairless. Lvs rounded, bluntly toothed, soon hairless, swaying in every breeze on thin flattened stalks; sucker lvs very like 1. Catkin scales deeply cut; March. Damp woods and heaths, fens, moors, mountains. T.

3* BLACK POPLAR Populus nigra. Spreading tree to 30 m, with rugged blackish bark and large bosses on trunk, rarely suckering; twigs and sticky buds hairless. Lvs ace-of-spades, shorter and broader on short shoots. Fl March-April, before lvs. By fresh water. B,F,G. The tall slender Lombardy Poplar var. italica is widely planted. **3a* Italian Poplar** P. × canadensis has a more fan-shaped crown, no bosses on trunk, more frequent suckers, and larger lvs with one or two pimples at base; female trees rare. Very widely planted. **3b* Balsam Poplar** P. gileadensis □ is like 3a with buds and young lvs balsam-fragrant and lvs without pimples, paler beneath, with hairs on veins. Less widely planted.

4* BOG MYRTLE or SWEET GALE Myrica gale (Bog-Myrtle Family, Myricaceae). Shrub to 2·5 m, fragrant from yellow dots on red-brown twigs and narrow oval lvs, downy beneath. Orange male and shorter reddish female catkins on different plants; Apr-May, before lvs. Wet heaths, bogs. T, western. **4a***, p.298.

5* WALNUT Juglans regia (Walnut Family, Juglandaceae). Spreading tree to 30 m; bark grey, smooth at first. Lvs pinnate, aromatic. Fls yellow-green, male in hanging catkins, female in short erect spikes; Apr-May. Fr a brown nut in a green case. Hedgerows, widely planted, often bird-sown.(B,F,G) from S E Europe.

Birch Family   Betulaceae

6* SILVER BIRCH Betula pendula. Erect tree to 30 m, with papery peeling white bark and dark bosses on trunk; young twigs hairless, with resinous dots. Lvs ace-of-spades, teeth of different depths. Catkins yellowish, male hanging, female erect; Apr-May, with lvs. Woods, scrub, heaths, moors. T. **6a* Downy Birch** B. pubescens □ is shorter and may be a shrub, with bark grey or brown, no bosses, young twigs downy and without dots, and lvs uniformly toothed. Usually wetter places, not on lime. Northern. 6 × 6a* is frequent.

7* DWARF BIRCH Betula nana. Semi-prostrate shrub to 1 m; twigs hairless. Lvs rounded, deeply toothed, downy when young. Catkins as 6 but smaller; June-July, after lvs. Moors, bogs, tundra. B,G,S. 7 × 6a occurs. **7a**, p.298.

8* ALDER Alnus glutinosa. Spreading tree or shrub to 20 m; bark dark brown, rugged, twigs hairless. Lvs roundish, toothed, blunt-tipped, almost hairless. Male catkins yellowish, female egg-shaped, purplish; Feb-March, before lvs. Fr cone-like. Fens, by fresh water. T. **8a* Grey Alder** A. incana □ has smooth bark, and young twigs and narrower pointed lvs downy at least when young. Drier places. T, but (B). 8 × 8a is found. **8b**, p.298.

See note on p. 30.

Hazel Family Corylaceae

1* HORNBEAM *Carpinus betulus.* Spreading tree to 25 m; bark smooth, grey, twigs slightly downy. Lvs pointed oval, toothed, scarcely downy. Catkins greenish; Apr-May with lvs. Fr a nut within *lfy bracts.* Woods; widely planted. T, south-eastern.

2* HAZEL *Corylus avellana.* Shrub to 6 m; bark red-brown, peeling. Lvs oval to rounded, with a small point, toothed, downy. Male fls in hanging catkins, pale yellow; female erect, bud-like, with bright red styles; *Jan-Mar,* before lvs. Fr a nut, enclosed in a lfy husk. Woods, scrub, hedges. T.

Beech Family Fagaceae

3* BEECH *Fagus sylvatica.* Spreading tree to 30 m; bark smooth, grey, buds red-brown. Lvs pointed oval, shallowly toothed, veins prominent at edge, silky white beneath at first. Fls greenish, male in a roundish *tassel,* female erect, in pairs; Apr-May, with lvs. Fr a warm brown nut ("mast") enclosed in a bristly husk. Woods, usually in pure stand; widely planted. T, southern.

4* SWEET CHESTNUT *Castanea sativa.* Spreading tree to 30 m; bark rugged, grey-brown. Lvs *elliptical,* toothed. Catkins yellowish, with a sickly fragrance, male fls at the top, female below; July. Fr a brown nut encased in a softly spiny cup. Widely planted, sometimes naturalised. (T) from S Europe.

5* EVERGREEN OAK or HOLM OAK *Quercus ilex.* Evergreen tree to 25 m; bark greyish, scaly, twigs grey downy. Lvs pointed oval, untoothed or with holly-like spines, grey downy beneath. Catkins greenish-yellow, male long, female short; May-June. Fr a nut (acorn) in a cup. (B), F.

6* PEDUNCULATE OAK *Quercus robur.* Spreading tree to 45 m; bark rugged, grey-brown, twigs and lvs almost hairless. Lvs oblong, *lobed,* the basal lobes usually overlapping the very short stalk. Catkins greenish-yellow, male longer than female; Apr-May, with lvs. Fr a nut (acorn) in a long-stalked scaly cup. Woods, usually in pure stand; widely planted. T. **6a* Sessile Oak or Durmast Oak** *Q. petraea* □ has lvs tapering to the unlobed base and acorns scarcely stalked. Less often planted. Southern. **6b* Turkey Oak** *Q. cerris* □ has downy young twigs, larger, more jaggedly lobed lvs and short-stalked acorns in bristly cups. (B), F, G. **6c White Oak** *Q. pubescens* is shorter or a shrub with thickly downy twigs, lvs downy beneath when young, and acorn cups downy. F, G. **6d* Red Oak** *Q. rubra* □ with dark red twigs and much larger lvs turning rich red in autumn is the most frequent of many planted N American oaks.

1* Hornbeam

3* Beech   **4* Sweet Chestnut**   **6* Pedunculate Oak**

1

2

3

5

6a

6

6b

6d

Elm Family   Ulmaceae

1* WYCH ELM *Ulmus glabra*. Spreading tree to 40 m, with rugged bark and no suckers; twigs roughly hairy when young. Lvs broad oval to elliptical, alternate, roughly hairy. Fls small, unstalked, with a *tuft of reddish stamens;* Mar-Apr, before lvs. Fr a pale green notched disc, the seed in the centre. Woods, hedges. T.   **1a* English Elm** *U. procera* □ suckers freely and has twigs always hairy, lvs smaller and rounder, sometimes almost hairless above, and fr (rare in B) with seed above middle. Rarely in woods. B, F.   **1b* Small-leaved Elm** *U. minor* □ suckers, and has hairless twigs, narrower lvs hairless beneath, and fr with seed above middle. Rarely in woods. B, F, G.   **1c Fluttering Elm** *U. laevis* has twigs and undersides of lvs hairless or softly downy, fls and fr long-stalked and fr with hairy margins. F, G, S, southern.

Plane Family   Platanaceae

2* LONDON PLANE *Platanus hybrida*. Spreading tree to 30 m, with smooth grey bark, constantly flaking off. Lvs alternate, sharply palmately lobed. Fls and fr in *roundish heads,* fls green. Widely planted, occasionally self-sown. (B, F, G), of unknown origin.

Maple Family Aceraceae

Trees or shrubs, with long-stalked, palmately-lobed, ivy-like opposite lvs. Fls small, greenish-yellow, 5-petalled, male and female separate on same tree. Fr a pair of keys (samara), each with a single long wing.

3* SYCAMORE *Acer pseudoplatanus*. Spreading tree to 30 m, with smooth grey bark, flaking when old. Lvs 5-lobed to about half-way, well toothed. Fls in *hanging* clusters; May. Fr with wings diverging at right angles. Woods, hedges, widely planted. (B), F, G, (S).   **3a* Norway Maple** *A. platanoides* □ has lvs with more sharply pointed lobes and fewer, sharper teeth, brighter yellow fls in erect clusters, appearing in April, before lvs, and fr as 4. T, but (B).   **3b Italian Maple** *A. opalus* □ has rather leathery, much less toothed lvs and fr with wings diverging more narrowly. F, G.

4* FIELD MAPLE *Acer campestre*. Small tree or shrub to 20 m, with roughish grey bark; twigs downy, later corky. Lvs bluntly 5-lobed, untoothed, much smaller than 3, red when young. Fls in *erect* clusters; May. Fr with wings diverging horizontally, usually downy. Woods, hedges. T, southern.   **4a Montpelier Maple** *A. monspessulanus* □ has lvs 3-lobed, rather leathery, shiny above, fls sometimes before lvs, and fr with wings almost parallel. F, G. eastern.

1a* English Elm   **3*** Sycamore

1* Wych Elm   **2*** London Plane   **4*** Field Maple

Horse-Chestnut Family   Hippocastanaceae

1* HORSE CHESTNUT *Aesculus hippocastanum.* Tree to 25 m, with grey bark, smooth at first, and sticky buds. Lvs opposite, palmate with *5-7 lflets.* Fls white, with a yellow to pink spot, 5-petalled, in conspicuous erect stalked clusters; May. Fr a large warm-brown nut, in a fleshy green, stoutly but softly spiny case. Widely planted, often self-sown. (B,F,G) from SE Europe.

Holly Family   Aquifoliaceae

2* HOLLY *Ilex aquifolium.* Evergreen tree or shrub to 10 m, with grey bark, smooth at first. Lvs oval, usually *spiny,* leathery, dark green. Fls 4-petalled, white, often purplish, in clusters, male and female on different trees; May-Aug. Fr a red berry. Woods, scrub, hedges. T, southern.

Lime Family   Tiliaceae

3* SMALL-LEAVED LIME *Tilia cordata.* Spreading tree to 30 m, with smooth dark brown bark and large bosses on trunk; young twigs more or less hairless. Lvs heart-shaped, greyish beneath, with faint side-veins and small tufts of reddish hairs. Fls 5-petalled, yellowish, *fragrant,* in small clusters, on stalks half-joined to an oblong bract; July. Fr a globular nut. Woods. T.   **3a* Large-leaved Lime** *T. platyphyllos* □ has no trunk bosses, young twigs usually downy, larger lvs with all veins prominent and uniformly grey downy beneath, and fewer larger fls to the cluster. Southern. **3b* Common Lime** *T.* × *vulgaris* is 3 × 3a, more like 3, but with larger lvs, more prominent veins and no red hairs. Widely planted, occasionally a natural hybrid.

Olive Family   Oleaceae

4* ASH *Fraxinus excelsior.* Spreading tree to 30 m, with grey bark, becoming rugged; buds *black.* Lvs opposite, *pinnate,* the lflets unstalked. Fls petalless, with tufts of purple-black stamens, becoming greenish; Apr-May, before lvs. Fr a key (samara) with a single wing. Woods, sometimes in pure stand, hedges. T. **4a Flowering Ash** *F. ornus* is much smaller, to 10 m, with lflets usually shortly stalked and fragrant whitish fls appearing with lvs. F,G.

Quassia Family   Simaroubaceae

5* TREE OF HEAVEN *Ailanthus altissima.* Tree to 20 m, with smooth grey bark; *suckering.* Lvs alternate, pinnate. Fls 5-petalled, greenish, in stalked clusters; *July.* Fr a key, reddish at first. Widely planted, frequently naturalised. (B,F,G) from China.

2* Holly

3* Small-leaved Lime

1* Horse Chestnut

4* Ash

Birthwort Family Aristolochiaceae

1* BIRTHWORT *Aristolochia clematitis*. Medium/tall hairless stinking perennial. Lvs bluntly heart-shaped, alternate, untoothed. Fls dull yellow, irregularly *tubular* with a flattened lip, swollen at the base, in small clusters at base of upper lvs; June-Sept. Waste places, often by water. (B),F,G, eastern.

2* ASARABACCA *Asarum europaeum*. Low creeping patch-forming downy perennial. Lvs kidney-shaped, shiny, long-stalked. Fls dull purple, solitary, *bell-shaped*, 3-lobed; Mar-Aug. Woods on lime-rich soils. (B),F,G, southern, scattered.

Mistletoe Family Loranthaceae

3* MISTLETOE *Viscum album*. Woody, regularly branched parasite on tree boughs (mainly apples and poplars, rarely conifers). Lvs elliptical, in pairs, yellowish, *leathery*. Fls inconspicuous, green, 4-petalled, stamens and styles on separate plants; Feb-Apr. Fr a sticky *white berry*; Nov-Feb. T, southern.

Sandalwood Family Santalaceae

4* BASTARD TOADFLAX *Thesium humifusum*. Short, often prostrate, parasitic perennial. Lvs linear, 1-veined, alternate. Fls dull white inside, yellowish-green outside, in stalked spikes, with three bracts but no sepals; petal tube 5-lobed, persistent and much shorter than fr; June-Aug. Dry grassland on lime. B,F,G, western. **4a** *T. ebracteatum* has only one bract. G. **4b** *T. alpinum* has purer white fls, the petal tube 4-lobed and 2-3 times as long as fr. F,G,S, mountains. **4c** *T. rostratum* is like 4b but has only one bract. G. **4d** *T. pyrenaicum* has petal-tube equalling fr. Dry places. F,G. **4e** *T. linophyllon* has runners and usually 3-veined lvs. Sandy places on acid soils. F,G, eastern. **4f** *T. bavarum* has dark green lanceolate 3-5-veined lvs. Mountains. S.

Hemp Family Cannabaceae

5* HOP *Humulus lupulus*. Tall, roughly hairy, square-stemmed climbing perennial, twining *clockwise*. Lvs deeply *palmately* lobed, toothed. Fls green, the branched clusters of males and the cone-like females on separate plants; July-Sept. Often in hedges, also cultivated. T,lowland.

6* HEMP *Cannabis sativa*. Tall, *strong smelling* annual. Lvs deeply 3-9-lobed, almost to the base. Fls green, male in a branching cluster, female, on separate plants, in stalked spikes; July-Sept. Casual on waste ground, also illegally sown as a source of marijuana. T, lowland.

Nettle Family Urticaceae

7* NETTLE *Urtica dioica*. Medium/tall perennial, armed with *stinging* hairs. Lvs heart-shaped, toothed, opposite, longer than their stalks. Fls green with yellow stamens, in catkins, stamens and styles on separate plants; June-Sept. Woods, waste ground, often near houses. T. **7a* Annual Nettle** *U. urens* □ is annual and smaller, with sting less powerful, lower lvs shorter than their stalks, and stamens and styles in separate fls on same plant. Cultivated ground.

8* PELLITORY OF THE WALL *Parietaria judaica*. Short/medium, much-branched, softly hairy perennial; stems reddish. Lvs untoothed, alternate, to 70 mm, not more than three times as long as their stalks. Fls green with yellow stamens, in small forked clusters, stamens and styles in separate fls; June-Oct. Walls, banks, rocks. B,F,G,western. **8a** *P. officinalis* is larger and scarcely branched, with lvs to 120 mm, at least four times as long as their stalks. Open places.F,G, eastern. **8b* Mind Your Own Business** *Soleirolia soleirolii* is much smaller, with tiny roundish lvs and solitary pink fls; May-Aug.

39

Dock Family   Polygonaceae

pp. 40-44. Frequent weeds of waste places and arable land, also wet places. Lvs alternate. Characteristically there is a sheath at the leaf base forming a tube around the stem (the *ochrea).* In bistorts *Polygonum* the usually five sepal-like petals of the small sepal-less fls persist and enclose the fr.

1* BISTORT *Polygonum bistorta.* Medium patch-forming, unbranched, almost hairless perennial. Lvs narrow *triangular,* their stalks winged in the upper part. Fls pink, in a dense spike, up to 15 mm broad; June-Oct. Meadows and woods, usually away from lime, often near water. B, F, G, (S).

2* AMPHIBIOUS BISTORT *Polygonum amphibium.* Short/medium creeping perennial with two forms. In water, hairless and floating, rooting at the lf-junctions, with lvs oblong, not tapering at the base. The terrestrial form □ is smaller, slightly hairy, the longer, thinner lvs with a rounded base. Both have pink fls in dense spikes; June-Sept. Near or in still or slow-flowing water or on waste ground. T, lowland.

3* REDSHANK *Polygonum persicaria.* Medium, sprawling, hairless, branched annual. Lvs lanceolate, usually *dark-spotted,* tapering to the base. Fls pink or whitish, in densely crowded spikes; June-Oct. Bare ground, often near water. T. **3a* Pale Persicaria** *P. lapathifolium* □ is larger and slightly hairy, with pale dots on lvs and fl-stalks, and stouter spikes of greenish-white or brick-red fls.

4* ALPINE BISTORT *Polygonum viviparum.* Short unbranched hairless perennial. Lvs narrow lanceolate, tapering at base. Fl spikes with pale pink or white fls at top and red *bulbils* at base; June-Aug. Mountain grassland and lowland in the north. B, G, S.

5* WATER-PEPPER *Polygonum hydropiper.* Medium hairless annual with a *burning taste*; ochreae not fringed. Lvs narrow lanceolate. Fls white, tinged pink or green, in a narrow, yellow-dotted nodding spike; July-Sept. Damp meadows, wet mud, shallow water. T, lowland. **5a* Tasteless Water-pepper** *P. mite* is tasteless with fringed ochreae and an erecter undotted spike of pinker fls. Southern. **5b* Small Water-pepper** *P. minus* agg. is like 5a, but has much narrower lvs, less than 10 mm broad, and a slenderer erect fl-spike.

6* KNOTGRASS *Polygonum aviculare.* Low hairless annual, with silvery, ragged few-veined ochreae. Lvs lanceolate, larger on main stem than on branches. Fls pink or white, 1-6 together at base of upper lvs; petals persistent, *enclosing fr*; June-Nov. Bare ground, including sea-shores. T. **6a* Sea Knotgrass** *P. maritimum* □ is a perennial with many-veined ochreae, down-rolled leaf-margins, and dead petals shorter than fr; July-Sept. Atlantic coasts. B (rare), F. **6b* Ray's Knotgrass** *P. oxyspermum* also has fr longer than dead petals. Coastal sand and shingle. Not east coasts. **6c*** *P. boreale* has broader lvs, longer lf-stalks emerging from the ochreae, and pink-edged petals. Coasts. B, S, northern. **6d*** *P. rurivagum* has long (10 mm) red-brown ochreae and much narrower lvs. Dry places on lime. Southern. **6e*** *P. arenastrum* is mat-forming, with lvs all similar, shortly oval. Open ground. Lowland. **6f** *P. patulum* has greenish fls and no lvs at top of fl-spike. F, G.

7* BUCKWHEAT *Fagopyrum esculentum.* Medium hairless annual, reddening. Lvs arrow-shaped. Fls white or pink in branched clusters; June-Sept. Cultivated and naturalised in open places. T, southern. **7a** *F. tataricum* is usually taller and less red with smaller fls in a looser head. A local weed in cultures of 7. F, G, S.

Dock Family (contd.)

1* BLACK BINDWEED *Bilderdykia convolvulus.* Tall, climbing annual, with broader lvs and totally different fls from the true bindweeds (p.184); twines *clockwise.* Lvs heart-shaped, powdery below. Fls greenish-pink, in loose spikes; July-Oct. Fr stalks 2 mm, jointed above the middle; fr dull-black, narrowly winged white. Cultivated and disturbed ground. T, lowland. **1a* Copse Bindweed** *B. dumetorum* □ is taller and slenderer with whiter fls, fr stalks up to 8 mm, jointed below the middle, and fr shiny black, broadly winged white. Often in hedges, in warm sunny places. T, southern.

2* JAPANESE KNOTWEED *Reynoutria japonica.* Tall, vigorous perennial, often forming dense thickets; stems stout, somewhat zigzag. Lvs broad triangular. Fls whiter in branched spikes; Aug-Oct. Waste ground. (T), from Japan. **2a* Giant Knotweed** *R. sachalinensis* □ is taller, reaching 3 m; lvs very long, to 300 mm, pointed oval with bluntly arrow-shaped base; fls in dense clusters, a shy flowerer. Garden escape. (B,F,G), local, from Sakhalin, N E Asia.

3* HIMALAYAN KNOTWEED *Polygonum polystachyum.* Tall stout patch-forming perennial. Lvs lanceolate, *tapering* at the base, often slightly hairy beneath. Fls pink or white with red stalks, in long branched spikes; July-Oct, a shy flowerer. Waste places, often by rivers. (B,F,G), from Assam.

4* ICELAND PURSLANE *Koenigia islandica.* Insignificant prostrate annual. Resembles 6, but has minute *broad* oval unstalked lvs and often reddish stems. Fls 3-petalled, green, short-stalked, in small clusters; July-Sept. Damp mud and bare ground. B(rare),S, northern.

Dock Family contd. on p. 44.

Mesembryanthemum Family   Aizoaceae

5* HOTTENTOT FIG *Carpobrotus edulis.* Prostrate perennial with long creeping woody stems, and stout *fleshy* lvs, *triangular* in section. Fls usually magenta with a yellow centre, sometimes all yellow; May-Aug. Cliffs and dunes. (B,F), western, from S Africa.

Purslane Family   Portulacaceae

6* BLINKS *Montia fontana.* Very variable, but always insignificant straggling annual/perennial; stems often reddish. Lvs *narrow* oval, tapering into the stalk. Fls tiny, white, 5-petalled, long-stalked, in loose clusters; April-Oct. Bare, usually wet places, avoiding lime. T.

7* SPRING BEAUTY *Montia perfoliata.* Short hairless annual. Lvs pointed oval, long-stalked at base, one pair completely *encircling* the stem. Fls white, the five petals *not* deeply notched; April-July. Acid sandy soils, often in quantity on disturbed ground. (B,F,G), northern, from N America. **7a*** *Portulaca oleracea* has spoon-shaped lvs crowded below yellow fls. Cultivated and as a weed. (B),F,G.

8* PINK PURSLANE *Montia sibirica.* Short annual, hairless and rather fleshy. Stem lvs unstalked, but *not* encircling the stem. Fls pink, the five petals deeply *notched*; April-July. Damp woods. (T), scattered, from N America and N E Asia.

2a* Giant Knotweed

1

1a

2

3

4

5

6

7

8

Dock Family (contd.)

DOCKS *Rumex* are mostly hairy perennials with alternate lvs and ochreae (see p.40). Fls small, green, in whorls (the lower accompanied by a small lf) in branched spikes, often turning red. Petals and sepals alike, six, the three inner larger, enlarging and becoming the 'valves' around the 3-sided nut. Fr characters are important for identification. Hybrids are frequent.

1* COMMON SORREL *Rumex acetosa.* Short/tall acid-tasting perennial. Lvs arrow-shaped, 2-6 times as long as wide, the lobes at the base pointing backwards, the upper clasping the stem; ochreae with fringed margin. Fl-head little branched, loose; May-Aug. Grassland and woods. T. **1a, 1b*, 1c**, p.298.

2* SHEEP'S SORREL *Rumex acetosella* agg. Slender low/short perennial. Lvs *arrow-shaped*, sometimes very narrow (**2a** □), the basal lobes spreading or pointing *forwards*, the upper stalked. Fl May-Aug. Dry, rather bare places, not on lime. T.

3* BROAD-LEAVED DOCK *Rumex obtusifolius.* Tall perennial with broad oblong lvs up to 25 cm long, hairy on veins beneath, the lower *heart-shaped* at the base. Fl-head lax; June-Oct. Fr valves triangular, toothed, one swollen. Very common on bare and disturbed ground. T. **3a* Curled Dock** R. *crispus* □ has narrower lvs with strongly wavy margins, not distinctly heart-shaped at the base; fl-head dense; valves more oval, not toothed, usually all three swollen. **3b* Northern Dock** R. *longifolius* has broad lvs, with strongly wavy margins, very dense fl-spikes, and large valves kidney-shaped, not toothed or swollen; June-July. Often by streams. T, northern. **3, 3a,** and **3b** hybridise. **3c*, 3d*, 3e*, 3f,** p.298.

4* CLUSTERED DOCK *Rumex conglomeratus.* Medium/tall perennial with *zig-zag* stem, and branches spreading. Lower lvs oblong with *rounded* base, long-stalked, the upper narrower, rarely red-fringed. Fls in whorls, Lfy nearly to the top, well spread out; June-Aug. Valves untoothed, oblong, all swollen. Woods, grassy and waste places. T, southern. **4a* Wood Dock** R. *sanguineus* has lvs rarely red-veined and only one valve swollen. Damp, usually shady places. 4a and 4b have stems straight, branches erect and only the lowest whorls lfy. **4b* Shore Dock** R. *rupestris* is stouter and shorter with greyish lvs. Shores, dune-slacks. B(rare),F, western.

5* FIDDLE DOCK *Rumex pulcher.* Short/medium, *spreading* perennial, with lvs often *waisted* like a fiddle. Fls as in 4; June-Aug. Fl-stems drooping, stiff and tangled in fr; valves toothed, all swollen. Dry, often waste places. T, southern, scattered.

6* WATER DOCK *Rumex hydrolapathum.* Very tall perennial reaching 2m; much branched, the branches erect. Lvs narrow, sometimes *over 1 m* long, 4-5 times as long as wide. Fl July-Sept. Valves triangular, not toothed, all swollen. Rivers and wet places. T. **6a* Scottish Dock** R. *aquaticus* is slenderer with broader, triangular lvs, less than three times as long as wide, and valves unswollen. T, scattered, rare in B. **6b*,** p.298.

7* MARSH DOCK *Rumex palustris.* Medium annual/perennial, much branched and turning *yellow-brown* in fr; stems rather wavy. Root lvs tapering, wavy-edged, c.6 times as long as wide. Fls in lfy whorls, well separated, but more crowded at the tips; June-Sept. Fr-stalks thick, stiff, shorter than valves; valves toothed, all swollen. Bare ground near water. T, southern. **7a* Golden Dock** R. *maritimus* turns golden-yellow in fr; fr-stalks slender and longer than the longer-toothed valves. Ponds, sometimes sea-shores. T.

8* MOUNTAIN SORREL *Oxyria digyna.* Short hairless perennial; stem almost lfless. Lvs fleshy, *kidney-shaped.* Fl June-Aug. Fr with *two* valves, not swollen. Damp rock ledges and by streams on mountains; to sea-level in the north. T.

3a* Curled Dock

45

Goosefoot Family   Chenopodiaceae

Goosefoots *Chenopodium* and Oraches *Atriplex* are annuals (except 1), often mealy, with alternate toothed lvs and spikes of inconspicuous green petalless fls, with joined sepals (except 3) and yellow anthers. There are few good diagnostic characters for the naked eye. In Goosefoots leaf-shape is important. Oraches are told from Goosefoots by their separate male and female fls, and by their sepals being enclosed by 2 triangular bracts. Bare ground, often by the sea.

1* GOOD KING HENRY *Chenopodium bonus-henricus.* Medium *perennial,* often reddish. Lvs large to 10 cm, mealy when young, *triangular,* almost untoothed. Fls in almost lfless spikes; May-Aug. Cf 6b, which has smoother lvs and stamens and styles in separate fls. Waste places. T, southern.

2* FAT HEN *Chenopodium album.* Medium/tall and deep green under a thick *white mealy* covering; stem often reddish. Lvs variable, lanceolate to diamond-shaped. Fl June-Oct. Abundant on disturbed ground. T.   **2a* Fig-leaved Goosefoot** *C. ficifolium* □ has large, deeply 3-lobed lower lvs, the middle lobe much the longest. B,F,G. **2b* Oak-leaved Goosefoot** *C. glaucum* is much smaller and often prostrate, mealy only beneath the narrow, wavy-edged lvs. **2c*, 2d*,** p. 299.

3* RED GOOSEFOOT *Chenopodium rubrum.* Variable, medium/tall, hairless, and often red-tinged. Lvs diamond-shaped, irregularly toothed. Sepals *not* joined together; Aug-Oct. Disturbed ground, manure heaps, common near the sea. T, southern. **3a* Small Red Goosefoot** *C. botryodes* is shorter and spreading, with less toothed lvs and fls persisting, covering fr. Drier parts of salt-marshes. B,F,G, scattered.

4* MAPLE-LEAVED GOOSEFOOT *Chenopodium hybridum.* Tall and hairless, with lvs usually triangular, tapering abruptly at the base, conspicuously though sparsely toothed. Fls in scattered lfless clusters; July-Oct. Disturbed ground. T, scattered. **4a* Nettle-leaved Goosefoot** *C. murale* is stouter with diamond-shaped lvs; fls more crowded. **4b* Upright Goosefoot** *C. urbicum* is almost hairless with smaller, less deeply toothed lvs, tapering gradually at the base. T, southern, rare in B.

5* MANY-SEEDED GOOSEFOOT *Chenopodium polyspermum.* Variable, short, spreading; stems square, often reddish. Lvs oval, *untoothed.* Fl July-Oct. Disturbed ground. T. **5a* Stinking Goosefoot** *C. vulvaria* is smaller and more prostrate and smells strongly of rotting fish. Also in salt-marshes. Southern, local.

6* COMMON ORACHE *Atriplex patula.* Very variable, medium/tall, much branched, usually mealy. Lvs lanceolate to triangular, the basal lobes pointing forwards, tapering into the stalk. Bracts toothed or not; July-Oct. Disturbed ground, sea-shores. T. **6a* Grass-leaved Orache** *A. littoralis* □ has almost linear lvs, the upper unstalked. Salt-marshes. **6b* Spear-leaved Orache** *A. hastata* □ agg. often reddens and has ascending branches, triangular lvs, the basal lobes at right angles to the stalk, and bracts always toothed. **6c, 6d,** p. 299.

7* FROSTED ORACHE *Atriplex laciniata.* Short/medium, sprawling, *silvery*; stems reddish, much branched, rather weak. Lvs diamond-shaped. Bracts 3-lobed; July-Sept. Dunes and beaches near high-tide mark. T, southern. **7a*** *A. rosea* is erecter with pointed bracts. Also inland. (B),F,G.

8* SEA BEET *Beta vulgaris* ssp. *maritima.* Sprawling, hairless, often red-tinged perennial with tall, upright shoots. Lvs dark green, *leathery,* untoothed. Fls as in Goosefoots (above); June-Sept. Wild ssp. of beets, mangolds, etc. Coasts. T, southern.

9 POLYCNEMUM *Polycnemum majus.* Short, often sprawling hairless annual. Lvs *sharp-tipped,* triangular in section. Fls greenish, petalless, solitary, at base of upper lvs; July-Oct. Dry places, bare ground. F,G, eastern. **9a,** p. 299.

2a* Fig-leaved Goosefoot

1

2

3   4   5

6a

6   6b   7   8   9

47

Goosefoot Family (contd.)

1* SEA PURSLANE *Halimione portulacoides.* Short/medium silvery *undershrub*; stems brownish. Lvs elliptical, untoothed, the lower opposite. Fls green, in short branched spikes, stamens and styles in separate fls; July-Oct. Fr unstalked. Banks of channels in salt-marshes. B,F,G, southern. **1a* Stalked Orache** *H. pedunculata* is a short annual, not woody, with all lvs alternate, and fr stalked. Drier parts of salt-marshes. T, southern, but extinct in B.

2* GLASSWORT *Salicornia europaea* agg. Low/medium fleshy annual, sometimes unbranched, often reddening or yellowing in fr. Lvs scale-like, opposite, *succulent,* translucent, fused to envelop the stem. Fls obscure, green with yellow anthers, usually in groups of three; Aug-Sept. Muddy salt-marshes, sometimes on sandy soils inland. T. **2a* Perennial Glasswort** *Arthrocnemum perenne* is a creeping woody perennial, forming tussocks up to 1 m across. Salt-marshes, gravelly shores. B,F, southern.

3* ANNUAL SEABLITE *Suaeda maritima.* Variable, short/medium, sometimes prostrate branched annual. Lvs cylindrical, alternate, fleshy with *pointed* tip and tapering base. Fls small, green, 1-3 in a cluster at base of upper lvs; stigmas 2; Aug-Oct. Coasts, especially salt-marshes. T, southern.

4* SHRUBBY SEABLITE *Suaeda vera.* Small greyish hairless *shrub* to 120 cm. Lvs cylindrical, rounded at base and *tip.* Fls as in 3; stigmas 3; June-Oct. Sand and shingle or among rocks above high-tide mark. B,F.

5* PRICKLY SALTWORT *Salsola kali.* Variable, short/medium, prickly, usually semi-prostrate annual, hairless or not; stem often striped. Lvs linear, unstalked, succulent, usually *sharp-tipped.* Fls green, solitary with lf-like bracts; fl persisting and covering the fr; July-Oct. Sandy coasts, in the south sometimes inland. T.

6 HAIRY SEABLITE *Bassia hirsuta.* Short/medium, often prostrate, slightly hairy annual. Lvs linear, fleshy. Fls resembling 3, but ripe fr *spiny.* Meadows near the sea. G. **6a** *Kochia laniflora* has thread-like lvs and spineless winged fr. Dry sandy places. G. **6b** *K. scoparia* is like 6a, but with broader 3-veined lvs. (F,G), from Asia. **6c** *Corispermum leptopterum* is like 6b with 1-petalled fls and flat seeds. (F,G),, from S Europe. **6d** *C. marschallii* is like 6c but softly-hairy and petal-less. G.

Amaranth Family   Amaranthaceae

7* GREEN AMARANTH *Amaranthus hybridus.* Medium/tall yellow-green hairless annual; much branched. Lvs pointed oval, alternate, untoothed. Fls tiny, 5-petalled, green or red, in very dense erect spikes, often branched, mixed with pointed bracts; July-Oct. Waste places. (B,F,G), from tropical America. **7a* Common Amaranth** *A. retroflexus* is shortly hairy, seeming greyish; fl-spike largely lfless, crowded and always green. From N America. **7b* White Amaranth** *A. albus* is hairless with whitish stems, sharp-tipped lvs and a lfy fl-spike. From N America. **7c* Love Lies Bleeding** *A. caudatus* □, with brilliant red-purple drooping tail-like fl-spikes, is a frequent garden escape. **7d** *A. blitoides, A. bouchonii, A. cruentus, A. graecizans, A. lividus* and other species also occur as casuals in the south of the area. They are a difficult group to identify.

7c* Love Lies Bleeding

Key to Chickweeds and Allies

(Pages 52-56)

(Fls white, 5-petalled).
A group of often confusingly similar plants of the Pink Family, Caryophyllaceae (pp.52-64), all with white or whitish flowers and separate sepals. The shape of the leaves and whether the petals are notched or not are two important clues to identification. Sometimes, as in the stitchworts, the petals are so deeply notched as to appear to be ten instead of five.

Plant

greyish: Thyme-leaved Sandwort 52; Marsh Stitchwort, Umbellate Chickweed, Upright Chickweed 54; Strapwort 56

stickily hairy: Sticky Mouse-ear 54; Corn Spurrey 56

hairs in lines on stem: Common Chickweed, Starwort Mouse-ear 54

hairless: Arctic Sandwort, Fine-leaved Sandwort 52; Sea Spurreys, Procumbent Pearlwort, Knawels, Strapwort, Smooth Rupturewort, Four-leaved Allseed 56

square-stemmed: Greater Stitchwort, Bog Stitchwort 54

stems woody: Curved Sandwort 52

Leaves

all alternate: Strapwort 56

all unstalked: Thyme-leaved Sandwort 52; Greater Stitchwort, Bog Stitchwort 54

linear: Spring Sandwort, Fine-leaved Sandwort 52; Corn Spurrey, Sea Spurreys, Pearlworts, Strapwort 56

curved: Curved Sandwort 52

whorled: Corn Spurrey 56

with stipules: Spurreys 56

tufted: Spring Sandwort 52; Knotted Pearlwort 56

fleshy: Arctic and Sea Sandworts 52; Stitchworts 54; Sea Spurreys, Sea Pearlwort 56

Flowers

tinged green or yellow: Sea Sandwort 52; Lesser Chickweed 54; Procumbent Pearlwort, Ruptureworts 56

pink: Spurreys 56

more than 10 mm: Greater Stitchwort, Water Chickweed, Field Mouse-ear 54

with notched petals: Stitchworts, Chickweeds, Mouse-ears 54

with no petals: Lesser Chickweed 54; Pearlworts, Knawels 56

petals shorter than sepals: Sandworts 52; Bog Stitchwort, Mouse-ears 54; Sea Spurreys, Pearlworts 56

Habitat

mountains: Sandworts 52; Mouse-ears 54; Pearlworts 56

walls: Sandworts 52; Umbellate Chickweed 54

lead-mines: Spring Sandwort 52

salt-marshes: Sea Sandwort 52; Stitchworts 54; Sea Spurreys 56

Pink Family   Caryophyllaceae

pp. 52-64. Lvs in opposite pairs, undivided, usually untoothed. Flg shoots repeatedly forked; petals 4-5, sepals may be separate (pp. 52-56) or fused into a tube (pp. 58-64). Fr dry. Sandworts have fls white (except 7), five *unnotched* petals and three styles.

1* THYME-LEAVED SANDWORT *Arenaria serpyllifolia.* Variable low/short, prostrate to bushy, hairy greyish annual. Lvs pointed oval, unstalked. Fls 5-8 mm, petals shorter than sepals; anthers yellowish; April-Nov. Dry, bare places and walls. T.

2* ARCTIC SANDWORT *Arenaria norvegica.* Low, *tufted* perennial, almost hairless. Lvs oval, rather fleshy, not clearly veined. Fls 8-12 mm, petals longer than sepals, anthers yellowish; June-Aug. Bare places, usually on lime. B, S, scattered, northern. **2a* Fringed Sandwort** *A. ciliata* is not tufted, and slightly hairy, with lvs not fleshy but clearly 1-veined. Mountains. T, rare in B (Ireland). **2b* Balearic Sandwort** *A. balearica* is mat-forming and has rounded lvs. Walls (B) from Mediterranean. **2c** *A. gothica* has narrower lvs. S (Gotland). **2d** *A. humifusa* has runners and much smaller, usually solitary fls with lilac anthers. S, northern.

3* THREE-VEINED SANDWORT *Moehringia trinervia.* Low/short, downy, rather straggling annual. Lvs pointed oval, with three or five *prominent veins.* Fls 6 mm, petals shorter than sepals; May-June. Woods on rich soils. T. **3a Mossy Sandwort** *M. muscosa* is hairless with linear, 1 to 3-veined lvs, and larger fls with four petals longer than sepals. Mountains, F, G.

4* SPRING SANDWORT *Minuartia verna.* Low mat-forming perennial; stems usually downy. Lvs stiff, linear, 3-veined in tufts. Fls 8-9 mm, petals longer than sepals, anthers pink; May-Sept. Dry rocky places, often on lime; old mine-tips. B, F, G, scattered. **4a* Mountain Sandwort** *M. rubella* is much smaller and tufted with almost lfless fl-stems; fls 5-8 mm, solitary, petals shorter than sepals; July-Aug. Mountain rock-ledges. B(rare), S. **4b** *M. biflora* is like 4a but with blunt 1-veined lvs and fls usually in pairs; petals longer than sepals; July. S. **4c** *M. setacea* is more tufted and hairless above. F, G, southern.

5* FINE-LEAVED SANDWORT *Minuartia hybrida.* Low, very slender, erect annual, usually much branched and hairless. Lvs linear, mostly near the base. Fls 6 mm, petals much shorter than narrow white-edged sepals, anthers pink; May-Sept. Dry stony and sandy places. B, F, G. **5a* Teesdale Sandwort** *M. stricta* is a tufted perennial, unbranched, but with several almost lfless fl-stems; fls 8 mm, usually three to a stem, petals equalling sepals; June-July. Wet places in mountains. B(rare), S, scattered. **5b** *M. viscosa* has stems stickily hairy and usually branched from above the middle. F, G, S, south-eastern. **5c** *M. rubra* has fls in a dense terminal cluster. G. **5d** *M. mediterranea* has fls on short crowded stalks. F, western, rare.

6* CURVED SANDWORT *Minuartia recurva.* Short, tufted perennial, slightly downy with woody stems and dense, somewhat *curved* lvs. Petals slightly longer than sepals; June-Oct. Mountains. Ireland (rare).

7* CYPHEL *Minuartia sedoides.* Low, cushion-forming, hairless perennial. Lvs linear, overlapping. Fls almost flush with the cushion, *yellow-green* often without petals, distinguishing it from Moss Campion (p.58); June-Aug. Bare, rocky places, usually in mountains. B, F, G.

8* SEA SANDWORT *Honkenya peploides.* Creeping, rather yellowish perennial; stems and oval lvs *fleshy.* Fls 6-10 mm, greenish-white, solitary among the lvs along the stems, petals not longer than sepals; May-Aug. Sand and shingle by sea. T.

Pink Family *(contd.)*

STITCHWORTS and CHICKWEEDS *Stellaria.* Fls white, petals five, notched or cleft, styles three.

1* GREATER STITCHWORT *Stellaria holostea.* Short, straggly perennial; stems square. Lvs narrow lanceolate, *unstalked.* Fls 20-30 mm, petals split to halfway; April-June. Woods, hedges, on heavier soils. T. **1a* Marsh Stitchwort** *S. palustris* is half the size and greyish; fls 12-18 mm, petals split to base; May-July. Wet, grassy places in lowlands. **1b* Lesser Stitchwort** *S. graminea* ☐ has fls 5-12 mm, petals equalling sepals; May-Aug. Rarely on lime. **1c, 1d, 1e,** p. 299.

2* COMMON CHICKWEED *Stellaria media.* Low/short, very variable, often prostrate annual, with a line of hairs down the weak round stem. Lvs *oval*, the lower long-stalked. Fls 8-10 mm, petals cleft to base, equalling sepals, or absent; all year. Bare and cultivated ground. T. **2a* Wood Stitchwort** *S. nemorum* ☐ is a short perennial with stem hairy all round and petals twice as long as sepals; May-July. Damp woods. Eastern. **2b* Greater Chickweed** *S. neglecta* ☐ is short and sometimes perennial, with fls 10 mm, petals slightly longer than sepals; Apr-July. Cf. 3. Damp shady places. Local. **2c* Lesser Chickweed** *S. pallida* ☐ has all lvs short-stalked, and fls 4-8 mm, always without petals; Mar-May. Sandy places. Southern. **2d* Bog Stitchwort** *S. alsine* ☐ has a square stem, unstalked lvs, fls 4-6 mm, and petals shorter than sepals; May-Aug. Wet places. **2e, 2f,** p. 299.

3* WATER CHICKWEED *Myosoton aquaticum.* Straggling medium perennial, downy above. Lvs pointed oval, the lower stalked. Fls 15 mm, petals deeply split, much longer than sepals; differs from 2b in having five styles; June-Oct. Wet places. T, southern.

4* UMBELLATE CHICKWEED *Holosteum umbellatum.* Low greyish annual. Lvs lanceolate. Fls in umbel-like heads, white; petals ragged, longer than sepals; Mar-May. Dry sandy places, walls. F, G, S, southern; extinct in B.

MOUSE-EARS *Cerastium.* Fls white, five notched petals, five styles (except 5d and 5e).

5* FIELD MOUSE-EAR *Cerastium arvense.* Low perennial, slightly hairy. Lvs narrow lanceolate, not tapering at base. Fls *12-20 mm*, petals twice as long as sepals; April-Aug. Dry open places on lime. T. **5a* Dusty Miller** *C. tomentosum* is densely white-hairy. Waysides. (T), from S E Europe. **5b* Alpine Mouse-ear** *C. alpinum* is shorter with long white hairs, lvs narrowed at base, bracts white-edged, and fls 18-25 mm; June-Aug. Mountain rocks. B, G, S. **5c* Arctic Mouse-ear** *C. arcticum,* as 5b but white hairs much shorter and bracts all green, lflike. **5d* Starwort Mouse-ear** *C. cerastoides* ☐ is hairless except for line of hairs down stem; fls 9-12 mm, styles 3; July-Aug. Mountains. B, S. **5e,** p. 299.

6* COMMON MOUSE-EAR *Cerastium fontanum.* Very variable, low/short perennial, slightly hairy, with lfy non-flowering shoots. Lvs lanceolate. Petals deeply notched, equalling sepals; lower bracts lflike, upper sometimes white-edged; Apr-Nov. Grassland, bare ground. T. **6a* Sea Mouse-ear** *C. diffusum* is a low annual with 3-7 styles and all bracts lflike; Mar-July. Dry places, especially near sea. **6b* Grey Mouse-ear** *C. brachypetalum* is annual with deeply-notched petals shorter than sepals, and all bracts lflike. Dry, open places on lime. T but (B), south-eastern. **6c* Little Mouse-ear** *C. semidecandrum* is a semi-prostrate annual, with slightly notched petals shorter than sepals and all bracts white-edged; Mar-May. Sandy ground. **6d* Dwarf Mouse-ear** *C. pumilum* resembles 6c but is erect and often reddish, with lower bracts lflike; Apr-June. Dry grassland on lime. Southern.

7* STICKY MOUSE-EAR *Cerastium glomeratum.* Low, yellowish, stickily hairy annual. Fls in *tight* heads, rarely opening fully, sepals very hairy; Apr-Oct. Bare ground. T.

8* UPRIGHT CHICKWEED *Moenchia erecta.* Low greyish hairless annual. Fls white with petals not notched, four styles and sepals with broad *white margins*; Apr-June. Bare sandy places. B, F, G.

Pink Family (contd.)

1* CORN SPURREY *Spergula arvensis*. Short/medium, stickily hairy annual. Lvs linear, blunt-tipped, in whorls, channelled on underside, with small stipule at base. Fls 4-8 mm, white, petals 5, not notched, styles 5; May-Sept. Disturbed ground, not on lime. T, lowland.

2* LESSER SEA SPURREY *Spergularia marina*. Low/short, rather fleshy annual. Lvs slightly pointed, with small stipule at base. Fls 5-8 mm, usually pink, with 5 pointed petals shorter than sepals; styles 5; May-Sept. Dry parts of salt-marshes. T, rare inland. **2a* Rock Sea Spurrey** *S. rupicola* is a stickily hairy perennial with shorter lvs and fls 8-10 mm, petals as long as sepals. Cliffs and rocks by sea. B, F, western. **2b* Greater Sea Spurrey** *S. media* □ is an almost hairless perennial; fls 9-12 mm, petals a paler, bluer pink, longer than sepals. Salt-marshes. T, inland in G. **2c* Sand Spurrey** *S. rubra* is a low/short, stickily hairy annual/biennial; lvs not fleshy, ending in a short, stiff hair-point; stipules silvery; fls 3-6 mm, pale pink. Sandy places, not on lime or salty soils. T. **2d, 2e, 2f,** p. 299.

3* KNOTTED PEARLWORT *Sagina nodosa*. Short, tufted perennial. Lvs linear, largest at base, forming stiff *clusters* ("knots") up stem. Fls white, 5-10 mm, petals 5, not notched, much larger than sepals; styles 5; July-Sept. Damp, sandy places, mainly on lime. T.

4* PROCUMBENT PEARLWORT *Sagina procumbens*. Low hairless tufted perennial, spreading from a non-flowering rosette. Lvs linear, ending in a minute bristle. Fls tiny, greenish-white, long-stalked, petals usually 4, much smaller than sepals, often absent; styles usually 4; May-Sept. Sepals spreading when fr ripe. Damp, bare places. T. **4a* Snow Pearlwort** *S. intermedia* agg. is compacter with petals 4-5, white, almost equalling sepals, styles 4-5, and sepals erect when fr ripe. B (rare), S, mountains. **4b* Alpine Pearlwort** *S. saginoides* resembles a large 4a with the habit of 4; fls on central rosette; July-Oct. **4** × **4b*** occurs. B, S, mountains. **4c* Annual Pearlwort** *S. apetala* is a variable low/short lax annual, not forming a rosette; petals usually absent, sepals sometimes erect in fr; Apr-Aug. Dry, sandy places. Southern. **4d* Sea Pearlwort** *S. maritima* is a dark green fleshy annual, like 4c but with lvs blunt, stubby, and sepals not spreading in fr. Open ground. T, coasts. **4e* Heath Pearlwort** *S. subulata* □ is slightly hairy with fls white, petals 5, equalling slightly sticky sepals; styles 5. Dry, sandy places. Scattered.

5* ANNUAL KNAWEL *Scleranthus annuus*. Low, rather spiky-looking annual. Lvs linear, *meeting round stem*. Fls in clusters, no petals, sepals 5, usually pointed, green, with narrow white margin; May-Oct. Dry, sandy places. T. **5a* Perennial Knawel** *S. perennis* is a stouter perennial; branches woody at base; sepals very blunt, with broad white margin. T, rare in B.

6* STRAPWORT *Corrigiola litoralis*. Low greyish annual with stems often reddish. Lvs linear, all *alternate*. Fls white, minute and crowded, petals 5, as long as often partly reddish sepals; June-Oct. Damp, sandy and gravelly places. T, southern, scattered, rare in B.

7* SMOOTH RUPTUREWORT *Herniaria glabra*. Variable prostrate bright green, more or less hairless annual/perennial. Fls *green*, in clusters at base of oval lvs, petals green, blunt; May-Oct. Fr pointed, longer than sepals. Dry, bare places, often on lime. T, southern, rare in B. **7a* Fringed Rupturewort** *H. ciliolata* is a slightly hairy perennial; fr blunt, shorter than sepals; June-Aug. Maritime sand and rocks. B (rare), F, south-western. **7b,** p. 299.

8* CORAL NECKLACE *Illecebrum verticillatum*. Prostrate hairless annual. Resembles 7 but has white fls with pointed petals; June-Oct. Damp, sandy ground, not on lime. B (rare), F, G, southern.

9* FOUR-LEAVED ALLSEED *Polycarpon tetraphyllum*. Low hairless branching annual. Lvs oval, apparently in *whorls* of four. Fls in clusters, white, tiny, petals shorter than sepals; June-Aug. Sandy places. B (rare), F, south-western. **9a*** *P. diphyllum* is unbranched, the often purplish lvs in pairs. Mainly coastal.

Pink Family *(contd.)*

CAMPIONS and CATCHFLIES *Silene* and *Lychnis* (pp.58-60) mostly have conspicuous fls with five petals, and sepals joined in a tube.

1* BLADDER CAMPION *Silene vulgaris.* Medium greyish perennial, usually hairless; all shoots flowering, woody at base. Lvs pointed oval, often wavy-edged. Fls white, petals deeply cleft, sepal-tube inflated to form a bladder; many fls in each cluster; styles 3; May-Sept. Arable, waste, and grassy places, especially on lime. T. **1a* Sea Campion** *S. maritima* □ is short, with non-flowering shoots and broader petals; clusters few-fld. Cliffs and shingle by sea, also on mountains.

2* NOTTINGHAM CATCHFLY *Silene nutans.* Variable medium unbranched perennial, downy below and sticky above. Lvs broader at tip. Fls white, drooping, opening and fragrant at night, all usually pointing one way; petals *rolled back,* very narrow and deeply cleft; May-Aug. Dry, undisturbed places. T. **2a* Italian Catchfly** *S. italica* is branched, with erect fls not all pointing one way; petals less recurved. (B, rare), F, local.

3 WHITE STICKY CATCHFLY *Silene viscosa.* Medium biennial/perennial, densely covered with *sticky hairs.* Lvs narrow oval, the lower with wavy margins. Fls white, 20 mm, in a long spike, petals deeply cleft, sepals hairy; June-July. Dry, grassy places. G, S, eastern. **3a** *S. tatarica* is almost hairless, with non-flowering shoots, narrower lvs, smaller fls and hairless sepals; Aug.

4* SPANISH CATCHFLY *Silene otites.* Medium perennial, stickily hairy near the base. Lvs broadest at tip, in rosettes. Fls small, 3-4 mm, *yellowish*, apparently in whorls, flhds very unlike other catchflies; stamens and styles often on different plants; June-Sept. Dry, sandy places. B(rare), F, G, south-eastern.

5* MOSS CAMPION *Silene acaulis.* Low, hairless, slightly woody perennial, forming characteristic bright green, moss-like *cushions,* often covered with the solitary, pink fls; June-Aug. Wild azalea (p.172) has duller green lvs and very different fls, and Cyphel (p.52) has less pointed lvs and yellowish fls. Bare places in mountains, and near sea-level in north. B, S.

6* SWEET WILLIAM CATCHFLY *Silene armeria.* Short hairless annual with all shoots flowering. Lvs lanceolate, greyish, the upper clasping the stem. Fls bright pink, clustered in rather flat-topped heads, petals only notched; June-Sept. Dry, often shady places, and as a garden escape. F, G, (B, S).

7 FLAXFIELD CATCHFLY *Silene linicola.* Medium annual; stem *rough,* branching. Lvs linear. Fls pink, with red-purple stripes, petals deeply cleft, sepals clearly veined, inflated, contracted at tip; June-July. Flax fields. F, G.

8 ROCK CATCHFLY *Silene rupestris.* Short, slender, hairless, *greyish* perennial. Lvs lanceolate. Fls white or pink, on long stalks, petals notched, sepal-tube 10-veined; June-Sept. Dry places, often in mountains. S; rare and southern in F, G.

9 NORTHERN CATCHFLY *Silene wahlbergella.* Short, unbranched, slightly hairy perennial with long, narrow lvs and solitary fls. Sepal-tube *inflated* and *longer* than the inconspicuous, purplish petals; June-Aug. Mountain meadows on lime. S. **9a** *S. furcata* is stickily hairy and branched, with smaller fls and less inflated sepal-tube.

10* FORKED CATCHFLY *Silene dichotoma.* Medium/tall annual, all stems flowering. Lvs lanceolate. Fls 15 mm, white, occasionally pink, petals *deeply cleft*; sepals hairy, not sticky; May-Aug. Waste places, (B casual), F, G.

1

1a

2

3

4

5

6

7

8

9

10

Pink Family *(contd.)*   CAMPIONS AND CATCHFLIES *(contd.)*

1* RED CAMPION *Silene dioica.* Medium/tall hairy perennial. Lvs pointed oval, the basal ones stalked. Fls bright *pink,* sometimes much paler as a result of hybridisation with 2□, giving a complete range of shades; petals cleft; calyx-teeth triangular, pointed; stamens and styles in separate fls, styles 5; Mar-Nov. Teeth on seed-pod rolled back. Rich soils especially in or near woods or shady hedges. T.

2* WHITE CAMPION *Silene alba.* Medium/tall, much branched, stickily-hairy perennial, resembling 1, but fls longer and *white,* calyx-teeth narrow lanceolate, blunt, and teeth on seed-pod upright; May-Oct. Hedges and bare places, often on cultivated ground. T.

3* NIGHT-FLOWERING CATCHFLY *Silene noctiflora.* Medium, stickily hairy annual. Lvs pointed oval. Fls pink, opening at night, and petals rolling inwards by mid-morning to show their *yellowish* undersides; styles *three,* in same fls as stamens. Teeth on seed-pod turned back. Arable fields. T, eastern.

4* RAGGED ROBIN *Lychnis flos-cuculi.* Medium/tall, rough-stemmed perennial. Lvs lanceolate. Fls bright pink, the characteristic irregularly and narrowly 4-lobed petals giving them a *ragged* appearance; May-Aug. Damp meadows and marshy places. T.

5* STICKY CATCHFLY *Lychnis viscaria.* Medium tufted perennial with lfy flowering shoots, *sticky* just below each lf-junction. Lvs lanceolate. Fls 20 mm, bright rosy red, petals notched, apparently *whorled,* in long spikes; May-Aug. Dry, rock places, rarely on lime. T, upland, rare in B.

6* ALPINE CATCHFLY *Lychnis alpina.* Short tufted perennial with lanceolate lvs and smooth stems ending in a compact cluster of fls, smaller (6-12 mm) and pinker than in 5, and *deeply* notched; June-July. Bare places in mountains. B(rare), S.

7* CORN COCKLE *Agrostemma githago.* Tall, hairy annual. Lvs narrow lanceolate. Fls pale red-purple, petals notched, sepals hairy, forming a tube with five *long,* narrow, spreading teeth, projecting well beyond the petals; May-Aug. Cornfields. T, southern, decreasing.

8* SOAPWORT *Saponaria officinalis.* Medium/tall, often rather straggling, hairless perennial, with runners. Lvs lanceolate. Fls soft pink, the petals not notched, apparently stalked, and *standing clear* of the scarcely inflated sepals; styles 2; June-Sept. Hedges and woods by streams, waysides. T, lowland, but (B). A double-fld form□ is a frequent escape in B.

1 × 2* Hybrid between Red and White Campions

8* Double-flowered Soapwort

Pink Family (contd.)

1* SMALL-FLOWERED CATCHFLY *Silene gallica.* Short/medium, stickily hairy annual. Lvs lanceolate, the lower broadest near tip. Fls 10-12 mm, mostly pointing in one direction; petals notched, white or pink, sometimes with a deep red spot (var. *quinquevulnera* □); June-Oct. Sandy places. B,F,G.

2* SAND CATCHFLY *Silene conica.* Short, stickily hairy, greyish annual. Lvs narrow lanceolate. Fls 4-5 mm, variably pink, the sepal-tube soon *swelling,* clearly veined but shorter than petals; May-June. Sandy places, mostly near the coast. B,F,G, southern.

3* BERRY CATCHFLY *Cucubalus baccifer.* Medium/tall downy perennial with rather brittle branches. Fls c18 mm, drooping, greenish-white, petals deeply cleft; July-Sept. Fr □ a round, *black berry.* Shady places. (B),F,G, eastern.

4 FASTIGIATE GYPSOPHILA *Gypsophila fastigiata.* Very variable low/tall perennial, hairless below, downy above. Lvs linear, 20-80 mm. Fls 5-8 mm, *white or lilac,* in a dense, *flat-topped* head, petals notched, less than twice as long as sepals; styles 2; June-Sept. Dry rocks. G,S, southern, scattered. **4a** *G. repens* is shorter and completely hairless, with lvs often curved and petals twice as long as sepals. Mountains. G, southern.

5 ANNUAL GYPSOPHILA *Gypsophila muralis.* Low/short, usually hairless annual. Lvs linear, 5-25 mm, greyish. Fls 4 mm, *pink,* in a *loose cluster,* petals slightly notched, twice as long as sepals; styles 2; June-Oct. Damp woods and meadows. F,G,S, southern.

6* COW BASIL *Vaccaria pyramidata.* Medium greyish hairless annual with a regularly branching stem. Lvs oval. Fls pink, 10-15 mm, long-stalked, with an inflated, *winged* sepal-tube; June-July. Arable fields, often on lime. (B, casual), F,G.

7* TUNIC FLOWER *Petrorhagia saxifraga* Short/medium, usually hairless perennial. Lvs linear. Fls pale pink or white, usually solitary, on long stalks, in a loose head, petals notched; June-Aug. Dry sandy places and walls. (B,rare),F,G, southern.

8 PROLIFEROUS PINK *Petrorhagia prolifera.* Short/medium hairless annual. Lvs linear, fused into a sheath at the base, about as long as wide. Fls pink, in *dense clusters,* each surrounded by several large, *brown, papery bracts*; fls opening one or two at a time; petals notched; May-Sept. Dry, open places, usually on lime. F,G,S, southern. **8a* Childling Pink** *P. nanteullii* is very similar with a slightly downy stem and longer leaf-sheaths, twice as long as wide. B(rare),F.

3* Fruit of Berry Catchfly

63

Pink Family *(contd.)*

PINKS *Dianthus* are perennials (except 7) with stiff, greyish linear lvs. Fls white, pink, or red, with sepals fused into an unribbed tube, surrounded by an epicalyx of 2-6 scales.

1* CHEDDAR PINK *Dianthus gratianopolitanus.* Short, hairless, densely tufted, with *long,* creeping sterile shoots. Lvs *rough-edged.* Fls 20-30 mm, solitary, fragrant, pale pink, petals shortly toothed; epicalyx with 2-4 short segments, pointed; May-July. Sunny rocks, often on lime. B(rare),F,G, scattered. **1a** *D. seguieri* is not greyish and has smaller, deeper pink fls, 2-4 together. G, southern.

2* CLOVE PINK *Dianthus caryophyllus.* Short/medium, hairless, tufted, with *smooth-edged* lvs; sterile shoots short. Fls 35-40 mm, sometimes in loose groups of up to five, fragrant, pink, petals *shortly toothed;* epicalyx as 1; July-Aug. Walls. (T) from S Europe. Also widely cultivated as Carnation.

3* WILD PINK *Dianthus plumarius.* Short, with the habit of 2 but *rough-edged* lvs. Fls 25-35 mm, fragrant, pink or white, the petals *cut to half-way* into narrow, feathery lobes; epicalyx with 2-4 pointed segments up to ⅓ as long as sepal tube; June-Aug. Walls. (T, southern) from S E Europe.

4 JERSEY PINK *Dianthus gallicus.* Short/medium, loosely tufted, *downy.* Lvs stiff, very short, not more than 15 mm. Fls 20-30 mm, mauve pink, the petals as in 3, but less deeply lobed; epicalyx as in 1; June-Aug. Coastal sand dunes. F.

5 LARGE PINK *Dianthus superbus.* Medium/tall, branching, hairless. Fls *30-50 mm,* pink, the petals as in 3 but divided *almost to the base;* epicalyx long, pointed; June-Sept. Dry, often shady places. F,G,S, south-eastern. **5a** *D. arenarius* has narrower lvs and white fls. G,S, eastern.

6* MAIDEN PINK *Dianthus deltoides.* Short/medium, loosely tufted, with *short,* creeping sterile shoots; fl-stems *roughly hairy.* Lvs short, roughly hairy at the edges. Fls 15-20 mm, pale pink, spotted, petals toothed; epicalyx usually with two long-pointed scales half as long as the sepal-tube; June-Sept. Dry grassy, usually sandy places. T.

7* DEPTFORD PINK *Dianthus armeria.* Medium stiff hairy *annual,* dark green, not greyish. Lvs shortly sheathed at the base. Fls 8-15 mm, bright pink or red, in dense 2-10-fld clusters, surrounded by *long, green, lfy bracts;* epicalyx with two scales as long as the softly hairy sepal-tube; June-Aug. Dry sandy places and tracksides. T, southern.

8 CARTHUSIAN PINK *Dianthus carthusianorum.* Variable, medium, hairless, with lf-sheaths several times longer than the diameter of the stem. Fls *c.* 20 mm, usually bright pink or red, in dense clusters surrounded by *short brown bracts;* epicalyx half as long as dark sepal-tube; May-Aug. Dry grassy places and open woods. F,G, eastern. **8a* Sweet William** *D. barbatus* □ is the familiar garden plant with very dense, many-fld flat heads and long-pointed epicalyx scales longer than sepal-tube. (T) from S Europe.

8a* Sweet William

Buttercup Family Ranunculaceae

pp. 66-76. Includes a very mixed bag of rather primitive plants, all with many stamens and normally five, usually very prominent, petals or petal-like sepals. The fls often contain small honey-lvs or nectaries, which secrete nectar.

1* STINKING HELLEBORE *Helleborus foetidus.* Tall foetid perennial with over-wintering stems. Lvs palmate, simply divided, all *on the stem,* the uppermost undivided. Fls *bell-shaped,* petal-less, sepals yellow-green, purple-tipped, in clusters; Jan-May. Dry woods and scrub, mainly on lime. B, F, G, southern.

2* GREEN HELLEBORE *Helleborus viridis.* Short perennial with *two root lvs* that die before winter; all lvs palmately divided, except in the south and east where forms with undivided lvs occur. Fls spreading, larger than 1, petal-less, the sepals dull green like the lvs; Feb-Apr. Damp woods in thick leaf-mould. B, F, G, southern.

3* LOVE-IN-A-MIST *Nigella damascena.* Short, branched annual. Lvs divided into *thread-like* segments. Fls solitary, with five pale blue sepals, *surrounded* by feathery lvs; June-July. Dry, open places. (B, casual, F), from Mediterranean. **3a****N. arvensis* is variable with fls often veined green and long-stalked, not surrounded by a web of lvs; June-Sept. Arable fields on lime. (B, casual), F, G.

4* WINTER ACONITE *Eranthis hyemalis.* Low hairless perennial. Lvs palmately lobed, all from roots and appearing after fl-stem has died down, except for three which form a *ruff* immediately below the solitary fl; sepals yellow, usually 6; Jan-Mar. Woods, naturalised from gardens. (B, F, G, scattered), from S Europe. Buttercup Family contd. on p. 67.

Water-lilies Nymphaeaceae

Hairless perennials of stagnant and slow-flowing fresh water. They root in the mud and have large floating lvs on long stalks, and conspicuous fls.

5* YELLOW WATER-LILY *Nuphar lutea.* Lvs oval, larger than 6, up to 40 cm, and some thin and *submerged.* Fls yellow, much smaller than 6, to 60 mm, held a few cm above the water surface; June-Sept. Still or slow-flowing water, sometimes deeper than 6. T, lowland. **5a* Least Water-Lily** *N. pumila* □ is smaller with fls to 30 mm and larger gaps between the petals; June-July. Hybridises with 5 and intermediates occur, sometimes in the absence of both parents. Stagnant water, mainly upland.

6* WHITE WATER-LILY *Nymphaea alba.* All lvs *floating,* almost circular, rather broader than the fls, the lobes at the base not overlapping. Fls 50-200 mm, white, fragrant; June-Sept. Still, shallow water. T, southern in S. **6a** *N. candida* □ is rather smaller with basal lobes of lvs overlapping, leaving little or no gap and deeper yellow stamens. F, G, S, scattered.

6a *Nymphaea candida*

Buttercup Family *(contd.)*

1* GLOBE FLOWER *Trollius europaeus.* Short/medium hairless perennial. Lvs palmate, deeply cut. Fls large, almost *spherical,* petal-less, the ten yellow sepals curving in at the top; May-Aug. Damp, grassy places, mainly upland. T.

2* MARSH MARIGOLD *Caltha palustris.* Short stout hairless perennial; dwarf and creeping on mountains. Lvs large, *kidney-shaped,* toothed, dark green, shiny, and often mottled paler on the upper surface. Fls variable, 10-50 mm, with five yellow sepals and no petals; Mar-Aug. Fr rather pod-like, clustered, conspicuous in summer. Wet places. T.

BUTTERCUPS *Ranunculus* mostly have shiny yellow fls (except Large White and Glacier, p.70) and both petals and sepals, distinguishing them from 1 and 2.

3* MEADOW BUTTERCUP *Ranunculus acris.* Medium/tall hairy perennial with characteristic 2-7-lobed lvs, rather rounded in outline, the end lobe *unstalked* (cf. 4 and 5). Fl-stalks unfurrowed, sepals *erect,* nectaries golden-yellow; Apr-Oct. Meadows, damp grassy places. T. **3a** *R. polyanthemos* is strongly branched, and less hairy, with 5-lobed lvs and narrow lf-segments; May-July. Damp woods on lime. F,G,S, eastern. **3b** *R. nemorosus* is like 3a but has 3-lobed lvs and broader lf-segments; beak on seed longer. Meadows. F,G,S (Oland). **3c Woolly Buttercup** *R. lanuginosus* is very hairy with lvs less deeply divided; nectaries orange-yellow; May-Aug. Damp woods. G.

4* BULBOUS BUTTERCUP *Ranunculus bulbosus.* Short, very hairy perennial with stem-base swollen. Lvs like 3 but less deeply lobed and end lobe *stalked.* Sepals *turned downwards,* fl-stalks furrowed; Mar-June. Drier grassland than 3 and 5, and prefers lime. T. **4a* Hairy Buttercup** *R. sardous* is a hairier annual with stem-base not swollen, smaller, often shiny lvs, and paler fls; May-Oct. Damp arable land, grassy places. T, southern. **4b** *R. illyricus* is medium and less hairy, with very narrow lf-segments. G,S (Gotland).

5* CREEPING BUTTERCUP *Ranunculus repens.* Short/medium, slightly hairy, creeping perennial, with rooting *runners.* Lvs rather triangular in outline, end lobe *stalked.* Resembles 4 more than 3, but sepals erect and fl-stalks not furrowed; May-Sept. Damp, often bare places. T. **5a Jersey Buttercup** *R. paludosus* is hairier and not creeping; seeds with long hooked beak. F. **5b** *R. montanus* agg. has no runners and stem lvs more divided. Mountains. G.

6* GREATER SPEARWORT *Ranunculus lingua.* Tall hairless perennial with long *runners.* Lvs to 25 cm long, *lanceolate,* toothed. Fls *more* than 20 mm, stalks not furrowed; June-Sept. Marshes and fens, becoming rarer. T, scattered.

7* LESSER SPEARWORT *Ranunculus flammula.* Very variable, usually short/medium, hairless perennial, sometimes creeping and then rooting at some lf-junctions of the often reddish stem, but with *no* runners. Lvs *lanceolate,* short- or unstalked, to 4 cm, more or less toothed. Fls *7-20 mm,* stalks furrowed; June-Oct. Wet places. T. **7a* Creeping Spearwort** *R. reptans* has runners and roots at every lf-junction; lvs long-stalked, broader; fls 5 mm; June-Aug. Hybridises with 7. Gravelly lake-shores. T but scattered, rare in B; mainly in mountains. **7b* Adderstongue Spearwort** *R. ophioglossifolius* is a short erect annual with broader lvs and fls 6-9 mm; May-Aug. Marshes. B (rare), F, south-western. **7c** *R. hyperboreus* is like 7a with broader, 5-lobed lvs and 3 petals. Mountains. S.

8* LESSER CELANDINE *Ranunculus ficaria.* Low/short hairless perennial with rather fleshy, dark green, heart-shaped lvs. In shady places may have bulbils at base of lf-stalk. Usually 8-12 narrow petals, whitening when old; Mar-May. Woods, hedge-banks, bare damp ground. T. **8a** *R. lapponicus* is creeping and rooting at lf-junctions, with 3-lobed lvs, 6-8 petals and down-turned sepals. Mountains. S. **8b** *R. cymbalaria* is smaller with 5-petalled fls. (S).

Buttercup Family *(contd.)*

BUTTERCUPS *Ranunculus (contd.)*

1* GOLDILOCKS BUTTERCUP *Ranunculus auricomus.* Short, slightly hairy perennial with lowest lvs only slightly lobed. Fls few, sepals purple-tinged, petals 0-5, often *distorted* and sometimes *absent,* otherwise resembling a slender Meadow Buttercup (p. 68); Apr-May. Woods, hedges. T. **1a, 1b,** p. 299.

2* CELERY-LEAVED BUTTERCUP *Ranunculus sceleratus.* Medium hairless annual with palmately lobed shiny lvs. Fls numerous, small, petals *no longer* than down-turned sepals; May-Sept. Fr head conical, with several hundred seeds. In or near slow or stagnant water. T, lowland. **2a,** p. 299.

3* CORN BUTTERCUP *Ranunculus arvensis.* Short/medium hairless annual with small (4-12 mm) pale yellow fls on unfurrowed stalks, and spreading sepals; May-July. Fr *spiny,* strongly ridged. Cornfield weed, often in rather damp places on lime. T, southern.

4* SMALL-FLOWERED BUTTERCUP *Ranunculus parviflorus.* Short hairy, usually sprawling annual, somewhat unbuttercup-like, though most resembling a small Hairy Buttercup (p. 68); stem lfy. Root-lvs circular. Fls tiny (3-6 mm), on furrowed stalks, petals no longer than down-turned sepals; May-July. Bare, dry places on lime. B, F, south-western.

5 LARGE WHITE BUTTERCUP *Ranunculus platanifolius.* Tall perennial. Lvs large, shallowly 5-7-lobed. Fls 20 mm, *white,* on long hairless stalks; May-Aug. Damp woods, mainly lowland. F, G. **5a,** p. 299.

6 GLACIER BUTTERCUP *Ranunculus glacialis.* Low hairless perennial with thick 3-5-lobed lvs. Fls large, solitary or few, white or pinkish, sepals with purple-brown hairs; June-Aug. Stony ground in mountains near snow-line, not on lime. S.

WATER CROWFOOTS *(Ranunculus* subgenus *Batrachium)* grow in all freshwater habitats from mud to swift water. Lvs palmately lobed when floating and/or much divided and feathery when submerged. Fls white, petals with yellow claw (except 8c). Most are very similar, and often best identified on habitat and lf-type.

7* COMMON WATER CROWFOOT *Ranunculus aquatilis.* Very variable annual/perennial, with submerged lvs branching in three planes and usually also *toothed* floating lvs. Fls 10-20 mm, nectaries circular. Fr-stalk shorter than lf-stalk; Apr-Sept. Still or slow water to 1 m deep. T. **7a* Pond Water Crowfoot** *R. peltatus* has more shallowly toothed floating lvs, longer (20-30 mm), pear-shaped nectaries and stalk longer than lf-stalk. Western. **7b* Brackish Water Crowfoot** *R. baudotii* is very like 7a but has more rigid lf-segments, and smaller (10-20 mm) fls sometimes with blue-tipped sepals. Brackish water. **7c* Thread-leaved Water Crowfoot** *R. trichophyllus* has no floating lvs and short, stiff, fan-like submerged lvs, repeatedly branching in three; fls 5-10 mm, nectaries half-moon-shaped. Sometimes in brackish water. **7d* Fan-leaved Water Crowfoot** *R. circinatus* is like 7c, but with lvs branching in two dimensions only, and larger (10-20 mm) fls. **7e* Chalk-stream Water Crowfoot** *R. penicillatus* usually lacks floating lvs and is larger and robust, with long trailing stems; fls 20-30 mm, nectaries pear-shaped; fr-stalk longer than lf-stalk. Fast streams, especially in limestone areas. **7f* River Water Crowfoot** *R. fluitans* is like 7e, but never has floating lvs; submerged lvs very dark green with long narrow thread-like segments. Rarely in limestone areas.

8* IVY-LEAVED CROWFOOT *Ranunculus hederaceus.* Creeping annual/perennial, with all lvs ivy-shaped. Fls 3-6 mm, petals equalling sepals; May-Sept. Shallow water, rarely in limestone areas. T, western. **8a* Round-leaved Crowfoot** *R. omiophyllus* has petals twice as long as sepals; fls 8-12 mm. B, F, western. **8b* Three-lobed Crowfoot** *R. tripartitus* is more delicate perennial, with both finely-divided submerged lvs and rounder floating lvs; sepals blue-tipped; Mar-June. B, F, G, western. **8c** *R. ololeucos* is like 8b but petals all white, more than twice as long as sometimes green-tipped sepals; May-July. Shallow water. F, G, western.

1

2

3

4

5

6

7

8

Buttercup Family (contd.)

MEADOW-RUES *Thalictrum* have fls in clusters with four inconspicuous petals and no sepals but long stamens which give the fls their colour and rather shaggy appearance. Lvs 2-3-pinnate.

1* COMMON MEADOW-RUE *Thalictrum flavum*. Tall, almost hairless perennial. End lflet *longer* than broad. Fls erect, densely crowded, with whitish petals and long yellow stamens. June-Aug. Wet meadows. T. **1a Small Meadow-rue** *T. simplex* is usually somewhat smaller, with lflets rather narrower, and fls drooping at first. Meadows. G, S, northern. **1b** *T. lucidum* is tufted with narrower lflets. G. **1c** *T. morisonii* has shiny stems, upper lflets narrow and untoothed and fls in more open clusters. G (rare), southern.

2* GREATER MEADOW-RUE *Thalictrum aquilegifolium*. Tall hairless perennial. Lflets about as long as broad. Fls larger than 1 and 3, greenish-white petals and very long, *broad* lilac-coloured stamens; June-July. Woods, meadows. (B), G, S, eastern.

3* LESSER MEADOW-RUE *Thalictrum minus*. Variable medium (in dry places) to tall (in wet places) perennial. End lflet roughly *as long as broad.* Fls yellowish or purplish, in loose clusters; June-Aug. Sand dunes, turf, rocks and mountain ledges, and by streams, mainly on lime. T, scattered.

4* ALPINE MEADOW-RUE *Thalictrum alpinum*. The smallest meadow-rue, a short delicate unbranched perennial. Lvs 2-trefoil, end lflet rounded. Fls purple, in an unbranched stalked spike; stamens violet with yellow anthers; May-July. Damp turf, ledges, usually in mountains. B, S, northern.

5* BANEBERRY *Actaea spicata*. Medium hairless perennial, strong smelling, rather umbellifer-like, with large, long-stalked 2-pinnate or 2-trifoliate root lvs. Fls white, in stalked spikes; petals short, usually 4, stamens long, *white*, conspicuous; May-July. Fr a large, shiny *black berry*. Woods and limestone pavements. T, eastern, rare in B. **5a** *A. erythrocarpa* has more deeply cut lvs and smaller red berries. S, northern.

6* TRAVELLER'S JOY *Clematis vitalba*. Tall clambering woody perennial, sometimes with immensely long stems. Lvs pinnate, often ending in twining tendrils. Fls fragrant, with four greenish petals and conspicuous stamens; July-Sept. Fr □ with long, grey, hairy *plumes,* in dense clusters, persisting through the winter, giving rise to the name Old Man's Beard. Hedges and scrub on lime. B, F, G, southern. **6a** *C. recta* □ looks very different, being erect and not woody; fls white. G (rare), southern.

7 ALPINE CLEMATIS *Clematis alpina*. A much smaller plant than 7, with 2-pinnate lvs and larger, solitary, showy *violet* fls; May-July. Fr with similar hairy plumes. Rocky mountain woods. G, S, rare.

6a *Clematis recta*

6 Traveller's Joy, fr.

Buttercup Family (contd.)

1* COLUMBINE *Aquilegia vulgaris.* Tall, branching perennial with 2-trifoliate lvs, the lflets rounded, dull green. Fls characteristic, each petal with a *long spur*, usually violet, but sometimes pink or white, especially when escaped from gardens; May-July. Woods and scrub, on lime. T.

2* MONKSHOOD *Aconitum napellus.* Tall hairless perennial, with lvs palmately divided almost to the midrib. Fls large, *violet or blue*, broadly hooded, in long spikes; May-Sept. Damp woods and by streams. B,F,G, southern. **2a** *A. firmum* has lvs less deeply divided and fewer fls. G. **2b** *A. variegatum* □ has fls blue, white or variegated, the hood taller, almost conical. Mountain woods. G, southern.

3 WOLFSBANE *Aconitum vulparia.* Tall hairless perennial with lvs palmately divided to midrib. Fls pale yellow, narrowly hooded, in branched spikes; June-Aug. Very shady, damp woods. F,G, eastern. **3a Northern Wolfsbane** *A. septentrionale* has fls hairy, usually violet, sometimes yellow, the hood broader at the base. S.

4* FORKING LARKSPUR *Consolida regalis.* Short, downy annual; widely branched. Lvs much divided into many very narrow segments. Fls deep purple, *long-spurred* in a loose spike; fl-stalks much longer than their lf-like bracts; June-Aug. Fr hairless. Arable fields on lime. T, but casual in B. **4a* Eastern Larkspur** *C. orientalis* is larger and more branched; fls blue in a dense spike; fl-stalks shorter than bracts; fr hairy. (B, rare, F,G), from S Europe. **4b* Larkspur** *C. ambigua* resembles 4a but is downy with fl-stalks as long as bracts. (B, casual), F.

5 YELLOW PHEASANT'S-EYE *Adonis vernalis.* Short perennial, stem scaly at base. Lvs 2-3-pinnate, feathery. Fls large, 40-80 mm across, *yellow*, with more than ten petals; Apr-May. Dry grassland. F,G,S, south-eastern.

6 LARGE PHEASANT'S-EYE *Adonis flammea.* Short/medium annual, like 7, but fls larger (25-30 mm) and sepals *hairy* at least at base and pressed against petals; June-July. Arable fields on lime. F,G.

7 SUMMER PHEASANT'S-EYE *Adonis aestivalis.* Short hairless annual with feathery 3-pinnate lvs, the lower stalked. Fls *red* (or yellow, var. *citrina*) 15-25 mm, with prominent black centres and black anthers; sepals hairless, touching with 5-8 spreading petals; June-July. Arable fields on lime. F, G. **7a* Pheasant's-Eye** *Adonis annua* has fls always red; June-Aug. Lower lvs unstalked. B,F,G, southern.

8* MOUSETAIL *Myosurus minimus.* Low hairless annual. Lvs linear, slightly fleshy, all in a basal tuft. Fls small, solitary, long-stalked, 5-7-petalled, greenish-yellow, with a greatly elongated plantain-like fr head, fancifully likened to a mouse's tail. Bare, usually damp ground. T.

2b *Aconitum variegatum*

Buttercup Family (contd.)

ANEMONES, *Anemone* and *Pulsatilla,* have conspicuous, solitary fls, opening in spring, with brightly coloured sepals and no petals. *Pulsatilla* usually has six, and *Anemone* more than six sepals.

1* PASQUE FLOWER *Pulsatilla vulgaris.* Low hairy perennial. Lvs 2-pinnate, feathery, covered with long hairs. Fls large, purple, *bell-shaped,* 50-80 mm, erect at first, later drooping, anthers bright yellow; Mar-May. Fr□ with long, silky plumes, sometimes persisting into late summer. Dry grassland on lime. T, southern.

2 PALE PASQUE FLOWER *Pulsatilla vernalis.* Short hairy perennial. Lvs much less divided than in 1, with broader lobes and fewer, shorter hairs. Fls similar to 1, but *white* inside and violet or pink outside, drooping at first, later erect; Apr-June. Stem lengthening in fr. Meadows, especially in mountains. F,G,S, eastern.

3 SMALL PASQUE FLOWER *Pulsatilla pratensis.* Low hairy perennial, much resembling 1, but fls smaller (40 mm), always *drooping;* dark purple in the north, paler southwards; May. Root lvs very feathery, stem lengthening in fr. Meadows. G,S.

4 EASTERN PASQUE FLOWER *Pulsatilla patens.* Differs from 1 in having broadly lobed root lvs, and *spreading* petals, so that fl is not bell-shaped; Apr-May. Meadows. S.

5* WOOD ANEMONE *Anemone nemorosa.* Low/short hairless perennial. Stem lvs 3, long-stalked, deeply palmately lobed, in a whorl just below the fl; root lvs similar, appearing after fls, few, sometimes absent. Fls solitary, white, 20-40 mm, sometimes pinkish, especially on the underside of the 6-12 hairless sepals; Mar-May. Deciduous woods, except on very acid soils, mountain meadows. T. **5a* Blue Anemone** *A. apennina*□ has blue-purple fls with 8-18 sepals; Mar-Apr. (B), F, G. **5b** *A. narcissiflora* □ has smaller fls 3-8 together in an umbel. Mountains. G (rare), southern.

6* YELLOW ANEMONE *Anemone ranunculoides.* Short hairy perennial. Lvs like 5, those on the stem very short-stalked. Fls bright *yellow,* buttercup-like, 15-20 mm, usually solitary and with 5-8 sepals, hairy beneath; Mar-May. Woods. T, but (B).

7 SNOWDROP WINDFLOWER *Anemone sylvestris.* Short/medium hairy perennial. Lvs palmately lobed. Fls white, to 70 mm, with only *five* sepals, hairy beneath, anthers yellow; Apr-June. Dry woods on lime. F,G,S, southern.

8 HEPATICA *Hepatica nobilis.* Low/short hairless perennial. Lvs *3-lobed,* untoothed, evergreen and purplish below. Fls solitary, 15-25 mm, usually blue-purple, occasionally pink or white, with 6-9 sepals and three sepal-like bracts immediately below them; anthers whitish; Mar-Apr. Woods and scrub on lime. F,G,S.

5a* Blue Anemone

5b *Anemone narcissiflora*

1* Pasque Flower

Barberry Family Berberidaceae

1* BARBERRY *Berberis vulgaris*. Deciduous shrub to 4 m; stem with 3-pointed *spines*. Lvs oval, sharply toothed. Fls yellow, in drooping spikes; May-June. Fr a bright red berry. Hedges, dry woods, often on lime. T, southern. **1a* Oregon Grape** *Mahonia aquifolium* ☐ is much shorter, with spineless stem, lvs pinnate and spine-toothed, fls fragrant, in erect spikes, and fr blue-black; Mar-May.

Fumitory Family Fumariaceae

Mostly rather floppy, hairless plants with lvs several times pinnately divided, and characteristic tubular, two-lipped, spurred fls. Fumitories *Fumaria* are mostly annual weeds of arable fields and waste places.

2* YELLOW CORYDALIS *Corydalis lutea*. Short perennial with lfy stems and lvs ending in a lflet. Fls bright *yellow*, in dense spikes opposite upper lvs; May-Sept. Fr drooping. Walls. (B,F,G), from S Europe. **2a*** *C. ochroleuca* ☐ has lf-stalks narrowly winged, cream-coloured fls, and an erect fr. Walls and rocks.

3* BULBOUS CORYDALIS *Corydalis solida*. Short perennial with only *1-3 lvs* on the stem and a scale below the lowest lf; lflets in threes. Fls 18-30 mm long, pink-purple, more than ten to each rather lax, *terminal* spike; bracts lobed; Apr-May. Woods, hedges. T but (B). **3a*** *C. bulbosa* is rather larger with no scale below the lowest lf; bracts not lobed, fl-spikes dense. **3b** *C. intermedia* has fls 10-15 mm long, only 2-8 to a spike, and unlobed bracts. F,G,S. **3c** *C. pumila* is like 3b with lobed bracts. G,S.

4* CLIMBING CORYDALIS *Corydalis claviculata*. Medium/tall, *climbing* annual, with lvs ending in a tendril. Fls 5-6 mm long, creamy, 6-8 in dense spikes opposite lvs; June-Sept. Rocky, often wooded places, not on lime. T, western.

5* COMMON FUMITORY *Fumaria officinalis* agg. Much the commonest species. Weak scrambling annual. Lf-segments flattened. Fls pink, tipped darker, the largest *7-9 mm* long, sepals more than ¼ the length of the petals; Apr-Oct. Bracts shorter than the fr-stalks. T.

6* WALL FUMITORY *Fumaria muralis*. Rather robuster than 5, with larger fls 9-12 mm, 12 on each spike; spikes shorter than their stalk; lower petals turned up at the edges. Fr stalks *not* curved back. Often on hedge-banks. T, western. **6a*** *F. bastardii* has fl-spikes longer than their stalks, fls 9-11 mm, 15-25 to a spike, and upper petals compressed, the lower with flat edges. B,F, western. **6b*** *F. martinii* is like 6a with fls 11-13 mm. B,F.

7* RAMPING FUMITORY *Fumaria capreolata*. Robust climbing annual, up to 1 m long. Fls white, tipped blackish-pink, the largest 10-14 mm, on short stalks, c 20 in each spike; upper petals compressed, lower with turned-up edges; May-Sept. Fr stalks *turned back*. Often in hedges. T, southern. **7a*** *F. purpurea* has purple fls, upper petal not compressed. B(rare). **7b*** *F. occidentalis* agg. has lower petal with broad flat margin. B (Cornwall, rare).

8* SMALL FUMITORY *Fumaria parviflora*. Slender annual. Lf-segments channelled. Fls *5-6 mm*, white or pinkish, tipped darker pink in *dense 20-fld* spikes on very short stalks; sepals less than ¼ the length of the petals; June-Sept. Bracts as long as fr-stalks. Usually on chalk. B,F,G, southern. **8a*** *F. densiflora* has fls 6-7 mm, always pink; bracts longer than fr-stalk. T, southern. **8b*** *F. vaillantii* agg. has lf-segments flat, fls always pink, 6-12 on each spike; and bracts slightly shorter than fr-stalks. T, southern. **8c** *F. schleicheri* is like 8b, but has deep pink fls and bracts much shorter than fr-stalks. F,G.

1a* Oregon Grape

Poppy Family   Papaveraceae

POPPIES *Papaver* are mostly weeds of cultivated ground with large, solitary, brightly-coloured fls, whose four petals have a rather creased, silky appearance. They have two sepals which fall off soon after the buds open. Juice usually white. Lvs 1-2-pinnate.

1* COMMON POPPY *Papaver rhoeas.* Medium, usually roughly hairy annual. Fls 70-100 mm, deep scarlet, often with a *dark centre*, anthers blue-black; fl-stalks with spreading hairs; June-Oct. Pod almost *round*, hairless. Arable fields and disturbed ground. T, southern.

2* PRICKLY POPPY *Papaver argemone.* Short, stiffly hairy annual with hairs closely pressed to stem. Fls 20-60 mm, generally smaller than 1 and 4, petals *pale scarlet*, not overlapping; May-July. Pod long and narrow, with numerous bristles. Arable fields, especially on sandy soils. T, southern.

3* ROUGH POPPY *Papaver hybridum.* Resembles 2 but with a *rounded* pod, densely covered with stiff yellow bristles. Fls darker, crimson; June-Aug. Arable fields, mainly on lime. T, southern.

4* LONG-HEADED POPPY *Papaver dubium.* Medium annual, with hairs closely pressed to stems; juice white. Fls paler than 1, 30-70 mm, *without* dark centres; anthers violet; June-Aug. Pod much *longer* than wide, hairless. Arable fields and disturbed ground. T. **4a* Babington's Poppy** *P. lecoqii* has juice turning yellow on exposure to air, petals more orange-red, and anthers brown or yellow. On lime. B,F.

5* OPIUM POPPY *Papaver somniferum.* Medium/tall, almost hairless perennial with large, wavy, greyish lvs, irregular in outline. Fls up to 180 mm, from *lilac* with dark centres to white; June-Aug. Arable fields or waste ground, mainly an escape from cultivation. (T, southern). Grown since early times as a source of opium.

6 ARCTIC POPPY *Papaver radicatum* agg. Low/short tufted hairy perennial with deeply lobed lvs; juice white or yellow. Fls yellow, sometimes very pale, 20-40 mm; June-Aug. Pod *softly hairy,* less than twice as long as wide. Bare, rocky ground in mountains. S.

7* YELLOW HORNED-POPPY *Glaucium flavum.* Medium/tall, hairless, rather greyish perennial, with wavy edged lvs clasping the branching stem. Root lvs more deeply lobed, all lvs rough to the touch. Fls 60-90 mm, yellow, variable; June-Sept. Pods *very long,* up to 300 mm, sickle-shaped, hairless. Sea shingle and waste places inland. T, western. **7a* Red Horned-Poppy** *G. corniculatum* □ has fls 30-50 mm, orange to red; stem and pod hairy. Bare and waste places. (T, casual) from S Europe.

8* WELSH POPPY *Meconopsis cambrica.* Medium, slightly hairy perennial with long-stalked, deeply lobed root lvs and similar, but short-stalked stem lvs. Fls usually *more than 50 mm,* yellow; June-Aug. Pod hairless, longer and narrower than in 6. Damp woods and rocks. B,F, upland and as a garden escape in B and G.

9* GREATER CELANDINE *Chelidonium majus.* Medium/tall, almost hairless perennial, its branched stems easily broken, revealing an orange juice. Lvs rather greyish, pinnately lobed. Fls *20-25 mm,* smaller than other poppies, in clusters of 3-8; Apr-Oct. Hedges, walls, waste places, often near houses. T.

8a* Red Horned Poppy

Key to Crucifers

The Crucifer or Cabbage Family, Cruciferae (pp. 84-98), is one of the most distinctive families, with its four petals arranged crosswise. Some other superficially similar four-petalled flowers are Greater Celandine (p.80), Tormentil and other potentillas (p.110-12), Willowherbs (p.150) and Bedstraws (pp.184-6). These are included in the Main Key on p.13. The shape of the fruits is one of the most important clues to crucifer identity.

Plant

foetid: Wall Rockets 84; Narrow-leaved Pepperwort, Field Pennycress, Swinecress 98

garlic-smelling: Garlic Mustard 90; Garlic Pennycress 98

grey or white: Alisons, Ball Mustard 84; Woad 86; Wild Cabbage, Rape, Black Mustard, Bastard Cabbage 88; Sea Kale. Stocks 92; Garden Rockcress, Tower Mustard 94; Alisons 96; Field Pepperwort, Perfoliate Pennycress 98

with bulbils: Coralroot Bittercress 90

with runners: Creeping Yellowcress 86; Cuckoo Flower, Large Bittercress 90; Sea Kale, Horse-radish, Dittander 92; Alpine Rockcress 94; Hoary Cress 98

Stems

woody below: Golden Alyssum 84; Wallflower 86; Wild and Wallflower Cabbages 88; Dame's Violet 90; Sea Kale, Stocks 92; Hoary Cress 98

leafless: Yellow Whitlow-grass 84; Isle of Man Cabbage 88; Shepherd's Cress, Rock and Common Whitlow-grasses, *Draba nivalis, D. fladnizensis* 96

angled: Wallflower, Common Watercress 86; Large Bittercress, Watercress 90

unbranched: Yellow Whitlow-grass, Ball Mustard 84; Cuckoo Flower, Coralroot Bittercress, Garlic Mustard 90; Hairy Rockcress 94; Whitlow-grasses 96

zig-zag: Wavy Bittercress 94

Flowers: fragrant: Wallflower 86; Dame's Violet 90; *Erysimum odoratum* 299.

yellow: all on pp.84-8; *Cardamine enneaphyllos* 90; Tower Mustard, Towercress 94; *Draba nemorosa* 96

white: Wild Radish 88; all on pp.90-2

pink/purple: Cuckoo Flower, Coralroot Bittercress, Dame's Violet, Honesties 90; Sea Rocket, Stocks, Early Scurvy-grass 92; Northern and Tall Rockcresses 94; Candytufts 96; Alpine Pennycress 98

Petals: notched: Hoary Alison, Common and Rock Whitlow-grasses 96

unequal: White Ball Mustard 92; Candytufts, Shepherd's Cress 96

absent: Narrow-leaved Bittercress 90; Hairy Bittercress 94; Narrow-leaved Pepperwort, Lesser Swinecress 98

veined: Bastard Cabbage, Wild Radish 88.

Sepals

erect (fls yellow): Yellow Whitlow-grass, Small Alison, Ball Mustard 84; Hedge and Treacle Mustards, Common Wintercress, Creeping Yellowcress 86; Wild and Wallflower Cabbages, Hairy Rocket, Wild Radish 88

spreading (fls yellow): Wall Rockets 84; Tall Rocket, Warty Cabbage, Yellowcresses 86; Wild Turnip, Black Mustard, Charlock, Bastard Cabbage 88

persistent: Small Alison 84

Seed-pods

long and narrow: Wall Rockets 84; Wallflower, Tall Rocket, Hedge and Treacle Mustards, Common Wintercress 86; all on pp. 88-90; Stocks 92; all on p. 94

oblong: Yellow Whitlow-grass, Gold of Pleasure 84; Woad, Great Yellowcress 86; Perennial Honesty 90; White Ball Mustard, Early Scurvy-grass 92; all on pp. 96-8

spherical: Alisons, Ball Mustard 84; Warty Cabbage 86; Austrian Yellowcress 299; Sea Kale, Horse-radish, Dittander, Common scurvy-grass 92; Common Whitlow-grass 96; Garden Cress, several Pepperworts 98

circular: Honesty 90; Wild Candytuft 96; Field Pennycress 98

2-lobed: Buckler Mustard 84; Swinecresses 98

notched: Wild Candytuft, Shepherd's Purse 96

wrinkled: Ball Mustard 94

warty: Warty Cabbage 96; Swinecress 98

jointed or segmented: Bastard Cabbage, Wild Radish 88; Sea Rocket, Sea Kale 92

flattened: Buckler Mustard 94; Honesties 90; Dittander 92; Northern and Tall Rockcresses 94; Wild Candytuft, Shepherd's Purse and Cress, Hutchinsia 96

triangular: Shepherd's Purse 96; Alpine, Perfoliate and Mountain Pennycress 98

twisted: Tower Mustard 94; Twisted Whitlow-grass 96

4-angled: Treacle Mustard, Common Wintercress 86; Black Mustard, Hairy Rocket 88

overtopping fls: London Rocket 86; Hairy Bittercress 94

spreading: Small Alison, Ball Mustard 84; Creeping Yellowcress 86; Wallflower Cabbage 88; Dame's Violet, Honesties, Watercress 90; Common Scurvy-grass 92; Shepherd's Purse, Wall Whitlow-grass 96; Austrian Yellowcress 299

closely pressed to stem: Hedge Mustard 86; Black Mustard 88; Stocks 92; Hairy Rockcress, Tower Mustard 94

hanging: Woad 86

Cabbage Family Cruciferae

Crucifers get their names from the characteristic cross made by their four, usually well separated petals; sepals 4; fls usually in stalked spikes. Seed pods are important characters and are either long and thin (siliquae), or less than three times as long as broad (siliculae) and of various shapes.

1* YELLOW WHITLOW-GRASS *Draba aizoides.* Low tufted hairless perennial with stiff narrow undivided lvs but a *lfless* stem. Petals yellow, much longer than sepals; Mar-May. Pod elliptical. Rocks and walls on lime. B(rare),F,G, southern, scattered. **1a** *D. alpina* agg. has broader lvs with a few branched hairs; June-July. S, northern. **1b** *D. crassifolia* is like 1a, but has unbranched hairs, still broader, thicker lvs, and smaller paler fls.

2* ANNUAL WALL ROCKET *Diplotaxis muralis.* Short/medium, hairless annual/biennial, much branched, with stems *lfless* or almost so. Root lvs stalked, deeply pinnately lobed. Fls 10-15 mm, deep yellow; June-Sept. Pods longer, exceeding their stalks, held at an angle to the stem. Walls, rocks, also a weed of bare and waste places. T, southern. **2a* Perennial Wall Rocket** *D. tenuifolia* is a taller perennial, woody at the base, with greyish lvs on the stem, fls 15-30 mm, and pods held stiffly erect, parallel to the stem, shorter than their stalks. B,F,G, southern. **2b** *D. viminea* is smaller, with smaller, paler fls. (G), from S Europe.

3* SMALL ALISON *Alyssum alyssoides.* Low/short, downy annual. Lvs small, lanceolate. Fls tiny, *pale* yellow, 3-4 mm, in dense spikes; style *minute;* Apr-June. Pod almost *round,* the hairy sepals persisting with it. Sandy fields, often on lime. T, eastern, but (B, rare).

4 MOUNTAIN ALISON *Alyssum montanum.* Low/short, hairy, often rather whitish perennial, with broader lvs and slightly larger fls than 3. Fls *bright* yellow, style longer than in 3; Apr-June. Pod round, sepals not persisting. Sandy places. F,G, southern. **4a* Golden Alyssum** *A. saxatile* is woody at base, with lvs often lobed and pods longer. Stony places on lime. (B),G(rare).

5* GOLD OF PLEASURE *Camelina sativa.* Short/medium, usually hairless annual. Lvs narrowly arrow-shaped, clasping the stem. Fls 3 mm, yellow; May-July. Pod *oblong,* becoming yellowish. A weed of arable and waste places. T, a rare casual in the north and west. **5a** *C. microcarpa* is hairy, the grey-green pods as long as wide. **5b** *C. alyssum* has a squashy, more rounded pod. Flax fields. **5c** *C. macrocarpa* is larger with larger pods. Flax fields. 5a-c are not in B.

6* BALL MUSTARD *Neslia paniculata.* Medium downy annual with lanceolate lvs, sometimes slightly toothed, the lower stalked, the upper clasping the stem. Fls yellow, 3-4 mm; June-Sept. Pod rounded, often rather flattened, *wrinkled.* Waste places. (B,casual,S),F,G, southern. **6a Mitre Cress** *Myagrum perfoliatum* is hairless and greyish, with more toothed lvs; pod larger, broad at the tip, tapering to the thicker stalk. (T, casual) from S Europe.

7 BUCKLER MUSTARD *Biscutella laevigata* agg. Short, slightly hairy perennial, with deeply lobed root lvs. Fls 5-10 mm, yellow; May-July. Pod much wider than long, flattened, with two almost *circular* lobes. Dry limestone rocks, occasionally in waste places. F,G.

Siliculae of

5a *Camelina microcarpa*

5b *C. alyssum*

5c *C. macrocarpa*

6a Mitre Cress

1

2

3

4

5

6

7

7

Cabbage Family (contd.)

1* WALLFLOWER *Cheiranthus cheiri.* Medium/tall perennial with narrow lanceolate untoothed lvs covered with *flattened* hairs. Fls *25 mm,* orange-yellow or orange-brown, fragrant; Mar-June. Pods long, flattened. Old walls, dry limestone rocks. (B,F,G), from the Mediterranean. **1a, 1b, 1c,** p.299.

2* HEDGE MUSTARD *Sisymbrium officinale.* Medium/tall, *stiff,* roughly hairy (occasionally hairless) annual with spreading branches and rosette lvs deeply pinnately lobed.Fls *3 mm,* pale yellow; May-Sept. Pods 6-20 mm, much shorter than 2a-2f, erect and closely *pressed* to stem, unlike 2a-2f. Waste places and hedge banks. T. **2a** *S. strictissimum* has lvs not or shallowly-toothed, not lobed; fls bright yellow, 4-6 mm, and pods hairless. (B),F,G. **2b* London Rocket** *S. irio* is often hairless, with lvs more narrowly lobed and pods hairless, overtopping fls when young. (T, rare in B). **2c* False London Rocket** *S. loeselii* is like 2b but hairy with larger fls and young pods not overtopping fls. **2d* Eastern Rocket** *S. orientale* has topmost lvs not lobed, fls 7 mm, and pods hairy at first. (T). **2e*** *S. austriacum* is biennial/perennial with some lvs not lobed, fls golden yellow, 7-10 mm and pods and their stalks twisted. (B,S),F,G. **2f*** *S. volgense* is like 2b with larger fls.

3* TALL ROCKET *Sisymbrium altissimum.* Tall, branched annual, hairless above. Root lvs pinnate, roughly hairy, soon dying; stem lvs unstalked, with very *narrow, deeply cut* lobes. Fls 10 mm, pale yellow; June-Aug. Pods long, to 100 mm. Waste places.(T), from E Europe. **3a* Flixweed** *Descurainia sophia* ☐ has lvs 2-pinnate with even narrower lobes, fls 3 mm, petals shorter than sepals, and pods 10-40 mm. T.

4* WOAD *Isatis tinctoria.* Medium/tall, mostly hairless biennial/perennial. Lower lvs narrow, downy; upper greyish, hairless, *arrow-shaped,* clasping the stem. Fls 4 mm, yellow; June-Aug. Pods dark brown, oblong, *hanging.* Dry places and rocks. (T, southern), from S Europe.

5* WARTY CABBAGE *Bunias orientalis.* Medium/tall, hairless perennial; stems warty. Lvs pinnately lobed, the topmost unlobed. Fls sulphur-yellow, 15 mm; May-Aug. Pods rounded, *warty,* shiny, long-stalked. Waste places. (T), from E Europe. **5a*** *B. erucago* is similar with pods 4-winged.

6* TREACLE MUSTARD *Erysimum cheiranthoides.* Short/tall, *square-stemmed* annual covered with flattened branched hairs. Rosette lvs lanceolate, *shallowly* and irregularly toothed, soon dying. Fls 6 mm, yellow, fl-stalks longer than sepals; June-Sept. Pods long, 10-30 mm, 4-angled. Waste places. T. **6a, 6b,** p.299.

7* COMMON WINTERCRESS *Barbarea vulgaris.* Medium/tall *hairless* perennial. Lvs with 2-5 pairs of coarse lobes, end lobe shorter than rest of lf; upper lvs undivided. Fls 7-9 mm, bright yellow; May-Aug. Pods long, 15-30 mm, stiffly *erect.* Damp places and roadsides. T. **7a* Small-flowered Wintercress** *B. stricta* has end lobe longer than rest of lf and 1-3 pairs of side lobes; fls 5-6 mm, pods 20-30 mm. **7b* Medium-flowered Wintercress** *B. intermedia* has all stem lvs lobed, and fls 6 mm. (T, southern) from Mediterranean. **7c* American Wintercress** *B. verna* has all lvs lobed, the lower with 6-10 pairs; fls 7-10 mm; Apr-July; pods 30-60 mm. (B,F,G, western), from Mediterranean.

8* GREAT YELLOWCRESS *Rorippa amphibia.* Tall hairless perennial, with runners. Lvs lanceolate, toothed or pinnately lobed. Fls 6 mm, bright yellow; June-Aug. Pods *oblong, long-stalked.* Wet places. T. **8a*,** p.299.

9* CREEPING YELLOWCRESS *Rorippa sylvestris.* Straggling, hairless perennial with runners and short/medium flg shoots. Lvs pinnately lobed, the upper sometimes unlobed. Fls 5 mm, yellow, petals longer than sepals; June-Sept. Pods long, 6-18 mm, equalling stalk. Damp and bare places. T. **9a* Marsh Yellowcress** *R. palustris* is erecter and stronger-stemmed; fls 3 mm, petals not longer than sepals; June-Oct; pod 4-9 mm, oblong, shorter than stalk. **9b* Iceland Yellowcress** *R. islandica* ☐ is like 9a but always prostrate and has pod longer than stalk. B,S. **9c** *R. pyrenaica* is downy at base, with upper lvs clasping the stem, smaller fls and shorter fr. F,G, southern.

1

2

3   3a   4

5   6   7   8   9   9b

Cabbage Family *(contd.)*

1, 2 and 8 are cultivated forms and their wild relatives, so much variability must be expected.

1* WILD CABBAGE *Brassica oleracea.* Medium/tall hairless perennial with a thick, straggling stem, woody at the base and clearly marked with old *lf scars.* Lvs *greyish,* pinnately lobed, somewhat clasping the stem. Fls 30-40 mm, in a long spike, the unopened *buds above* the fls; May-Sept. Sea-cliffs in B and F, western; also (T), widely cultivated as cabbage, cauliflower, etc.

2* WILD TURNIP *Brassica rapa.* Tall slender annual/biennial. Root lvs pinnately lobed, roughly hairy, bright green, stem lvs clasping. Fls 20 mm, in a more or less flat-topped spike, the *fls above* the unopened buds; Apr-Aug. Pods long. Cultivated, and on bare ground, especially by streams. (T). **2a* Rape** *B. napus* has greyish lvs and fls dull yellow in a spike like 1, but less elongated. Fls buff in var. *napobrassica* (Swede). **2b*** *B. juncea* is hairless, with greyish lvs, the lower very large and compound, the end lflet much larger than the others; fls level with the buds. (B,G), from Asia.

3* BLACK MUSTARD *Brassica nigra.* Tall, rather greyish, hairless annual. Fls yellow, 15 mm; June-Aug. All lvs *stalked,* the lowest pinnately lobed, bristly and bright green. Pods long, pressed against stem. Waste places, sea cliffs, stream banks, woods; also cultivated. T. **3a* Hoary mustard** *Hirschfeldia incana* is whitish with coarse hairs below, and has paler fls. Waste places. (B,G),F.

4* CHARLOCK *Sinapis arvensis.* Medium/tall, roughly hairy annual. Lower lvs large, irregularly lobed and toothed, the upper narrow, with pointed teeth, *unstalked* but not clasping the stem. Fls 15-20 mm, yellow; Apr-Oct. Pods long, markedly beaded. Arable fields, waste places, most often on lime. T. **4a* White Mustard** *S. alba* □ has all lvs deeply pinnately-lobed, sepals longer. Often cultivated and as a weed. (T, southern), from Mediterranean.

5* HAIRY ROCKET *Erucastrum gallicum.* Short/medium, densely hairy annual. Lvs pinnately lobed, the lower stalked, the upper not clasping the stem. Fls pale yellow, 10-15 mm, sepals erect; May-Sept. Pods *less than 40 mm long,* beaded. Waste places, dry rocks. T, southern, but (B,rare). **5a*** *E. nasturtiifolium* has upper lvs not clasping, fls bright yellow and sepals spreading.

6* WALLFLOWER CABBAGE *Rhynchosinapis cheiranthos.* Medium biennial with base of lfy stem hairy. Lvs deeply pinnately lobed. Fls yellow, 20-25 mm; June-Aug. Pods long, *more than 40 mm.* Dry and waste places, rarely on lime. (B),F,G, south-western. **6a* Isle of Man Cabbage** *R. monensis* is entirely hairless, with an almost lfless stem. Sandy places by coast. B, western. **6b* Lundy Cabbage** *R. wrightii* is densely hairy. Cliffs. B,rare (Lundy Is.).

7* BASTARD CABBAGE *Rapistrum rugosum.* Medium hairy annual whose stalked lvs have three pairs of lobes. Fls pale yellow, 10 mm; May-Sept. Pods with *two segments,* the upper large and rounded, with an abrupt beak, the lower narrower, closely pressed to the stem. Waste ground. (B,F,G), from the Mediterranean. **7a*** *R. perenne* is a perennial, stiffly hairy below, with six pairs of lf-lobes; fls bright yellow; and upper pod-segment narrowing gradually into beak.

8* WILD RADISH *Raphanus raphanistrum.* Medium, roughly hairy annual, slightly resembling 4 but with the teeth of the upper lvs blunt. Fls larger (25-30 mm), pale yellow or white with lilac veins; May-Sept. Pods long, markedly *beaded,* and breaking easily at the joints. Bare and waste ground. T. **8a* Sea Radish,** ssp. *maritimus,* is taller and bushier, with fls always yellow, and pods not breaking easily. Coasts. B,F,G. **8b* Garden Radish** *R. sativus* has an edible tuberous root, fls white or lilac, and pods not beaded.

4a* White Mustard

Cabbage Family (contd.)

BITTERCRESSES *Cardamine* have pinnate lvs (except 1a) and long pods.

1* CUCKOO FLOWER *Cardamine pratensis* agg. Medium hairless perennial, sometimes tufted, with root lvs forming a *rosette;* stem lvs dissimilar. Fls variable, 12-30 mm, lilac or white, anthers yellow; Apr-June. Pods to 40 mm. Damp places. T. **1a** *C. bellidifolia* □ is dwarf and tufted, with all lvs undivided, few or none on stems, and very small (7-10 mm) white fls. S, mountains.

2* LARGE BITTERCRESS *Cardamine amara.* Medium hairless perennial with basal lvs *not* in a rosette; stem lvs with *broader* lflets than 1. Fls white, 12 mm, anthers violet; Apr-June. Damp or wet places, often in woods. T. **2a*** *C. raphanifolia*□ is taller with a large, rounded terminal lflet, deep lilac fls and yellow anthers. Streamsides. (B, rare), from S Europe.

3* CORALROOT BITTERCRESS *Cardamine bulbifera.* Medium hairless perennial with *no root lvs;* upper stem lvs narrow lanceolate, with purple-brown *bulbils* at their base. Fls 12-18 mm, lilac to purple; Apr-June. Pods 20-35 mm. Woods, often of beech, on lime. T, southern. **3a** *C. heptaphylla* has all lvs pinnate and no bulbils; fls *c.* 20 mm, often white, pods up to 80 mm. F,G, southern. **3b** *C. pentaphyllos* is like 3a, but with lvs palmate or trefoil. **3a × 3b** occurs. Mountain woods. G, southern. **3c** *C. enneaphyllos* □ has 2-trefoil lvs, no bulbils and drooping pale yellow or white fls. Mountain woods. G, southern.

4* NARROW-LEAVED BITTERCRESS *Cardamine impatiens.* Medium hairless annual/biennial. Lvs with up to 18 *3-lobed lflets;* lf-stalks clasping the stem. Fls white, but petals inconspicuous or absent; anthers greenish-yellow; May-Aug. Pods explosive when ripe. Damp, shady, often rocky places, on lime in the north. T, mainly western. **4a** *C. parviflora* is a short annual, each lf with fewer, unlobed lflets; petals always present. F,G,S, scattered.

5* GARLIC MUSTARD *Alliaria petiolata.* Medium hairless biennial with long-stalked lvs, heart-shaped at the base, smelling of *garlic* when crushed. Fls white; Apr-Aug. Pods long. Hedges and open woods. T.

6* DAME'S VIOLET *Hesperis matronalis.* Medium/tall hairy perennial, with toothed lanceolate short-stalked lvs. Fls 15-20 mm, white or violet, very *fragrant;* May-Aug. Pods long, to 100 mm, curving upwards. Hedges, roadsides, a frequent garden escape. (B,G,S), F.

7 PERENNIAL HONESTY *Lunaria rediviva.* Tall perennial with a hairy stem. Lvs pointed oval, sharply toothed, the upper stalked. Fls fragrant, purple, *c.* 20 mm; May-July. Pods long, very *flattened,* elliptical with pointed tip. Damp woods, mainly on lime. T but (B), eastern. **7a* Honesty** *L. annua* is biennial with unstalked, coarsely toothed upper lvs, and larger, scentless fls, 25-30 mm; Apr-June. Pods almost round. A frequent garden escape. (T), from S E Europe.

8* WATERCRESS *Nasturtium officinale.* Short/medium *creeping* hairless perennial, with pinnate lvs, green in autumn. Fls white; May-Oct. Pods less than 18 mm long, with two rows of seeds inside. In and beside shallow, flowing water, often on lime. T, southern. **8a* One-rowed Watercress** *N. microphyllum* is extremely like 8 but with lvs turning purplish-brown in autumn, and pods more than 16 mm long, containing only one row of seeds; fls two weeks later. Less often on lime. **8 × 8a*** is intermediate, with lvs turning purplish-brown, but deformed pods. Occurs with parents, and often cultivated.

2a* *C. raphanifolia*

3a *Cardamine enneaphyllos*   **1a** *C. bellidifolia*   **4a** *C. parviflora*

Cabbage Family (contd.)

1* SEA ROCKET *Cakile maritima* agg. Short/medium greyish hairless *succulent* annual. Lvs linear to pinnately lobed. Fls lilac, sometimes very pale: June-Sept. Pods egg-shaped, angled. Sandy coasts. T.

2* SEA KALE *Crambe maritima.* Stout medium greyish hairless perennial, making large clumps, with thick stems, woody at the base. Lvs large, rounded, fleshy, with *crinkly* lobes or teeth, but narrower up the stem. Fls 10-15 mm, white, in a rather flat-topped head; June-Aug. Pods globular. Sea shores, on sand and shingle. T.

3* WHITE BALL MUSTARD *Calepina irregularis.* Medium/tall hairless annual. Stem lvs *clasping,* toothed, arrow-shaped. Fls 2-4 mm, with unequal petals; May-June. Pods more or less spherical, beaked. Waste places, mainly on lime; not a coastal plant. (B, rare), F, G.

4* HOARY STOCK *Matthiola incana.* The garden plant, a medium/tall, grey, stickily hairy annual/perennial; stem lfless and usually woody at the base. Lvs narrow lanceolate, untoothed. Fls *30 mm, purple,* sometimes very pale, fragrant; May-July. Pods long, stout, hairy. Sea cliffs. B(rare), F, western. **4a* Sea Stock** *M. sinuata* □ is a less bushy, whitish biennial, lfy and not woody at the base; root lvs lobed or wavy-edged. Pods stickily hairy. Sand dunes.

5* HORSE-RADISH *Armoracia rusticana.* Tall hairless perennial with *large* (to 50 cm), very long-stalked, wavy-edged, toothed root lvs and *narrow,* short-stalked stem lvs. Fls 8-9 mm, white, densely clustered on long-stalked lfy spikes; May-Aug. Pods almost spherical. Cultivated, and naturalised in damp places and waysides. (T, southern) from E Europe.

6* DITTANDER *Lepidium latifolium.* Tall greyish hairless perennial resembling 5 but with smaller shorter-stalked lvs, not wavy-edged. Fls 2-3 mm, sepals *white-edged;* July-Aug. Pods rounded, rather flattened. Damp places, usually near the coast. T, southern.

7* COMMON SCURVY-GRASS *Cochlearia officinalis* agg. Very variable, low/medium hairless perennial with *fleshy,* heart-shaped lvs, the upper loosely clasping the stem. Fls white, 8-10 mm; Apr-Aug. Pod *swollen,* almost spherical. Dry saltmarshes and banks, occasionally inland. T. **7a* English Scurvy-grass** *C. anglica* □ has upper lvs clearly clasping the stem, lowest lvs not heart-shaped, tapering to the base, and usually larger fls; Apr-July. Hybridises with 7. Muddy shores. B, G, S. **7b** *C. aestuaria* has pod with blunt or notched tip. F. **7c** *C. fenestrata* has pod 3-4 times as long as broad. S. **7d** *Kernera saxatilis* has lvs broad to narrow lanceolate, often well toothed, not fleshy, the upper sometimes clasping the stem. Mountains, on lime. G, southern, rare.

8* EARLY SCURVY-GRASS *Cochlearia danica.* Very variable, low/short overwintering annual, differing from 7 especially in its oval pods, often lilac fls, and *stalked* upper lvs; Jan-Sept. Usually sandy places by the sea. T. **8a Greenland Scurvy-grass** *C. groenlandica* is perennial with usually lfless stems. Iceland.

4a* Sea Stock

7a* English Scurvy-grass

Cabbage Family *(contd.)*

1* HAIRY BITTERCRESS *Cardamine hirsuta*. Low/short, rather hairy annual, with a compact rosette of pinnate lvs, each with 3-7 pairs of lflets. Fls less than 5 mm, with four stamens, less obvious than *long, erect* pods; Feb-Nov. Bare open ground. T.   **1a* Wavy Bittercress** *C. flexuosa* is usually biennial and has wavy stems, a loose rosette of fewer lvs, each with 13-15 lflets, six stamens and pods not overtopping stem; Apr-Sept. Damp woods and rocks. T.

2* NORTHERN ROCKCRESS *Cardaminopsis petraea*. Short, slightly hairy perennial with *lobed* lower lvs and narrow, toothed stem lvs. Fls 5-8 mm, white, sometimes lilac; June-Aug. Pods long, strongly *flattened*. Rocks and river shingle in mountains. B, G, S.

3 TALL ROCKCRESS *Cardaminopsis arenosa*. Annual/perennial, taller than 2 and all lvs more deeply lobed. Fls larger than 2, white in the north, lilac further south; Apr-June. Pods like 2. Sandy soils, usually lime-rich. F, G, S. **3a** *C. halleri* is smaller with runners and rounded lvs; Apr-May; pod stalks longer. Mountains, naturalised by roadsides. F, G, scattered. **3b** *Arabidopsis suecica* has lvs less deeply lobed, fls always white and pods not flattened. Gravel. S.

4* THALE CRESS *Arabidopsis thaliana*. Short/medium annual with roughly hairy, *elliptical* toothed root lvs, forming a rosette; stem lvs narrower, not stalked or clasping the stem. Fls 3 mm, white; Mar-Oct. Pods long, only slightly flattened. Dry sandy soils and walls. T. **4a** *Sisymbrium supinum* is larger, with pinnately cut lvs, fls slightly larger and pods not flattened. F, G, S. **4b** *Braya linearis* is more tufted, with untoothed linear lvs. Arctic gravels on lime. S.

5* HAIRY ROCKCRESS *Arabis hirsuta* agg. Short/medium hairy biennial/perennial with both stem lvs and long pods *tightly pressed* against stem; stem lvs tapering at base; root lvs in a rosette, long oval, only slightly toothed. Fls white, 3-5 mm, numerous; May-Aug. Dry grassland, dunes, rocks, walls, on lime. T. **5a* Bristol Rockcress** *A. stricta* is a low/short perennial, with stem sparsely lfy and root lvs more deeply toothed; fls creamy; Mar-May. B, western, rare.

6* TOWER MUSTARD *Arabis glabra*. Tall, almost hairless, *grey-green* biennial with softly hairy, toothed, short-lived root lvs and arrow-shaped stem lvs *clasping* the stem. Fls 5-6 mm, creamy; May-July. Pods long, numerous, closely pressed to stem. Dry banks, roadsides, open woods. T, eastern. **6a** *A. pauciflora* has hairless root lvs which persist after flowering, and guitar-shaped stem lvs; pods fewer. Dry rocky meadows and woods, on lime. F, G, south-western.

7* ALPINE ROCKCRESS *Arabis alpina*. Low/short creeping hairy perennial, with long *runners,* forming conspicuous grey-green *mats.* Stem lvs deeply toothed, bases bluntly arrow-shaped. Fls 6-10 mm, white, petals spreading; May-July. Mountain rocks. B (rare), S. **7a* Garden Rockcress** *A. caucasica* is greyish-white; stem lvs with fewer shallower teeth, sharply arrow-shaped at base; fls 15 mm; Mar-May. Garden escape, on rocks and by roads, on lime. (T), from S Europe.

8* TOWERCRESS *Arabis turrita*. Medium, softly hairy biennial/perennial with an often reddish stem. Root lvs long-stalked, densely grey-hairy. Fls yellowish-white; Apr-July. Pods very *long,* up to 150 mm, all *twisted* to one side and downwards. Rocks, walls. (B, F, G), southern, from S Europe.

9 ANNUAL ROCKCRESS *Arabis recta*. Short, softly hairy annual with *untoothed* lvs, the oval root ones soon withering, the stem ones clasping with arrow-shaped points. Fl white, less than 5 mm, petals erect; Apr-June. Pods long, 10-35 mm, spreading. Mountain rocks. F, G.

Cabbage Family (contd.)

1* WILD CANDYTUFT *Iberis amara*. Short, slightly downy annual with more or less deeply toothed, hairless lvs scattered up the stem. Fls 6-8 mm, white or mauve, petals *unequal;* May-Sept. Pods rounded, less than 5 mm long. Bare ground and arable fields, on lime. B'F,G, southern. **1a** *I. intermedia* is larger with narrower, pointed, untoothed upper lvs and pods more than 5 mm. F,G, southern. **1b** *I. umbellata* □ has purplish fls in a flat-topped head. Garden escape. (B,F,G), from the Mediterranean.

2* SHEPHERD'S PURSE *Capsella bursa-pastoris* agg. Very variable, low/ medium annual/biennial, hairy or hairless. Lvs lanceolate, pinnately lobed, deeply toothed or untoothed, mainly in a basal rosette. Fls 2-3 mm, white; all year. Pod an inverted, notched *triangle*. Extremely common weed. T. **2a*** *C. rubella* □ has reddish petals, as long as sepals, and pods with rounded lobes. (B, rare), F.

3* SHEPHERD'S CRESS *Teesdalia nudicaulis*. Low/short, hairless annual with an almost *lfless* stem, and a rosette of pinately lobed root lvs. Fls 2 mm, petals unequal, not notched; June-Oct. Open sandy places, not on lime. T, mainly southern.

4* HOARY ALISON *Berteroa incana*. Medium hairy annual, grey with hairs. Lvs lanceolate, untoothed. Fls 5-8 mm, white, petals deeply *cleft;* June-Oct. Pods oval. Casual on waste ground. (B,F,S),G, eastern.

5* SWEET ALISON *Lobularia maritima*. Short, densely hairy, greyish perennial, with *narrow* untoothed lvs. Fls 5-6 mm, fragrant, white, petals not notched; June-Sept. A garden escape, especially near the sea. (T), from S Europe.

6* WALL WHITLOW-GRASS *Draba muralis*. Short hairy annual with sparsely lfy stem. Lvs rounded, toothed, *partly clasping* the stem. Fls 2-3 mm, white, petals not notched; Apr-June. Pods elliptical, 5 mm, spreading. Rocks and walls, mainly on lime. T, scattered. **6a** *D. nemorosa* has narrower lvs, smaller pale yellow fls and longer fr stalks. G,S. **6b* Twisted Whitlow-grass** *D. incana* □ is a taller biennial/perennial with lfier stems, narrower lvs, larger fls, notched petals and longer pods twisted when ripe; May-July. Rocks, dunes, mountains. B,G,S, northern. **6c** *D. daurica* agg. is like 6b, but with creamy fls and straight pods. S. 6d-6f are all low perennials with stems lfless or almost so; June-July. Mountain rocks. **6d* Rock Whitlow-grass** *D. norvegica* agg. □ has lvs not always toothed and petals notched. B,S. **6e** *D. nivalis* s smaller than 6c, with stems always lfless and lvs densely hairy, usually untoothed. S. **6f** *D. fladnizensis* is like 6d, with stem hairless and lvs sparsely hairy. S.

7* COMMON WHITLOW-GRASS *Erophila verna* agg. Very variable, low, slightly hairy annual with a lfless stem. Lvs lanceolate, in a basal rosette. Fls white, petals *cleft to base,* 3-6 mm; Mar-May. Pods elliptical or rounded, on long stalks. Bare, often sandy ground, rocks. T.

8* HUTCHINSIA *Hornungia petraea*. Low hairless annual with a rosette of pinnate lvs and a lfy stem. Fls minute, greenish-white, petals not notched; Mar-May. Pod elliptical' 2 mm. Bare sandy ground, rocks, mainly on lime. T, southern, scattered. **8a** *Hymenolobus procumbens* is taller, with undivided stem lvs and fls slightly larger; pods kidney-shaped, flat, to 5 mm. F,G, scattered.

1b *Iberis umbellata*　　**2a*** *Capsella rubella*　　**6b*** Twisted Whitlow-grass　　**6d*** Rock Whitlow-grass

Cabbage Family (contd.)

1* HOARY CRESS *Cardaria draba.* Medium, usually hairless perennial with *greyish,* pointed, toothed lvs, clasping the stem. Fls 5-6 mm, white, stalked, in flat-topped, *umbel-like* clusters; May-June. Pods kidney-shaped, beaked. Roadsides and disturbed ground, forming clumps. T, southern.

2* FIELD PEPPERWORT *Lepidium campestre.* Medium greyish annual/biennial, with many spreading hairs. Lvs lanceolate, the upper rather triangular and clasping the stem. Fls 2-3 mm, white, petals *longer* than sepals, anthers yellow; May-Aug. Pods roundish, flattened, notched, rough, *as long* as their stalks. Dry banks, disturbed ground. T. **2a* Smith's Pepperwort** *L. heterophyllum* is an often shorter, more branched perennial, with longer petals, violet anthers, and smooth pods. Also in grassland. T, south-western. **2b* Poor Man's Pepperwort** *L. virginicum* has downward directed hairs, toothed lvs not clasping stem, and pods shorter than stalks. (T) from N America. **2c*** *L. divaricatum* has toothed lvs not clasping stem lvs; petals equalling or shorter than sepals. (B,G,S) from Africa. **2d*, 2e,** p.300.

3* GARDEN CRESS *Lepidium sativum.* Short hairless strong-smelling annual, the cress used for salads. Lvs *linear,* not clasping the stem, the lower pinnate. Fls sometimes reddish, 4-5 mm, petals longer than sepals; June-July. Pod roundish, flattened *longer* than its stalk. Disturbed ground. (T), from W Asia.

4* NARROW-LEAVED PEPPERWORT *Lepidium ruderale.* Short/medium, almost hairless annual/biennial, smelling strongly like 3, with pinnate root lvs and narrow undivided stem lvs. Fls green, usually petalless; May-Sept. Pods oval, flattened, *shorter* than their stalks. Waste places, commonest near coasts. T. **4a*** *L. densiflorum* is not cress-scented and has hairs pointing upwards, stem lvs slightly toothed, fls sometimes with minute petals and pods as long as stalks. (T) from N America. **4b* Least Pepperwort** *L. neglectum* is like 4a with spreading hairs, undivided stem lvs and a round pod. **4c** *L. bonariense* is hairy, all lvs pinnate. (B,G,S).

5* FIELD PENNYCRESS *Thlaspi arvense.* Short/medium hairless, foetid annual with broad lanceolate toothed lvs clasping the stem, the lower not in a rosette. Fls 4-6 mm, white, anthers yellow; May-Aug. Pods rounded, with *broad,* almost translucent *wings,* the very short style persisting in a deep notch. Waste ground. T. **5a* Garlic Pennycress** *T. alliaceum* smells of garlic and has narrower, less winged pods; Apr-June. (B),F,G.

6* ALPINE PENNYCRESS *Thlaspi alpestre.* Very variable short hairless perennial with non-flowering rosettes. All lvs *untoothed,* somewhat clasping the stem. Fls white or mauve, 4-8 mm, anthers violet; Apr-July. Pods heart-shaped, winged, the style usually projecting beyond the notch. Limestone rocks and lead-mine debris, and in mountain woods away from lime. T, scattered, eastern.

7* PERFOLIATE PENNYCRESS *Thlaspi perfoliatum* is low/short and rather greyish, with all lf rosettes flowering. Lvs lanceolate, deeply clasping stem. Fls white, anthers yellow; Apr-May. Pods heart-shaped, winged. Stony and waste ground, on lime. T, southern, rare in B.

8 MOUNTAIN PENNYCRESS *Thlaspi montanum.* Short hairless mat-forming perennial, with *sparsely lfy* stems. Root lvs long-stalked. Fls 10-12 mm, anthers yellow; Apr-June. Pods heart-shaped, flattened, broadly winged, style projecting beyond notch. Rocky grassland, on lime. F,G, southern.

9* SWINECRESS *Coronopus squamatus.* Prostrate annual/biennial. Lvs 1-2-pinnate, lflets narrow. Fls 2-3 mm, white, in small clusters opposite lvs; June-Sept. Pods longer than stalks, rounded, clearly 2-lobed, warty, *not* notched, with prominent style. Characteristic weed of paths and *trodden* ground. T, southern.

10* LESSER SWINECRESS *Coronopus didymus.* More delicate than 9, often *semi-erect,* with a pungent smell and more feathery lvs. Fls 1-2 mm, white or petalless; June-Sept. Pods shorter than stalks, more obviously 2-lobed than 9, and *notched,* with obscure style. Waste places. T, southern.

1

2

3

4

5

6

7

8

9

10

Mignonette Family   Resedaceae

1* WELD *Reseda luteola.* Tall hairless unbranched biennial. Lvs lanceolate, *untoothed,* wavy-edged. Fls 4-5 mm, yellow-green, with *four* sepals and petals; June-Sept. Pods globular, erect. Disturbed ground, often on lime. T, southern. **1a** *Sesamoides canescens* has 6-petalled fls. F, western, rare.

2* WILD MIGNONETTE *Reseda lutea.* Medium branched biennial/perennial, the stem clothed with small *pinnate* lvs. Fls 6 mm, yellow-green, with *six* sepals and petals; June-Sept. Pods oblong, erect. Disturbed ground, often on lime. T, southern. **2a* Corn Mignonette** *R. phyteuma* is a short downy annual with sometimes unlobed lvs, fls whitish and pods drooping. (B, F, G). from S Europe. **2b* White Mignonette** *R. alba* is medium, the white fls with five sepals and petals. (B, F, G), from S Europe.

Sundew Family   Droseraceae

Insectivorous plants, whose lvs are covered with long sticky hairs which curve inwards to trap insects.

3* COMMON SUNDEW *Drosera rotundifolia.* Low/short perennial with a lfless fl-stem more than twice as long as the neat rosette of *round* red lvs with long stalks. Fls 5 mm, white, 5-petalled, in a shortly stalked spike, arising from the rosette; June-Aug. Hybridises with 4a. Wet heaths and moors, sphagnum bogs, never on lime. T.

4* OBLONG-LEAVED SUNDEW *Drosera intermedia.* Differs from 3 in its long *narrow* lvs, tapering to their shorter hairless stalks. Fl-stem arising from *below* rosette and curving upwards. Habitat similar. T, rather scattered.   **4a* Great Sundew** *D. anglica* □ has larger longer lvs and fl-stem arising from rosette. Sometimes on lime. Northern.

Stonecrop Family   Crassulaceae see p. 102

5* NAVELWORT *Umbilicus rupestris.* Low/medium hairless perennial, the *round* fleshy lvs with a *"navel"* in the centre, above the lf-stalk; stem lvs less rounded. Fls bell-shaped, toothed, greenish or pinkish-white; June-Aug. Rocks, banks, walls, not on lime. B, F, western.

6* ROSEROOT *Rhodiola rosea.* Short hairless grey perennial, often purple-tinged. Lvs thick, stiff, *succulent,* more or less oval, often hiding the stem. Fls yellow, 4-petalled, in rather flat-topped heads, stamens and styles on separate plants; anthers purple; May-June. Fr orange. Mountain rocks, sea-cliffs. B, S, northern, and rare in F.

Saxifrage Family   Saxifragaceae see p. 104

7* OPPOSITE-LEAVED GOLDEN SAXIFRAGE *Chrysosplenium oppositifolium.* Low creeping perennial, slightly hairy; stem square. Lvs roundish, bluntly toothed, opposite; root lvs as long as stalks. Fls 3-4 mm, greenish, *without* petals, sepals 4, stamens 8, anthers bright yellow; March-July. Wet places. T, but western in S. **7a* Alternate-leaved Golden Saxifrage** *C. alternifolium* □ has a triangular stem, alternate, broader, more toothed lvs, root lvs on long stalks, and fls 5-6 mm. Eastern. **7b** *C. tetrandrum* is like 7a, but completely hairless and with 4 stamens. S, northern.

Grass of Parnassus Family   Parnassiaceae

8* GRASS OF PARNASSUS *Parnassia palustris.* Short hairless tufted perennial with a rosette of heart-shaped, untoothed, stalked lvs and a stem with a *single* lf and a *solitary* white fl, 15-30 mm, with five clearly veined petals; June-Sept. Wet, marshy places. T.

2

4a

4

3

7a

8

5

6

7

Stonecrop Family Crassulaceae

See also p. 100. Mostly hairless perennials with untoothed, fleshy, un- or short-stalked lvs. Fls star-like, usually conspicuous, with five petals and sepals. Characteristic of dry rocky places and walls.

1 * ORPINE Sedum telephium. Very variable, short/tall slightly greyish perennial, with clusters of erect, often red-tinged lfy stems. Lvs up to 80 mm long, oblong, *flattened*, slightly toothed, alternate. Fls pinkish-red, whitish, greenish or yellowish, petals spreading, in flattened heads; July-Sept. Shady places and woods. T. **1a * Caucasian Stonecrop** S. spurium □ is shorter and creeping, forming mats, with erect, downy flg stems; lvs to 30 mm, more toothed, and larger fls, usually pinkish-red, rarely white, petals erect. Garden escape. (T), from Caucasus.

2 * BITING STONECROP Sedum acre. Low creeping evergreen with *peppery* taste. Lvs 3-6 mm, cylindrical, not spreading, yellowish, broadest at base. Fls 12 mm, bright yellow; May-July. Dry, bare places, often on lime. T. **2a * Tasteless Stonecrop** S. sexangulare is not peppery, has lvs 3-9 mm, broadest at middle, and paler 8-10 mm fls; July-Sept. T, eastern, but (B).

3 * WHITE STONECROP Sedum album. Bright green, often red-tinged, mat-forming evergreen, creeping with short, erect flg stems. Lvs *6-12 mm*, rarely to 25 mm, alternate, cylindrical, rather flattened on the upper surface. Fls 6-9 mm, *white*, in a dense, much-branched, *flat-topped* head; June-Aug. Rocks, walls. T, southern.

4 * REFLEXED STONECROP Sedum reflexum. Short creeping evergreen, with short, erect flg shoots and low lfy non-flg ones. Lvs 8-20 mm, *cylindrical*, not persisting when dead. Fls 15 mm, *yellow*, usually with seven petals and sepals; June-Aug. Rocks, walls. T, southern, but (B). **4a * Rock Stonecrop** S. forsteranum has narrower flattened lvs, crowded into terminal rosettes on the non-flg shoots; dead lvs persistent; fls 12 mm. Sometimes in damp places. B, F, G, western.

5 ALPINE STONECROP Sedum alpestre. Low evergreen, somewhat resembling 2 but not peppery. Lvs rather flattened, parallel-sided. Fls 6-8 mm, dull yellow; June-July. Mountain rocks, not on lime. F, G.

6 * ENGLISH STONECROP Sedum anglicum. Low mat-forming evergreen, greyish but soon reddening. Lvs *3-5 mm*, cylindrical, alternate. Fls 12 mm, *white* (pink below), the flhead much less branched than in 3; June-Sept. B, F, S, western, rarely on lime. **6a * Thick-leaved Stonecrop** S. dasyphyllum is smaller and tufted with opposite, slightly sticky or downy lvs, and fls 5-6 mm. (B), F, G, southern.

7 * HAIRY STONECROP Sedum villosum. Low, reddish, *downy* biennial. Lvs 6-12 mm, alternate, flattened on the upper surface. Fls 6 mm, *pink*, on long stalks; stamens 10; June-Aug. Wet places and by streams, in mountains. T, scattered. **7a** S. rubens is annual with lvs 10-20 mm and unstalked fls; stamens 5; May-July. (F, G), from S Europe. **7b** S. hirsutum is very hairy; fls usually white. Dry rocks. F. **7c** S. andegavense is a hairless annual with more rounded lvs; fls usually white. F. **7d** S. hispanicum has 6 white petals with pink veins. (G).

8 ANNUAL STONECROP Sedum annuum. Low hairless annual/biennial. Lvs flattened on both surfaces, 6 mm. Fls *yellow*, short-stalked; June-Aug. G, S, only on mountains in the south.

9 * MOSSY STONECROP Crassula tillaea. Very low moss-like annual, with tiny crowded broad opposite lvs pressed against stem. Fls white, 1-2 mm, petals usually three, shorter than sepals; June-Sept. Damp sandy places, wet in winter. B, F, G, western. **9a * Water Tillaea** C. aquatica has narrower, less crowded lvs and short-stalked 4-petalled fls. B (rare), G, S. **9b** C. vaillantii is like 9a with long- stalked fls. F.

10 HEN-AND-CHICKENS HOUSELEEK Jovibarba sobolifera. Succulent perennial with lvs in roundish rosettes, often tipped red. Fls pale yellow, bell-shaped, 6-petalled; July-Aug. Dry, often sandy grassland. G. **10a**, p. 300.

Saxifrages   Saxifragaceae

See also p.100. Rather short perennials (except 6). Lvs alternate (except 9). Stamens 10, petals, sepals, styles 5. Fr a capsule in two parts, often joined at the base.

1* STARRY SAXIFRAGE *Saxifraga stellaris*. Low/short, sparsely hairy, with lfless stem and a *rosette* of very short-stalked, *toothed* lvs. Fls white, in a lax head, 10-15 mm, with conspicuous red anthers; sepals turned down; June-Aug. The commonest white-fld mountain saxifrage; in wet places. T. **1a** *S. foliosa* has bulbils. S.

2* ST PATRICK'S CABBAGE *Saxifraga spathularis*. Loosely tufted, with short, lfless fl-stem. Lvs spoon-shaped, mostly in rosettes, hairless, with pointed teeth and a narrow *translucent* margin; stalks long, flattened. Petals 4-5 mm, white with several *red spots*, sepals turned down; June-July. Rocks, not on lime. Ireland. **2a*, 2b*, 2c*, 2d*, 2e***, and hybrids, p.300.

3* ARCTIC SAXIFRAGE *Saxifraga nivalis*. Low, with basal rosette of small, thick lvs, 15-30 mm, rounded, coarsely toothed, *purple* beneath, and sticky at the edges. Fl-stalks lfless, very hairy. Fls *whitish*, unstalked, in a *crowded* head, petals unspotted; sepals spreading; July-Aug. B,S, mountains. **3a**, p.300.

4* RUE-LEAVED SAXIFRAGE *Saxifraga tridactylites*. Low, erect, *stickily hairy* annual, often reddish. Upper lvs spoon-shaped, 3-5-lobed, the lower untoothed; all lvs on stem. Fls *c.* 5 mm, white, with notched petals. Walls, bare places, lowland, mainly on lime. T, southern. **4a, 4b**, p.300

5* YELLOW SAXIFRAGE *Saxifraga aizoides*. Low/short, very *lfy*, slightly hairy. Lvs 10-20 mm, narrow, *unstalked*, fleshy. Fls in loose head, petals *yellow*, well spaced; June-Sept. Streams and wet places in mountains. B,G,S. **5a* Marsh Saxifrage** *S. hirculus* has stem lfless, red-hairy; lvs in rosette, fls solitary and petals overlapping. Bogs. T, scattered, rare in B.

6* MEADOW SAXIFRAGE *Saxifraga granulata*. Short/medium, downy, with *bulbils* at lf-bases, but no barren shoots. Lvs kidney-shaped, shallowly lobed, mainly basal. Fls 20-30 mm, white, not spotted, in loose clusters; Apr-June. Grassy places, lowland. T.

7* HIGHLAND SAXIFRAGE *Saxifraga rivularis*. Very low, hairless, with runners from bulbils at stem base; stem lfy. Lvs stalked, with *3-5 blunt lobes*. Fls few, 6-10 mm, dowdy, white; July-Aug. Wet mountain rocks. B(rare), S. **7a* Drooping Saxifrage** *S. cernua* is larger; lvs with 3-7 pointed lobes; fls larger, solitary, replaced by bulbils at least in fr.

8* MOSSY SAXIFRAGE *Saxifraga hypnoides*. Low/short, hairy, with many long, lfy, *barren* shoots forming a mat. Lvs long-hairy, usually *3-lobed*, clustered at the end of each stem. Fls 10-15 mm, white, buds nodding; bulbils often present; May-July. Wet, often grassy places in hills. B,F,S, western. **8a* Tufted Saxifrage** *S. cespitosa* is tufted with shorter shoots; lvs hairier, hairs shorter; fls smaller, greenish, on shorter stem. Rocks. B(rare), S. **8b* Irish Saxifrage** *S. rosacea* agg. may have habit of 8 or 8a, otherwise as 8 with erect fl-buds, broader lf-lobes and no bulbils. B(Ireland, rare), F, G.

9* PURPLE SAXIFRAGE *Saxifraga oppositifolia*. Creeping, with trailing stems covered with small, unstalked, *opposite* lvs, often lime-encrusted. Fls 10-20 mm, *purple*, almost unstalked; Apr-May, July. Damp places in mountains, often on lime. B,S.

10 LIVELONG SAXIFRAGE *Saxifraga paniculata*. Short, tufted with long, oblong, finely-toothed lvs, lime-encrusted, in hemispherical rosettes. Fl-stem long, mostly lfless, with a *loose panicle* of white fls; June-Aug. Mountain rocks. F,G. **10a** *S. cotyledon* is larger with broader, pointed lvs. Fl-stem much denser, with flg branches from the base. G,S. **11** (Hawkweed Saxifrage), p.300.

Rose Family   Rosaceae

Lvs alternate, usually pinnate, with stipules at base of lf-stalk. Fls usually open with five conspicuous petals, except for those on this page; often with an epicalyx – an outer ring of sepals below the true sepals.

1* MEADOWSWEET *Filipendula ulmaria.* Tall hairless perennial with lfy stems. Lvs long-stalked, pinnate, with 2-5 pairs of toothed lflets, each more than 20 mm, and small lflets in between, *stipules* green above, downy and pale beneath. Fls in *dense clusters,* creamy, fragrant, with 5 or 6 petals, 2-5 mm; June-Sept. Marshes, fens, swamps, wet meadows and woods. T. **1a* Dropwort** *F. vulgaris* □ is short/medium, with sparsely lfy stem, the shorter lvs with up to 20 pairs of lflets, each less than 20 mm. Fl-head rather flattened, less dense, petals 5-9 mm. Dry grassland, on lime. T, southern.

2 GOATSBEARD SPIRAEA *Aruncus dioicus.* Tall perennial with very large lvs, like 1 but *without* stipules, the flheads with many long, *finger-like* branches, held more or less horizontal; June-Aug. Mountain woods. F, G.

3* AGRIMONY *Agrimonia eupatoria.* Medium downy perennial with pinnate lvs, not at all sticky. Fls 5-8 mm, *yellow,* in a spike; June-Aug. Fr grooved, covered with small erect *hooks* at its apex. Dry grassy places, often on lime. T. **3a* Fragrant Agrimony** *A. procera* is aromatic and altogether larger with lvs stickily hairy beneath; fr hardly grooved, some of the hooks bent back. Rarely on lime. Southern. **3b* Bastard Agrimony** *Aremonia agrimonioides* is shorter with lfy clusters of fls, 7-10 mm; May-June. Mountain woods, waysides. (B), G.

4* GREAT BURNET *Sanguisorba officinalis.* Medium/tall hairless perennial. Lvs pinnate, with 3-7 pairs of lflets, 20-40 mm long. Fls tiny, on dense *oblong* heads, 10-20 mm long, stamens and styles in the same fl; petals absent, sepals deep red; June-Sept. Damp grassland. T, southern.

5* SALAD BURNET *Sanguisorba minor.* Short, rather greyish, almost hairless perennial, similar to 4, but with 4-12 pairs of lflets all less than 20 mm. Fls in *round* heads, the *upper with red styles,* the lower with yellow stamens; sepals green, no petals; May-Sept. Dry, grassy places, mainly on lime. T, southern. **5a* Pirri-pirri Bur** *Acaena anserinifolia* □ is a creeping under-shrub, with fls whitish, stamens and styles in the same fl, and fr softly spiny. Not on lime. (B), from Australia.

6* LADY'S MANTLE *Alchemilla vulgaris* agg. A variable collection of similar microspecies: low/medium, usually rather densely hairy, but sometimes hairless perennials. Lvs palmately lobed, *not* cut to the base, *green* on both sides. Fls 3-5 mm, in loose clusters with an epicalyx but no petals, sepals green, anthers yellow; May-Sept. Grassy places. T.

7* ALPINE LADY'S MANTLE *Alchemilla alpina.* Smaller and hairier than 6, with lvs divided *to the base* into 5-7 narrow lobes, *silvery grey* with hairs beneath. Fls pale green, 3 mm; June-Aug. Mountain grassland. T. **7a*** *A. conjuncta* has lvs divided to about ⅔ into 7-9 broader segments. River shingle. T, but (B, northern, rare), from the Alps.

8* PARSLEY PIERT *Aphanes arvensis* agg. More or less prostrate hairy, pale green annual. Lvs 3-lobed, toothed. Fls minute, green, in clusters *opposite* the lvs, surrounded by the toothed lf-like stipules; Apr-Oct. Bare ground. T, southern.

1a* Dropwort

5a* Pirri-pirri Bur

1

2

3

4

5

6

7

8

Rose Family *(contd.)*

ROSES *Rosa* are shrubs with thin stems armed with thorns. Lvs pinnate, with stipules. Fls large, white, pink or red, with large, floppy petals, easily detached. Fr a berry-like hip.

1* DOG ROSE *Rosa canina* agg. Variable; stems arching, to 3 m. Thorns *curved*. Lvs with 2-3 pairs of toothed lflets, *hairless* or sometimes downy beneath. Fls 45-50 mm, pink or white, styles not joined; June-July. Sepals with narrow lobes, falling before the hips turn red. Hedges, scrub. T, commoner in the south. **1a*** *R. stylosa* has styles joined in a column. B, F, G, western.

2* BURNET ROSE *Rosa pimpinellifolia*. Suckering stems with *straight* thorns and stiff bristles, forming short bushes to 50 cm. Lvs with 3-5 pairs of small, rounded lflets. Fls 20-40 mm, cream, sometimes pink; sepals not lobed; May-July. Ripe hips dark purple or *black*. Dry, open places, often on coastal dunes. T, rare in S. **2a Provence Rose** *R. gallica* □ has larger, more leathery, bluish-green lflets, larger deep pink fls and hips red. F, G.

3* FIELD ROSE *Rosa arvensis*. Scrambling, forming bushes not more than 1 m tall, unless on other shrubs; stems green, thorns curved. Lflets hairless, 2-3 pairs on each lf. Fls 30-50 mm, *always* white, the styles *joined* into a column; sepals scarcely lobed, soon falling; July-August. Hips red. Woods, hedges, scrub. B, F, G, southern.

4* DOWNY ROSE *Rosa tomentosa* agg. To 2 m; resembling 1 but lflets *densely downy;* thorns may be straight. Fls 30-40 mm, deep pink, sometimes pale; June-July. Sepals pinnate, sometimes persisting till red hips fall or decay. Hedges, scrub, especially in the hills. T, northern. **4a* Sweet Briar** *R. rubiginosa* agg. has lflets densely coated with sweet-smelling brown sticky hairs. Scrub, usually on lime. T, southern.

5* BRAMBLE *Rubus fruticosus* agg. Very variable scrambling shrub armed with easily detached thorns; over 2000 microspecies have been described. Rather woody, with long biennial, usually angled stems, rooting where they touch the ground. Lvs prickly, with 3-5 lflets. Fls 20-30 mm, white or pink; May-Nov. Fr (blackberry) with several fleshy segments, red at first, *purplish-black* when ripe. Woods, scrub, open and waste ground. T. **5a* Dewberry** *R. caesius* has weak, round, less prickly stems; lvs always with three lflets; fls always white; fr bluish, with a waxy bloom, fleshier, with fewer and larger segments. Also in damp places, preferring lime. T, southern.

6* RASPBERRY *Rubus idaeus*. Unbranched *erect* perennial with woody, biennial stems, armed with weak prickles. Lvs with 3-7 lflets. Fls smaller than in 5, the petals *erect,* well spaced, *always* white, rather inconspicuous; May-Aug. Fr *red* when ripe. Shady places. T, and widely cultivated.

7* STONE BRAMBLE *Rubus saxatilis*. Resembles 5 but smaller, slenderer, prostrate, and spineless; stems annual. Lvs with three lflets. Fls smaller than 5 or 6, 8-10 mm, the narrow petals *no longer than* the sepals; June-Aug. Fr red when ripe, the segments fewer and larger. Shady, rocky places, often on limestone. T, north-eastern.

8* CLOUDBERRY *Rubus chamaemorus*. Low/short creeping downy perennial with *rounded*, palmately lobed lvs. Fls always white, conspicuous, solitary, stamens and styles on separate plants; June-Aug, but a shy flowerer in B. Fr *orange* when ripe. Upland bogs and damp moors. B, G, S, northern.

9 ARCTIC BRAMBLE *Rubus arcticus*. Short spineless creeping perennial. Lvs with three lflets, smaller than 7. All stems flg; fls 15-25 mm, *bright red,* petals sometimes toothed; June-July. Moors. S, probably extinct in B.

2a Provence Rose

Rose Family (contd.)

1* WILD STRAWBERRY *Fragaria vesca*. Low/short perennial, with long *runners,* rooting at intervals. Lvs trefoil, bright green, with pointed teeth, the lflets paler beneath and hairy with *flattened* silky hairs, less than 80 mm long; hairs on stalks *not* spreading. Fls 12-18 mm, white, on a short stalk no longer than the lvs; Apr-July. Fr the familiar strawberry, but much smaller, covered all over with seeds, the sepals bent back from it. Dry, grassy places and woods, often on lime in the north. T. **1a* Hautbois Strawberry** *F. moschata* □ is larger, usually without runners; lflets more than 60 mm; fl-stalks longer than lvs; fls 15-25 mm; fr without seeds at base. T, but (B), eastern. **1b** *F. viridis* is smaller with short runners, downy or hairless lflets and sepals pressed against fr; fr not easily detached from stalk. F,G,S. **1c* Garden Strawberry** *F.* × *ananassa* □ has much larger fls and fr. (T), a frequent escape.

2* BARREN STRAWBERRY *Potentilla sterilis*. Low hairy perennial, usually with runners, and with *spreading* hairs on stem and lvs. Lvs smaller than 1, blunt-toothed, and bluish; lflets 5-25 mm, toothed. Fls 10-15 mm, white, with gaps between the slightly notched petals; Feb-May. Fr dry, *not* strawberry-like. Dry grassland and open woods. T, southern. **2a White Cinquefoil** *P. alba* has lvs with five very narrow, scarcely toothed lobes, green above and silvery below. Fls 15-20 mm; Apr-May. F, G, south-eastern. **2b Pink Barren Strawberry** *P. micrantha* is shorter, with no runners and often pinkish fls, 6-8 mm. G. **2c** *P. montana* has lflets untoothed in lower part, green above and grey beneath, and larger fls; May-June. Woods. F.

3* ROCK CINQUEFOIL *Potentilla rupestris*. Short/medium hairy perennial without runners. Root lvs *pinnate* with rather rounded lflets, and simpler, usually trefoil lvs up the stem. Fls 15-25 mm, *white;* May-June. Rocky slopes. T, scattered, rare in B.

4* MARSH CINQUEFOIL *Potentilla palustris*. Short/medium hairless perennial with pinnate lvs. Fls *star-shaped,* erect; petals narrow, deep purple, shorter than the broader spreading, *maroon* sepals, 10-15 mm long; May-July. Lime-free wet places. T.

5* WATER AVENS *Geum rivale*. Medium downy perennial with pinnate root lvs, becoming trefoil up the stem, and small stipules, 5 mm, at base of stem lvs. Fls nodding, *bell-shaped,* petals dull pink, 8-15 mm; sepals purple; Apr-Sept. Fr with persistent hairy style, hooked. Damp, often shady places. T. **5 × 6*** *G. × intermedium* is found where the parents grow together.

6* HERB BENNET *Geum urbanum*. Medium downy perennial. Lvs pinnate, the smaller with lf-like stipules up stems. Fls 8-15 mm, *yellow,* petals erect, sepals green; style as in 5; May-Sept. Woods and shady places on fertile soils. T. **6a Alpine Avens** *G. montanum* is low/short with fls 25-40 mm, solitary, 6-petalled, on lfless stems; styles in fr long, feathery, not hooked. Rocky mountain pastures. G. **6b** *G. hispidum* is hairier with smaller fls and stipules. S.

7* MOUNTAIN AVENS *Dryas octopetala*. Low creeping downy perennial with long stems. Lvs *oblong*, blunt-toothed, 5-20 mm, green above, *grey* beneath. Fls 20-40 mm, white, with *8 or more* petals, and many conspicuous yellow stamens; May-July. Fr with long, feathery styles. Mountain rocks and to sea-level in the north, on lime. B,S.

1* Wild Strawberry   1a* Hautbois Strawberry   1c* Garden Strawberry   2* Barren Strawberry

1

1a

3

2

4

5

6

7

Rose Family (contd.)

CINQUEFOILS *Potentilla* (pp.110-12) are usually creeping perennials with compound lvs, conspicuous yellow (p.112) or white (p.110) fls, five petals (except 2), numerous stamens, and an epicalyx.

1* SHRUBBY CINQUEFOIL *Potentilla fruticosa*. Small, branched, downy *shrub* up to 1 m high. Lvs deciduous, pinnate; lflets untoothed, greyish, hairy, narrow, usually five. Fls *c.* 20 mm, in loose clusters; May-July. Wet hollows, river banks, upland rocks. T, scattered, rare in B.

2* TORMENTIL *Potentilla erecta*. Creeping and patch-forming, with short, downy flg stems, not rooting at lf-junctions. Root lvs in rosette, trefoil, soon withering; stem lvs unstalked, appearing to have five lflets from conspicuous toothed lf-like stipules. Fls 7-11 mm, with *four* petals and sepals, numerous in loose heads; May-Sept. Moors and grassy places, not on lime. T. **2a* Trailing Tormentil** *P. anglica* □ has stems longer, rooting at lf-junctions, some lvs with five lflets, rosette persistent, and short-stalked stem lvs with untoothed stipules; fls 14-18 mm, solitary, petals 4-5. Often in woods. T, southern. **2b* Creeping Cinquefoil** *P. reptans* □ has stems to 1 m, rooting at lf-junctions, all lvs with 5-7 lflets, stem lvs stalked with untoothed stipules; fls 17-25 mm, solitary, petals 5. Waste places, bare ground, hedges, also on lime. The hybrids 2 × 2a, 2 × 2b and 2a × 2b are frequent.

3* SPRING CINQUEFOIL *Potentilla tabernaemontani*. Low/short, rather hairy, mat-forming, with creeping, *rooting* branches. Root lvs with 5-7 lflets, each 10-40 mm, and narrow stipules; upper lvs trefoil, unstalked. Fls 10-20 mm; epicalyx segments blunt, lanceolate; Apr-June. Dry, often rocky limestone grassland. T, southern. **3a* Alpine Cinquefoil** *P. crantzii* is not mat-forming, rarely rooting; fls 10-25 mm, often orange-spotted; epicalyx segments oblong; June-July. Mountain rocks. T. **3b** *P. heptaphylla* is taller and more erect, not mat-forming; lflets 5-7; fls 10-15 mm, numerous. G,S. **3c** *P. thuringiaca* is medium, erect, very hairy, with 5-9 lflets each up to 60 mm or more; epicalyx segments pointed. G,S. **3d** *P. pusilla* has star-shaped hairs and smaller fls. G. **3e** *P. aurea* is like 3a, but has appressed silky hairs and end tooth of lflets always smaller than neighbours. Not on lime. G, southern. **3f** *P. hyparctica* is like 3a with unspotted petals. S, Arctic.

4* HOARY CINQUEFOIL *Potentilla argentea* agg. Short/medium, with woolly, usually erect flg stems. Lvs with five lflets, dark green above, silvery *white* with dense hairs beneath; lflets pinnately lobed, except at base, margins inrolled. Fls 10-15 mm, style rounded at tip; June-Sept. Dry grassy places. T, southern. **4a** *P. collina* has 5-7 lflets, white or grey-green beneath; fls 8-10 mm, style club-shaped. F,G,S. **4b** *P. inclinata* has lvs grey beneath. F,G. **4c** *P. nivea* agg.□ usually has trefoil lvs. S.

5* SULPHUR CINQUEFOIL *Potentilla recta*. Short/medium, hairy, *stiffly* erect, lower lvs with 5-7 lflets to 100 mm. Fls 20-25 mm, petals longer than sepals, pale yellow, in clusters; June-Sept. Dry grassy and waste places. (B,S),F,G, southern. **5a* Norwegian Cinquefoil** *P. norvegica* is shorter with trefoil lvs; fls 10-15 mm, brighter yellow, petals no longer than sepals, which enlarge in fr. (B,F),G,S. **5b*** *P. intermedia* is like 5a with 5 lflets. (T, eastern).

6 GREY CINQUEFOIL *Potentilla cinerea*. Mat-forming with short, woody stems, as 4 but all creeping, *densely* covered with grey hairs. Lvs with 3-5 lflets, grey-green above, green beneath; stipules very narrow. Fls 10-15 mm, few; Apr-May. Dry grassy and rocky places, mainly upland. F,G,S, south-eastern.

7* SILVERWEED *Potentilla anserina*. Creeping with long runners, rooting at lf-junctions. Lvs *pinnate,* often silvery, at least beneath, lflets toothed. Fls 15-20 mm, *solitary*, petals twice as long as sepals; May-Aug. Damp grassy places. T. **7a, 7b,** p.300.

8* SIBBALDIA *Sibbaldia procumbens*. Very low, tufted and covered with stiff hairs. Lvs trefoil, lflets 3-toothed at tip. Fls *5 mm*, few; petals shorter than sepals, sometimes absent; July-Aug. Mountain grassland. T.

1

2a 2b

4c

4

2

3

5

6

7

8

Rose Family (contd.)

1* CRAB APPLE *Malus sylvestris.* Small tree to 10 m, rather spiny. Lvs toothed, pointed oval, hairless when mature. Fls 30-40 mm, white or pink, in loose clusters; May. Fr the familiar green apple, sometimes turning red. Woods and hedges. T. **1a* Cultivated Apple** *M. domestica* is unarmed; mature lvs downy. (T).

2 WILD PEAR *Pyrus pyraster.* Deciduous tree to 20 m, usually *spiny;* twigs grey to brown. Lvs roundish to elliptical, slightly toothed, hairless when mature. Fls white, 5-petalled, in clusters; Apr. Fr roundish to pear-shaped, not fleshy, yellow, brown or black, surmounted by dead sepals. Woods, scrub. ?B, F, G. **2a* Cultivated Pear** *P. communis* may not be spiny and has twigs red-brown and fr fleshy and sweet-tasting. Widely cultivated, sometimes naturalised in hedges. (T). **2b, 2c*,** p. 301.

3* WHITEBEAM *Sorbus aria* agg. Very variable tree, to 25 m. Lvs to 120 mm, more or less pointed oval, widest below the middle, *silvery white* with short hairs beneath, the teeth pointing to the tip. Fls 10-15 mm, white, in dense, rather flat heads; May-June. Fr 8-15 mm long, scarlet. Woods and scrub on lime. B, F, G. **3a* Rowan** *S. aucuparia* □ usually only reaches 15 m; lvs pinnate with 5-7 pairs of lflets, green on both sides; fls 8-10 mm, fr 6-9 mm long. Woods, heaths, moors, rocky places, often upland. Rarely on lime. **3b*, 3c*, 3d*,** p. 301.

4* HAWTHORN *Crataegus monogyna.* Hairless *thorny* deciduous shrub or small tree, 2-10 m. Lvs deeply *3-5-lobed,* more than half-way to midrib. Fls 8-15 mm, white, in broad dense flattened clusters; one style; May-June. Fr a *crimson* berry (haw). Hedges, scrub. T. **4a* Midland Hawthorn** *C. laevigata* □ has lvs lobed less than half-way to midrib; styles 2. More tolerant of shade. T, southern. Hybrid **4** × **4a** is abundant and variable. **4b,** p. 301.

5* BLACKTHORN *Prunus spinosa.* Deciduous thorny shrub, 1-4 m, much branched, with black or dark brown twigs. Lvs 20-40 mm, rather *oval,* toothed. Fls 10-15 mm, white, solitary, in short dense spikes, usually appearing *before* the lvs; Mar-May. Fr a *bluish-black* berry (sloe). Scrub, hedges. T. **5a* Cherry-Plum** *P. cerasifera* is an almost hairless thornless shrub or tree to 8 m; twigs glossy, green; fls 15-20 mm, appearing with lvs; Feb-Mar. Widely planted. (B, F, G), from S E Europe. **5b* Wild Plum** *P. domestica* is a small thornless tree with larger lvs; fls 15-25 mm in small clusters, appearing with lvs; Apr-May. Woods, hedges. (T), from Caucasus.

6* WILD CHERRY *Prunus avium.* Deciduous tree to 25 m. Lvs 60-150 mm, pale, often reddish, sparingly *downy* beneath, dull above, on long stalks. Fls 15-30 mm, white, few in loose clusters appearing with lvs; Apr-May. Fr a bright *red* bitter cherry, 10 mm across. Woods and hedges. T. **6a*, 6b, 6c,** p.301.

7* BIRD CHERRY *Prunus padus.* Deciduous tree to 15 m with lvs 50-100 mm, almost hairless, its long crowded *drooping* fl-spikes quite unlike 6; fls 10-15 mm; May. Fr *black.* Woods, hedges, mainly upland. T. **7a*,** p. 301.

8* WILD COTONEASTER *Cotoneaster integerrimus.* Deciduous shrub rarely more than 1 m, *not* thorny, all but the youngest twigs hairless. Lvs 15-40 mm, oval, *untoothed,* grey with dense long hairs beneath. Fls pink in groups of 2-4; Apr-June. Fr a small red berry. Limestone rocks, stony ground, mainly upland. T, eastern, rare in B. **8a** *C. niger* has black fr. S. **8b* Small-leaved Cotoneaster** *C. microphyllus* □ is evergreen with lvs 5-8 mm long and usually solitary white fls. Bird sown. (B) from Asia. **8c, 8d,** p. 301. Other cotoneasters, see p. 307.

9 AMELANCHIER *Amelanchier ovalis.* Deciduous tree to 12 m, with toothed hairless oval lvs, 30-70 mm Fls white in short erect spikes, petals 10-13 mm, *narrow;* Apr. Fr a small purple-black berry. Woods, scrub. F, G, southern. Other amelanchiers, see p. 307.

3a* Rowan

4a* Midland Hawthorn

Key to Peaflowers

(Pages 118-30)

The members of the Peaflower Family, Leguminosae (pp. 118-30), have the most distinctive flower-shape in the whole British flora. Some of the smaller species, however, such as the clovers, have to be fairly closely examined for this distinctive shape to become apparent. At first sight a clover appears to have a single globular flower, but in fact it is a tight head of small peaflowers. The family includes several shrubs and undershrubs.

Trees, shrubs or undershrubs

All on p. 120; Rest-harrow, Large Yellow Rest-harrow 126

Stems

ridged or angled: Broom, False Acacia 120; Bithynian Vetch 122; Spring, Sea and Tuberous Peas, Vetchlings 124

winged: Winged Broom 120; Everlasting and Marsh Peas, Hairy and Bitter Vetchlings 124

spiny: Gorse, Petty Whin, False Acacia 120; Rest-harrow 126

Leaves

grasslike: Grass Vetchling 124

palmate: Lupins 120

trefoil: Broom, Laburnum 120; pp. 126, 128 and 130

pinnate: Bladder Senna, False Acacia 120; all on pp. 118 and 122; Spring and Sea Peas, Meadow and Bitter Vetchling 124; Sainfoin, Kidney Vetch 126; Horseshoe Vetch, Birdsfoot 128

lanceolate: Broom, Dyer's Greenweed, Petty Whin 120

1 pair: Everlasting and Tuberous Peas, Meadow and Hairy Vetchlings 124

as spines: Gorse 120

as tendrils: Yellow Vetchling 124

with dark spot: Spotted Medick 128

Tendrils present

p. 122; Everlasting, Marsh and Tuberous Peas; Hairy, Meadow and Yellow Vetchlings 124

116

Flowers

yellow: Wild Lentil, Yellow Alpine Milk-vetch, Yellow Milk-vetch 118; all on p. 120; *Vicia pisiformis,* Bush and Yellow Vetches 122; Meadow and Yellow Vetchlings 124; Large Yellow Rest-harrow, Kidney Vetch, Classical Fenugreek, Lucerne 126; all on p. 128; Fenugreek, Mountain Clover 130

white/greenish: Goat's Rue, Wild Liquorice, Yellow Milk-vetch 118; Tree and Garden Lupins, False Acacia 120; Wood Vetch 122; Meadow and Hairy Vetchlings, Sea and Tuberous Peas 124; Red, White, Crimson, Haresfoot and Suffocated Clovers 130

pink/purple: Purple Milk-vetch, Goat's Rue 118; Wild Lupin 120; pp. 122 and 124; Sainfoin, Rest-harrow and Small Rest-harrow, Kidney Vetch 126; Birdsfoot 128; p. 130

blue/violet: Purple Milk-vetch 118; Wild Lupin 120; Tufted and Bush Vetches 122; Spring and Sea Peas, Hairy and Bitter Vetchlings 124; Lucerne 126

parti-coloured: Crown Vetch 118; Wild Lupin 120; Tufted, Fine-leaved, Danzig, Yellow and Bithynian Vetches 122; Large Yellow Rest-harrow, Kidney Vetch, Classical Fenugreek, Lucerne 126

in spikes: Milk-vetches *(Astragalus)* 118; p. 120; pp. 122, 124 and 126

in clusters: Scorpion Senna 120; all on p. 118 except Milk-vetches *(Astragalus)*; Kidney Vetch 126; all on pp. 128 and 130

solitary: Hairy Tare, Common, Yellow and Bithynian Vetches 122; Vetchlings 124; Classical Fenugreek 126; Dragon's Teeth, Spotted Medick 128

Pods

inflated: Bladder Senna 120

angled: Dragon's Teeth 128

hanging: Small Rest-harrow 126

curved: Classical Fenugreek, Lucerne 126; Horseshoe Vetch, Black Medick 128

buried: Burrowing Clover 130

warty: Sainfoin 126

spirally twisted: Lucerne 126; Spotted Medick 128

spreading: Birdsfoot Trefoil, Birdsfoot 128

Pea Family   Leguminosae

pp. 118-30. Has highly distinctive 5-petalled flowers, the broad and often erect "standard" at the top, the two narrower "wings" at the sides, and the two lowest joined as the "keel" which conceals the stamens and styles. Fruit a usually elongated pod.

MILK-VETCHES *Astragalus*. Lvs pinnate, usually with a terminal lflet. Fls in spikes or heads at base of lvs, keel *blunt*-tipped; calyx-teeth short.

1* PURPLE MILK-VETCH *Astragalus danicus*. Low/short downy perennial. Fls *violet;* May-July. Pods whitely hairy. Mountain and lime-rich grassland, dunes. T. **1a* Alpine Milk-vetch** *A. alpinus* □ has much paler fls and brown hairs on pods. Mountain rocks and grassland. T, but rare in B. **1b** *A. arenarius* has fewer, narrower lflets and paler fls on shorter stalks. G, S. **1c** *A. norvegicus* □ has fewer lflets and paler violet fls. S.

2 WILD LENTIL *Astragalus cicer*. Medium/tall hairy perennial. Fls pale *yellow,* sepal-tube and pod black-haired; June-July. Grassy places. F, G.

3 YELLOW ALPINE MILK-VETCH *Astragalus frigidus*. Short perennial, almost hairless. Lvs greyish. Fls yellowish-white; teeth of reddish sepal-tube tipped with black hairs; July-Aug. *Mountain* grassland. F, G, S, northern. **3a** *A. penduliflorus* has narrower lflets, yellower fls and sepal-tube green. S.

4* WILD LIQUORICE *Astragalus glycyphyllos*. Medium/tall straggling perennial, slightly hairy; stems *zigzag*. Fls greenish-cream; June-Aug. Grassland, scrub, open woods. T.

5* YELLOW MILK-VETCH *Oxytropis campestris*. Short hairy perennial. Lvs pinnate, with terminal lflet. Fls in rounded heads on lfless stems, creamy white or pale *yellow*, often tinged purple, keel finely *pointed;* June-July. Mountain rocks and grassland, rarely by the sea. B (rare), S. **5a Hairy Milk-vetch** *O. pilosa* is very hairy and has pale yellow fls on stems lfy at base. Grassland, rocks. F, G, eastern.

6* MOUNTAIN MILK-VETCH *Oxytropis halleri*. Like 5 but fls *purple*. B (rare), S. **6a Northern Milk-vetch** *O. lapponica* □ is slenderer and has smaller fls on stems lfy at base. S. **6b** *O. deflexa* is like 6a with whitish fls. S, Arctic, rare.

7* CROWN VETCH *Coronilla varia*. Medium/tall straggling hairless perennial. Lvs pinnate, with a terminal lflet. Fls in rounded heads, *particoloured* (pink, lilac, keel tipped purple); June-Aug. Pod 4-angled. Grassy places. (B, S), F, G.

8* GOAT'S RUE *Galega officinalis*. Medium/tall *erect* hairless perennial. Lvs pinnate, with a terminal lflet. Fls in heads, white or pinkish-lilac. 5 *bristle-like* calyx-teeth; July-Sept. Pod rounded. Damp and waste ground. (B), F, G.

6a Northern Milk-vetch

1c *Astragalus norvegicus*

Pea Family *(contd.)*

1* GORSE *Ulex europaeus*. Evergreen shrub to 2½ m, hairy. Lvs as rigid furrowed *spines*. Fls rich golden-yellow, in lfy stalked spikes, almond-scented, wings longer than keel and standard longer still; sepals yellowish, with spreading hairs as long as petals; all year. Heaths, grassland. B,F,G, western. **1a* Dwarf Gorse** *U. minor* is smaller, often prostrate, with shorter, less stout spines, smaller and paler fls, wings as long as keel, and sepals nearly as long as petals; July-Nov. B,F. **1b* Western Gorse** *U. gallii* ☐ is intermediate between 1 and 1a, with harder yellow fls. July-Nov. B,F.

2* BROOM *Cytisus scoparius*. Tall deciduous *spineless shrub,* sometimes prostrate by the sea; almost hairless; stems ridged. Lvs lanceolate and *trefoil.* Fls yellow in lfy stalked spikes; Apr-June. Heaths, open woods. T. 2a and 2b are shorter, with clover-like lvs. **2a Clustered Broom** *Chamaecytisus supinus* has fls in terminal clusters. On lime. F,G. **2b,** p. 300.

3* DYER'S GREENWEED *Genista tinctoria*. Medium deciduous *spineless* undershrub, sometimes prostrate, slightly hairy. Lvs *lanceolate.* Fls yellow, standard equalling keel, in lfy stalked spikes; June-Aug. Pods hairless. Grassland, heaths, open woods. T. **3a* Hairy Greenweed** *G. pilosa* is half-evergreen and more often prostrate, with small dark green oval lvs, smaller fls and downy pods; May-June. Drier habitats. **3b Winged Broom** *Chamaespartium sagittale* has young stems broadly winged. F,G. **3c Black Broom** *Lembotropis nigricans* has trefoil lvs and fls in longer lfless spikes, wings shorter than keel. Dry places. G, eastern.

4* PETTY WHIN *Genista anglica*. Slender short/medium undershrub, almost hairless, *spiny* (rarely spineless). Lvs small, pointed oval. Fls yellow in lfy stalked spikes, standard shorter than keel; Apr-June. Heaths, moors. T, western. **4a German Greenweed** *G. germanica* has branched spines and smaller hairy flowers with standard much shorter than keel. F,G,S, eastern.

5* LABURNUM *Laburnum anagyroides*. Deciduous *tree* to 7 m; bark smooth. Lvs trefoil. Fls yellow in drooping stalked spikes; May-June. Woods and scrub in mountains; widely planted. (B),F,G.

6* BLADDER SENNA *Colutea arborescens*. Deciduous shrub to 3 m, hairy. Lvs *pinnate.* Fls in stalked spikes, deep yellow, sometimes marked red; June-Aug. Pods papery, inflated. Open woods, scrub, also naturalised on waste ground in N and W. (B),F,G.

7* TREE LUPIN *Lupinus arboreus*. Evergreen shrub to 3 m, hairy. Lvs *palmate.* Fls in a stalked spike, *yellow* or white, sometimes tinged mauve, scented. Sandy soils. (B), from California. May-Aug. **7a* Sweet Lupin** *L. luteus* is a hairy annual widely grown for fodder. (B),F,G.

8* WILD LUPIN *Lupinus nootkatensis*. Medium/tall perennial, hairy. Lvs *palmate*, lflets 6-8. Fls in stalked spikes, *blue* and purple; May-July. River shingle, moors. (B,S), from N America. **8a* Garden Lupin** *L. polyphyllus* has larger lvs with 10-17 lflets and longer spikes of sometimes also pink, white or yellow fls. Widespread garden escape. (T), from N America. **8b* Annual Lupin** *L. angustifolia,* smaller, slenderer and always annual, is widely grown as a crop, as are the shorter Bitter Blue Lupin *L. micranthus,* covered with brown hairs, and White Lupin *L. albus,* whose white fls are tipped blue.

9* FALSE ACACIA *Robinia pseudacacia*. Deciduous tree to 25 m, suckering; bark grey-brown, deeply furrowed; shoots *thorny.* Lvs pinnate. Fls in drooping stalked spikes, white; June. Sandy soils. (T), from N America.

10 SCORPION SENNA *Coronilla emerus*. Deciduous shrub to 3-4 m. Lvs pinnate, *greyish.* Fls in heads, pale yellow often tipped red; Apr-June. Pods 50-100 mm. Scrub. F,G,S. **10a Small Scorpion Vetch** *C. vaginalis* is shorter, with shorter rounder green lflets and pods less than 35 mm. G. **10b** *C. minima* is like 10a, with lflets unstalked. F. **10c Scorpion Vetch** *C. coronata* is shorter and not woody. Dry places on lime. G. 10a-10c have much shorter pods.

Pea Family (contd.)

VETCHES *Vicia.* Mostly climbing or clambering plants. Lvs pinnate ending (except 4) in a tendril, usually branched. Fls usually in heads, on stalks from base of lvs. Pods long, more or less flattened. All except 8 differ from Peas *Lathyrus* (p. 124) in not having stems winged or angled.

1* TUFTED VETCH *Vicia cracca.* Clambering perennial to 1½ m, slightly downy. Lflets 8-12 pairs. Fls *10-40* in one-sided spikes, *blue*-violet, 10-12 mm; June-Aug. Pods brown. Hedges, bushy places. T.   **1a* Fine-leaved Vetch** *V. tenuifolia* has narrower lflets and larger, often white-winged fls. T, eastern, but (B).   **1b* Danzig Vetch** *V. cassubica* has shorter lflets and 4-15 pinker fls with whitish wings and keel. T but (B).   **1c* Fodder Vetch** *V. villosa* □ is a variable annual with 6-8 pairs of lflets, often larger fls, sometimes with white or yellow wings, and sepal-tube swollen at base; June-Nov. Bare and waste ground. T but (B).

2* WOOD VETCH *Vicia sylvatica.* Clambering hairless perennial to 2 m. Lflets 5-12 pairs. Fls *5-20* in one-sided spikes, white or *pale* lilac, with purple veins, 12-20 mm; June-Aug. Pods black. Open woods, rarely on sea cliffs and shingle. T.   **2a** *V. dumetorum* has blue-purple fls.   **2b** *V. pisiformis* □ has yellow fls. Both 2a and 2b have 3-5 pairs of lflets and brown pods. F, G, S, eastern.

3* BUSH VETCH *Vicia sepium.* Medium/tall clambering downy perennial. Lflets 3-9 pairs. Fls 2-6 in a short spike, *blue*-purple, rarely yellow, 12-15 mm; Apr-Nov. Pods black. Hedges, bushy places. T.   **3a*** *V. pannonica* has narrower lflets and longer, sometimes yellow fls. Grassy places. F, (B, G).

4* UPRIGHT VETCH *Vicia orobus.* Medium *erect* perennial, slightly downy; stems *unwinged.* Lvs ending in a minute *point,* not a tendril. Fls pink-lilac; May-June. Pods yellowish-brown. Scrub, rocks. T, western.

5* COMMON VETCH *Vicia sativa.* Variable short/medium clambering downy annual. Lflets 3-8 pairs, varying from almost linear to oval, tendrils sometimes unbranched; usually a *dark spot* on toothed stipules at base. Fls 1-2, reddish-purple, *10-30 mm;* Apr-Sept. Pods downy or hairless, black to yellow-brown. Bare or grassy ground; large forms are cultivated. T.   **5a* Spring Vetch** *V. lathyroides* □ is a prostrate downy annual, with 2-4 pairs of narrow lflets, tendrils always unbranched, untoothed unspotted stipules, solitary deep lilac 5-8 mm fls and black hairless pods; Apr-May. Sandy ground, especially near sea.

6* HAIRY TARE *Vicia hirsuta.* Slender short annual, almost hairless. Lflets 4-10 pairs, tendrils sometimes unbranched. Fls 1-8 in spike, pale *lilac, 4-5 mm;* sepal-teeth longer than tube; May-Aug. Pods downy. Grassy places. T.   **6a* Smooth Tare** *V. tetrasperma* □ has 3-6 lflet pairs, 1-2 larger (4-8 mm) deep lilac fls and hairless pods.   **6b* Slender Tare** *V. tenuissima* □ has 2-5 lflet pairs, 2-5 still larger (6-9 mm) deep lilac fls, and pods downy or hairless; June-Aug. B, F, G, southern. 6a and 6b both have calyx-teeth shorter than tube and tendrils usually unbranched.

7* YELLOW VETCH *Vicia lutea.* More or less prostrate downy annual. Lflets 3-10 pairs, tendrils branched or not. Fls 1-3, *yellowish-white,* often tinged purple, 20-35 mm; June-Sept. Pods usually downy. B, F, G.

8* BITHYNIAN VETCH *Vicia bithynica.* Medium clambering hairy annual; stems 4-angled. Lflets 2-3 pairs, large, *broad.* Fls 1-3, purple, wings and keel *creamy white,* 16-20 mm; May-June. Pods brown or yellow, downy. Grassy places, hedges. B, F, western.   **8a** *V. narbonensis* has broader lflets, longer, dark purple fls and black pods. Woods. F, (G).

2b *Vicia pisiformis*

1

1c

2

3

4

5a

6b

6a

8

5

6

7

Pea Family (contd.)

PEAS *Lathyrus.* Differ from all Vetches (p. 122), except Bithynian Vetch, in their winged or angled stems, but cf 4a. Most have fewer lflets, but some have more. Pods usually brown.

1* BROAD-LEAVED EVERLASTING PEA *Lathyrus latifolius.* Climbing perennial to 3 m, hairless or downy; stems winged. Lflets *one* pair, large, *broad.* Fls 5-15, bright magenta-pink, 20-30 mm; July-Sept. Scrub, widespread garden escape. (B), F, G. **1a* Narrow-leaved Everlasting Pea** *L. sylvestris* □ has narrower lvs and stipules, and smaller muddy pink fls; June-Aug. Woods, scrub, hedges. T. **1b** *L. heterophyllus* has 2-3 pairs of lflets and smaller fls. F, G, S.

2* MEADOW VETCHLING *Lathyrus pratensis.* Short/tall clambering perennial, hairless or downy; stems angled. Lflets one, occasionally two, pairs. Fls 4-12, *yellow;* May-Aug. Pods black. Grassy places. T.

3 SPRING PEA *Lathyrus vernus.* Short *erect* hairless perennial, stems angled. Lflets 2-4 pairs, *no* tendril. Fls 3-10, red-purple, fading blue, 13-20 mm; April-June. Pods brown. Woods, scrub. F, G, S. **3a* Black Pea** *L. niger* □ is sparsely downy, with 3-6 lflet pairs, smaller bluish-purple fls and black pods; May-July. T, rare in B.

4* BITTER VETCHLING *Lathyrus montanus.* Short erect perennial, hairless; stems *winged.* Lflets 2-4 pairs, sometimes very narrow (var. *tenuifolius* □), lvs ending in a *point* not a tendril. Fls 2-6, red-purple fading blue, 10-16 mm; Apr-July. Pods red-brown. Woods, scrub, heaths. T. **4a** *L. pannonicus* has stems not or narrowly winged, lflets very narrow, fls creamy tinged red-purple, and pale brown pods. F, G, southern.

5* SEA PEA *Lathyrus japonicus.* Prostrate fleshy perennial, hairless or downy; stems angled. Lflets 2-5 pairs, greyish, sometimes no tendril. Fls 2-12, purple fading blue, 14-22 mm; June-Aug. Shingle and dunes *by sea.* B, G, S.

6* MARSH PEA *Lathyrus palustris.* Medium/tall clambering perennial, slightly downy. Lflets 2-5 pairs, tendril branched. Fls 2-8, bluish-purple, 12-20 mm; June-July. Pods black. *Marshes,* fens. T, northern.

7* TUBEROUS PEA *Lathyrus tuberosus.* Medium/tall clambering perennial, more or less hairless; stems angled. Lflets one pair. Fls 2-7, bright *crimson,* slightly fragrant, 12-20 mm; June-July. Grassy and cultivated ground. (B), F, G.

8* HAIRY VETCHLING *Lathyrus hirsutus.* Medium/tall clambering annual, slightly downy; stems winged. Lflets one pair, tendrils branched. Fls 1-3, pale red-purple, fading blue, keel *creamy,* 7-15 mm; June-Aug. Pods brown, downy. Grassy and bare places. (B), F, G, eastern. **8a** *L. sphaericus* has stems angled, solitary orange-red fls and hairless pods. F, G, S, southern.

9* GRASS VETCHLING *Lathyrus nissolia.* Short/tall erect annual, hairless or slightly downy. Unique among peaflowers of region for long narrow *grasslike* 'lvs', (actually modified stems) with no tendrils, very hard to detect among the grass. Fls 1-2, long-stalked, crimson; May-July. Pods pale brown. Grassy places. B, F, G.

10* YELLOW VETCHLING *Lathyrus aphaca.* Medium/tall hairless clambering annual. Unique among peaflowers for its lvs having become unbranched *tendrils* direct from the stems and its stipules developed into *triangular* greyish apparent leaves, *joined* around the stem. Fls usually solitary, yellow; May-Aug. Dry grassy places. B, F, G.

3a* Black Pea

4* Bitter Vetchling (var. *tenuifolius*)

Pea Family (contd.)

1* SAINFOIN *Onobrychis viciifolia.* Short/medium erect, rarely prostrate, downy perennial. Lvs *pinnate.* Fls in stalked spikes, bright *pink;* June-Sept. Pods warty, pitted. Dry grassy and waste places; widely cultivated. (T), from E Europe. **1a** *O. arenaria* has lflets narrower and smaller paler fls with purple veins. F, G, southern. **1b** *Hedysarum hedysaroides* has purplish fls and pods constricted between seeds. G.

2* REST-HARROW *Ononis repens.* Short/medium semi-erect hairy undershrub; stem hairy all round, sometimes with soft *spines.* Lvs *trefoil* or oval, blunt or slightly notched. Fls in lfy stalked spikes, pink, wings equalling keel; July-Sept. Dry grassy places. T. **2a* Spiny Rest-harrow** *O. spinosa* □ is erect with sharper spines, stem with two lines of hairs, lflets never notched and sometimes pointed, and reddish-pink fls with wings shorter than keel. **2b** *O. arvensis* is taller, with fls in pairs. G, S.

3 LARGE YELLOW REST-HARROW *Ononis natrix.* Medium erect undershrub, stickily hairy. Lvs trefoil, variable. Fls in branched spikes, *deep yellow,* often veined red or purple; May-Aug. Pods short. Dry open places. F, G, southern. **3a** *O. pusilla* has broader lflets and fls half the size. F.

4* SMALL REST-HARROW *Ononis reclinata.* Low semi-prostrate annual, stickily hairy, with shiny dots. Lvs trefoil. Fls in lfy spikes, pink, 5-10 mm; May-June. Pods hang straight down when ripe. Sandy turf and cliffs by sea. B (rare), F.

5* KIDNEY VETCH *Anthyllis vulneraria.* Very variable, low/short, semi-prostrate/erect, annual/perennial, silkily hairy. Lvs *pinnate,* the lowest sometimes lanceolate. Fls in heads, sometimes paired, yellow, orange, red, purple, whitish or particoloured; two lfy bracts at base; Apr-Sept. Dry grassland, often by sea and on mountains. T.

6* RIBBED MELILOT *Melilotus officinalis.* Medium/tall hairless biennial. Lvs trefoil. Fls in stalked *spikes,* yellow, keel shorter than wings and standard; June-Sept. Pods hairless, brown when ripe. Bare and waste ground. T, but (B). **6a* Tall Melilot** *M. altissima* has deeper yellow fls, keel equalling wings and standard, and pods downy, black when ripe. **6b** *M. dentata* has smaller fls, wings shorter than standard, longer than keel, and hairless net-veined pods, blackish-brown when ripe. Saline soils. G, S. **6c* Small Melilot** *M. indica* is a short/medium annual, with smaller fls, wings and keel both shorter than standard, and pods hairless, net-veined, olive-green when ripe. (B), F, G, southern.

7* WHITE MELILOT *Melilotus alba.* Tall hairless biennial. Lvs trefoil. Fls in stalked spikes, *white;* July-Oct. Pods hairless, net-veined, brownish when ripe. Bare and waste ground; grown as Bokhara Clover. T, but (B).

8 CLASSICAL FENUGREEK *Trigonella foenum-graecum.* Short/medium annual, almost hairless. Lvs trefoil, toothed. Fls *1-2* at base of upper lvs, yellowish-white, tinged purple at base; Apr-May. Pods hairless, slightly curved, beaked. Bare and waste ground; widely cultivated. F, G, southern. **8a** *T. monspeliaca* is hairy, with smaller lvs and fls; fls yellower, not purple at base, 4-14 in a spike. F.

9* LUCERNE *Medicago sativa.* Medium hairless perennial. Lvs *trefoil.* Fls in short spikes, of varying shades of violet (spp. *sativa),* yellow (**9a** ssp. *falcata* □), green and livid black (hybrids between the two); June-Oct. Pods sometimes curved or spiral. Grassy, waste and cultivated places; widely grown as Alfalfa. T.

5* Kidney Vetch, colour varieties

127

Pea Family (contd.)

1* BIRDSFOOT TREFOIL *Lotus corniculatus*. More or less prostrate perennial, downy or hairless. Lvs with five lflets, lowest pair bent back so that lvs appear *trefoil*. Fls 2-7 in a head, yellow, often tinged orange or red, 10-16 mm; sepal-teeth erect in bud, an obtuse angle between the two upper; May-Sept. Pods straight, in a head resembling a bird's foot, 15-30 mm. Grassy places. T. **1a* Narrow-leaved Birdsfoot Trefoil** *L. tenuis* □ is slenderer, with narrower lflets, fewer and smaller yellow fls in each head, and the rear sepal-teeth converging; June-Aug. **1b* Greater Birdsfoot Trefoil** *L. uliginosus* is larger and erect, medium/tall, with broader lflets greyish beneath, 5-12 fls in each head and sepal-teeth spreading in bud, the two upper at an acute angle; June-Aug. Damp grassland, marshes.

2* HAIRY BIRDSFOOT TREFOIL *Lotus subbiflorus*. Prostrate *hairy* annual. Lvs as 1 (above). Fls 2-4 in a head, on stalks much longer than lvs, orange-yellow, 5-10 mm; June-Sept. Pods 6-12 mm. Grassy places. B, F. **2a* Slender Birdsfoot Trefoil** *L. angustissimus* is less hairy, with 1-2 fls in a head, on stalks usually shorter than lvs, yellow, and much longer (15-30 mm) pods; May-Aug.

3* DRAGON'S TEETH *Tetragonolobus maritimus*. Prostrate patch-forming perennial, hairless or downy. Lvs trefoil. Fls *solitary*, pale yellow, sometimes red-veined, 25-30 mm; May-Sept. Pods 4-angled, dark brown, 30-60 mm. Grassy places, often damp. T, but (B, rare).

4* HORSESHOE VETCH *Hippocrepis comosa*. Prostrate downy perennial. Lvs *pinnate*, ending in a lflet. Fls 5-12 in a head, yellow, 6-10 mm; May-July. Pods curved, with horseshoe-like segments, 15-30 mm. Dry grassland. B, F, G.

5* BIRDSFOOT *Ornithopus perpusillus*. Prostrate downy annual. Lvs *pinnate*, ending in a lflet. Fls 3-8 in a head, yellowish-white or pink, veined red, 3-5 mm; bracts pinnate, lf-like; May-Aug. Pods beaded, 10-18 mm, in a head resembling a bird's foot. Bare and grassy places. T, southern. **5a* Orange Birdsfoot** *O. pinnatus* □ has 1-5 orange-yellow fls with no lf-like bract. B(rare), F.

6* BLACK MEDICK *Medicago lupulina*. Low, often prostrate annual, usually downy. Lvs trefoil, lflets with a minute *point*. Fls yellow, 10-50 in a short-stalked rounded head, 3-8 mm; Apr-Oct. Pods curved, *black* when ripe, not covered by dead fls. Bare and grassy places; widely cultivated as Nonsuch. T. **6a* Lesser Trefoil** *Trifolium dubium* □ is almost hairless; lflets very slightly notched, not pointed, the middle one longer-stalked; fls 3-15 in the head; pods straight, covered by dead brown fls; May-Sept. Cf. 8. **6b* Slender Trefoil** *Trifolium micranthum* □, is like 6a with all lflets unstalked, fls 1-6 in a head, with long hairlike stalks, 2-3 mm. Southern.

7* SPOTTED MEDICK *Medicago arabica*. Prostrate annual, hairless or slightly downy. Lvs trefoil, usually with a *dark spot* on each lflet; stipules toothed. Fls 1-4 in a head, yellow, 5-7 mm; Apr-Sept. Pods spiny, spirally coiled, faintly netted. Bare and grassy places. B, F, G, southern. **7a* Toothed Medick** *M. polymorpha* is smaller and hairier, with unspotted lvs, stipules more deeply toothed, and strongly netted but more shortly spiral pods; May-Aug. **7b* Bur Medick** *M. minima* is smaller still, and downy, with unspotted lvs, untoothed pointed stipules, and two rows of hooked spines on the pods; Apr-July. T.

8* HOP TREFOIL *Trifolium campestre*. Low/short annual, hairy. Lvs trefoil, lflets with *no* point, the middle one longer-stalked. Fls *20-30* in rounded stalked 10-15 mm heads, yellow; May-Sept. Pods covered by pale brown dead fls. Grassy places. T. **8a* Large Hop Trefoil** *T. aureum* is a stouter biennial, with narrower lflets, all unstalked, larger, more golden-yellow, unstalked fls, and pods with a hooked beak; June-Aug. T, but (B). **8b** *T. spadiceum* is slenderer and almost hairless, with all lflets unstalked, and much larger, often paired heads of larger golden-yellow fls, turning dark brown when dead; June-Aug. Not on lime. F, G, S, eastern. **8c** *T. patens* is less hairy, with larger lvs and longer fls. Damp places. F. **8d** *T. badium* has fls turning bright chestnut brown. G.

Pea Family (contd.)

CLOVERS: *Trifolium*. pp.128-30. Lvs trefoil. Fls small, usually in a dense rounded head, wings longer than keel, sepal-tube 5-toothed. Pods covered by dead fls. Grassy places.

1* RED CLOVER *Trifolium pratense*. Variable erect low/tall hairy perennial. Lflets often with a whitish crescent; stipules triangular, bristle-pointed. Flheads globular or egg-shaped, sometimes paired, usually unstalked, with two lvs closely beneath, 20-40 mm; fls of varying shades of *pink-purple,* rarely creamy or white; May-Oct. Widely cultivated. T. **1a* Zigzag Clover** *T. medium* is usually short/ medium, with stems often slightly zigzag, stipules narrow, not bristle-pointed, and flheads longer-stalked, fls a uniform hard reddish-purple; June-Aug. Also in open woods and scrub. **1b,** p.301.

2 MOUNTAIN CLOVER *Trifolium montanum*. Short/medium hairy perennial. Flheads often paired, well stalked, 15-30 mm; fls *yellowish-white,* rarely pink, turning brownish; May-July. Especially in mountains. F,G,S. **2a* Sulphur Clover** *T. ochroleucon* ☐ has two lvs closely below the larger (20-40 mm) un- or short-stalked flheads. Lowlands. B,F,G. **2b,** p.301.

3* WHITE CLOVER *Trifolium repens*. Creeping perennial, more or less hairless, stems rooting at lf-junctions. Lflets usually with a whitish mark. Flheads globular, long-stalked, 15-30 mm; fls *white,* pink, or rarely purple, turning brown, scented, angles between sepal-teeth acute; May-Oct. Widely cultivated as Wild White, Dutch or Kentish Clover. T. **3a* Western Clover** *T. occidentale* has unmarked lvs, almost circular lflets, red stipules and small heads of scentless fls; Mar-May. Near the sea. B,F, western. **3b* Alsike Clover** *T. hybridum* ☐ is short/medium and erect, with unmarked lvs and shorter-stalked heads of more often pink fls, angles between sepal-teeth blunt; June-Sept. Often on waste ground; also widely cultivated. (B,S),F,G. **3c** *T. michelianum* is annual with pink fls. Wet places. F, rare.

4* STRAWBERRY CLOVER *Trifolium fragiferum*. Creeping hairy perennial; stems often rooting at lf-junctions. Flheads globular, 10-14 mm, *swelling* in fr to appear like miniature pale pink strawberries; fls pink; June-Sept. T. **4a* Sea Clover** *T. squamosum* is a short, often erect annual with two lvs closely beneath the unstalked flheads and fr heads not swollen but with conspicuously star-like spreading sepal-teeth; June-Oct. Damp grassland, especially by the sea. B,F. **4b*,** p.301.

5* KNOTTED CLOVER *Trifolium striatum*. Low, more or less erect downy annual. Lvs downy on *both* sides. Flheads egg-shaped, unstalked at base of lvs, 10-15 mm; fls pink, sepal-teeth swollen in fr, with erect teeth; June-July. Dry places. T, southern. **5a*, 5b*, 5c*,** p.301.

6* HARESFOOT CLOVER *Trifolium arvense*. Low/short hairy annual/biennial. Lflets narrow. Flheads *egg-shaped,* stalked, 10-20 mm; fls whitish or pale pink, much shorter than sepal-tube; June-Sept. Scarce on lime. T. **6a* Upright Clover** *T. strictum* ☐ has larger but still narrower lflets, and smaller (7-10 mm), more globular, longer-stalked heads of larger bright pink fls, much longer than calyx; June. B(rare),F.

7* CRIMSON CLOVER *Trifolium incarnatum*. Short/medium hairy annual. Flhds elongated, 10-40 mm; fls *bright* crimson, pink or creamy white, sepal-teeth spreading in fr; June-July. Widely cultivated. Pale-fld subsp. *molinerii,* Long-headed Clover, is native on cliff-tops by the sea. B,F(G,S), southern. **7a,** p.301

8* FENUGREEK *Trifolium ornithopodioides*. Prostrate hairless annual. Fls 2-4 in short-stalked heads, white or pale pink, 6-8 mm; June-July. Pods much longer than sepal-tube. Bare dry places, often damp in winter. B,F,G, southern. **8a* Burrowing Clover** *T. subterraneum* ☐ has larger (8-14 mm) fls and fr heads buried in the ground; May-June.

9* SUFFOCATED CLOVER *Trifolium suffocatum*. Prostrate hairless annual. Flheads unstalked, tightly packed, 5-6 mm; fls white, almost *hidden* by green calyx-teeth, recurved in fr; July-Aug. Dry places. B,F.

Wood-Sorrel Family Oxalidaceae

Downy. Lvs trefoil, often closing up at night. Fls 5-petalled, cup-shaped. Fr a capsule.

1* WOOD-SORREL *Oxalis acetosella*. Low creeping perennial. Fls *white,* usually veined mauve, sometimes tinged purple; Apr-May. Woods, shady places, also on mountains. T.

2* YELLOW OXALIS *Oxalis corniculata*. Low/short creeping annual/perennial, stems rooting at lf-junctions. Lvs often purplish, stipules tiny. Fls *yellow,* 4-7 mm; May-Oct. Fr stalks bent right back. Bare, especially cultivated, ground. T, southern, but (B). **2a* Upright Yellow Oxalis** *O. europaea* is erect, not rooting at lf-junctions, with lvs rarely purplish, no stipules and fr stalks not bent back. (T) from N America. **2b***, p. 302.

3* PINK OXALIS *Oxalis articulata*. Low/short tufted perennial. Lvs spotted pale orange beneath. Fls in a cluster, *rose-pink;* May-Sept. Garden escape. (B,F), from S America. **3a*, 3b***, p. 302.

4* BERMUDA BUTTERCUP *Oxalis pes-caprae*. Low/short tufted perennial. Lflets sometimes with a pale brown blotch. Fls in an umbel-like cluster, *yellow,* 20-25 mm, Mar-June. Naturalised weed. (B, rare, F), from S Africa.

Geranium Family Geraniaceae: see p. 134

5* COMMON STORKSBILL *Erodium cicutarium*. Low/medium annual/biennial, hairy, sometimes stickily so; often strong-smelling. Lvs *2-pinnate,* stipules lanceolate. Fls in umbel-like heads with five, easily shed, often unequal petals, purplish-pink to white, sometimes with a blackish spot at base of two upper petals, 4-11 mm; Apr-Sept. Fr with a long twisted beak. Bare and grassy places, frequent by the sea. T. **5a* Musk Storksbill** *E. moschatum* smells musky, is usually stickily hairy, and has 1-pinnate lvs, broader stipules and larger (15 mm) always purplish-pink fls with no blotches. **5b**, p. 302.

6* SEA STORKSBILL *Erodium maritimum*. Low, often prostrate, downy annual/biennial. Lvs oval, toothed, often in a rosette. Fls usually solitary, pale pink or white, the five petals dropping very soon; May-Aug. Fr with a long twisted beak. Bare ground, short turf, usually near the sea. B, F.

Flax Family Linaceae

Slender, hairless. Lvs undivided, untoothed; no stipules. Fls 5-petalled (except 10), open. Fr a capsule.

7* PERENNIAL FLAX *Linum perenne* agg. Short/medium hairless perennial. Lvs more or less linear, alternate, 1-veined or obscurely 3-veined. Fls in a cluster, on erect stalks, pale to bright blue, 25 mm; anthers and stigmas unequal; inner sepals often blunt, much shorter than fr; May-Aug. Dry grassland. B, F, G. **7a* Pale Flax** *L. bienne* □ has 1-3 veined lvs, pale bluish-lilac fls that drop early, and all sepals pointed, nearly as long as the more pointed fr. B, F. **7b* Common Flax** *L. usitatissimum* □ is annual with lvs always 3-veined and sepals like 7a; June-Oct. Widely cultivated. (T). **7c, 7d, 7e,** p. 302.

8* PURGING FLAX *Linum catharticum*. Low annual. Lvs oblong, opposite, 1-veined. Fls in a loose cluster, white, 4-6 mm; May-Sept. Grassland, fens, especially on lime. T.

9 YELLOW FLAX *L. flavum*. Variable medium hairless perennial, sometimes woody at base. Lvs lanceolate, the lower broader at tip. Fls yellow, *c* 30 mm, in a cluster. Dry grassy places. G.

10* ALLSEED *Radiola linoides*. A tiny (rarely above 5 cm) annual, with repeatedly branching thread-like stems. Lvs oblong, opposite, 1-veined. Fls 4-petalled, white, 1 mm, equalling sepals; July-Aug. Bare damp places, not on lime. T, western.

9 Yellow Flax

1

2

3

4

5

7b

7

7a

6

8

10

Geranium Family Geraniaceae

pp. 132-34. Hairy or downy. Lvs deeply palmately lobed or cut, with stipules. Fls 5-petalled, sepals often bristle-tipped, stamens prominent. Fr ends in long straight pointed beak, whence "cranesbill".

1* MEADOW CRANESBILL *Geranium pratense*. Medium/tall perennial. Lvs cut nearly to base. Fls *blue*, 25-30 mm; petals not notched; June-Sept. Fl stalks bent down after flg, becoming erect when fr ripe. Grassy places, usually on lime. T. **1a* Wood Cranesbill** *G. sylvaticum* ☐ has less deeply cut lvs, generally smaller, more reddish-mauve fls and fl stalks always erect. Also in woods and on mountains. Northern.

2* BLOODY CRANESBILL *Geranium sanguineum*. Short/medium tufted perennial. Lvs cut nearly to base. Fls *bright* purple-crimson, very rarely pink, 25-30 mm. Dry grassy places, usually on lime. T.

3* FRENCH CRANESBILL *Geranium endressii*. Medium perennial. Lvs cut to more than half-way. Fls *pink*, unveined, 25 mm; petals not notched; June-Aug. Garden escape. (B,F), from S W France. The hybrid **3 × 3a***, with dark-veined, slightly notched pink petals, is more frequent in B than either. **3a* Pencilled Cranesbill** *G. versicolor* ☐ has paler pink fls, pencilled with violet veins, and notched petals. From Italy and Balkans. **3b* Knotted Cranesbill** *G. nodosum* has stems swollen at lf-junctions, smaller purple-veined mauve fls and petals deeply notched. (B,F,G), from S Europe. **3c* Dusky Cranesbill** *G. phaeum* has less deeply cut lvs, smaller drooping maroon fls and unnotched wavy-edged petals. Damp and shady places; garden escape in the West. (B),F,G.

4 MARSH CRANESBILL *Geranium palustre*. Medium/tall perennial. Lvs cut to more than half-way. Fls pale *purple* or lilac, 30 mm; petals not or very slightly notched; July-Aug. Fr stalks bent down. Damp grassy places. F,G,S, eastern.

5* HEDGEROW CRANESBILL *Geranium pyrenaicum*. Medium perennial. Lvs cut to about half-way. Fls pink-purple to lilac, often *mauve-pink*, 15 mm; petals well notched; May-Sept. Fr stalks bent down. Grassy places, open woods. T, southern, but (B). **5a** *G. bohemicum* agg. ☐ is annual/biennial, with smaller, bright violet-blue fls, less deeply notched petals and erect fr stalks. Eastern. **5b,** p. 302.

6* HERB ROBERT *Geranium robertianum*. Medium annual/biennial, strong-smelling, often reddening. Lvs triangular, 1-2-pinnate. Fls *pink*, 20 mm; petals scarcely notched, pollen orange; Apr-Nov. Fr slightly wrinkled. Shady places, rocks, shingle. T. **6a* Little Robin** *G. purpureum* ☐ has much smaller fls, yellow pollen and strongly wrinkled fr. B,F.

7* SHINING CRANESBILL *Geranium lucidum*. Short annual, almost hairless, often reddening. Lvs *glossy,* rounded, cut to half-way. Fls pink, 15 mm; petals not notched; May-Aug. Fr wrinkled. Shady rocks and hedge-banks. T.

8* DOVESFOOT CRANESBILL *Geranium molle*. Short semi-prostrate annual, with long hairs on stems. Root lvs cut to less than half-way, stem lvs to half-way or more, all rounded. Fls dark or pale pinkish-purple, 10 mm; petals well notched; Apr-Sept. Fr hairless, usually wrinkled. Bare and grassy places. T. **8a* Round-leaved Cranesbill** *G. rotundifolium* ☐ is taller, with rounder, scarcely cut, wavy-edged lvs, pink fls, unnotched petals and downy unwrinkled fr. **8b* Small-flowered Cranesbill** *G. pusillum* ☐ has only short hairs on stem, stem lvs more narrowly cut to more than half-way, smaller lilac fls, half the stamens without anthers, and fr downy, not wrinkled.

9* CUT-LEAVED CRANESBILL *Geranium dissectum*. Short/medium straggling annual/biennial. Lvs cut almost to base. Fls pink-purple, 8 mm; petals notched; May-Sept. Fr downy. Bare and sparsely grassy places. T. **9a* Long-stalked Cranesbill** *G. columbinum* ☐ sometimes reddens and has larger longer-stalked fls, petals not notched and fr hairless. Dry grassy places.

8a* Round-leaved Cranesbill

8b* Small-flowered Cranesbill

Spurge Family Euphorbiaceae

SPURGES *Euphorbia.* Stems with milky juice. Lvs undivided, usually alternate and untoothed. Fls in usually broadening umbel-like clusters, yellowish-green; no petals or sepals, but a solitary 3-styled female fl, surrounded by several 1-anthered male fls is cupped by a calyx-like bract, with 4-5 glands on its lip, and 2-3 small lflike bracts at its base. Fr rounded, stalked.

1* WOOD SPURGE *Euphorbia amygdaloides.* Medium/tall unbranched perennial, downy, often tinged red. Lvs lanceolate. Fls yellow, glands horned, bracts joined together; *Apr-May. Woods,* scrub. B,F,G. **1a* Cypress Spurge** *E. cyparissias* □ is shorter and hairless, with much narrower lvs, turning bright red, and bracts not joined; May-Aug. Bare and grassy places. T, but (B). **1b* Leafy Spurge** *E. esula* grows in large patches, has bracts not joined, and may have glands either long- or short-horned; June-Aug. Grassy and waste places. T, but (B). **1c** *E. lucida* is like 1b with lvs shiny above. G, rare.

2* SEA SPURGE *Euphorbia paralias.* Short/medium greyish hairless perennial; unbranched. Lvs fleshy, oval, midrib *not* prominent beneath, overlapping up the stem. Fls with short-horned glands; June-Sept. Sands by the *sea.* B,F,G, southern. **2a** *E. seguieriana* has narrower, more pointed, less fleshy lvs and glands not horned; Apr-Aug. Woods, grassy places, on lime. F,G.

3* PORTLAND SPURGE *Euphorbia portlandica.* Short greyish hairless perennial; branched from the base, often reddening. Lvs oval, *midrib* prominent below. Fls with long-horned glands; Apr-Sept. Sands and short turf by the sea. B,F.

4* CAPER SPURGE *Euphorbia lathyrus.* Tall greyish hairless biennial. Lvs narrow, opposite. Fls with horned glands; June-Aug. Fr large, *3-sided,* resembling a green caper. Bare and cultivated ground. B(rare),F,G.

5* IRISH SPURGE *Euphorbia hyberna.* Medium little-branched perennial, almost hairless; often in clumps. Lvs untoothed, downy beneath, reddening. Fls with yellow bracts and *kidney-shaped* yellow glands; Apr-July. Fr with long and short warts. Damp or shady places. B,F, southern. **5a, 5b, 5c*, 5d,** p.302.

6* SUN SPURGE *Euphorbia helioscopia.* Short, usually unbranched, hairless annual. Lvs oval, finely toothed. Fls in a broad umbel, bracts often yellowish, glands green, oval; Apr-Nov and through mild winters. Fr smooth. Common weed of cultivation. T.

7* PETTY SPURGE *Euphorbia peplus.* Low/short, often branched hairless annual. Lvs oval, untoothed. Fls green, glands horned; Apr-Nov and through mild winters. Fr with wavy ridges. Common weed of cultivation. T. **7a* Broad-leaved Spurge** *E. platyphyllos* □ is taller and may be downy, with longer lvs finely toothed at the tip, larger yellower fls with rounded glands and fr with rounded warts; June-Sept. B,F,G. **7b* Upright Spurge** *E. serrulata* differs from 7a in being slenderer and always hairless, with stems often red and long warts on fr. Open woods. T, rare in B.

8* DWARF SPURGE *Euphorbia exigua.* Low slender greyish annual, hairless. Lvs linear. Fls yellowish, glands horned; May-Oct. Fr smooth. Weed of cultivation, especially cornfields. T.

9* PURPLE SPURGE *Euphorbia peplis.* Prostrate fleshy hairless annual; stems crimson. Lvs oblong, greyish, opposite. Fls stalked, along stems, 1-2 mm; July-Sept. Sandy shores. B(rare),F, southern.

10* DOG'S MERCURY *Mercurialis perennis.* Low/short unbranched downy foetid perennial. Lvs broad lanceolate, opposite. Fls in stalked tassels, petalless, green, 4-5 mm, stamens and styles on separate plants; Feb-May. Woods, shady places, often carpeting the ground; also on mountains. T. **10a* Annual Mercury** *M. annua* is paler, almost hairless and often branched, with female (styled) tassels scarcely stalked and sometimes on same plant as male; May-Nov and through mild winters. A weed of bare ground. **10b** *M. ovata* has very short fl-stalks. G, eastern, rare.

137

Balsam Family   Balsaminaceae

BALSAMS *Impatiens*. Hairless annuals, stems fleshy. Lvs oval, stalked, slightly toothed. Fls in long-stalked spikes from base of lvs, 5-petalled, with a broad lip, a small hood and a spur, often curved. Fr cylindrical, exploding when ripe.

1* HIMALAYAN BALSAM *Impatiens glandulifera*. Tall, stems often reddish. Lvs opposite or in threes, with small red teeth. Fls large, pale to dark pink-purple, with short bent spur; July-Oct. Bare places, especially by streams. (T).

2* TOUCH-ME-NOT BALSAM *Impatiens noli-tangere*. Medium/tall, lvs alternate. Fls *yellow*, sometimes spotted red-brown. spur curved downwards; July-Sept. Damp shady places. T.   **2a* Orange Balsam** *I. capensis* □   has red-spotted orange fls with spur bent into a crook; July-Sept. By rivers and canals. (B, F), from N America.

3* SMALL BALSAM *Impatiens parviflora*. Short/medium, lvs alternate. Fls *pale* yellow, spur short and straight, 3-5 mm; June-Sept. Shady or bare places. (T), from C Asia.

Milkwort Family   Polygalaceae

Hairless perennials. Lvs unstalked, usually alternate, no stipules. Fls 5-sepalled, the two inner large and petal-like on either side of the three true petals, which are joined at the base. Stamens also petal-like, in two Y-shaped bundles of four. Fr flat, heart-shaped, slightly winged.

4* COMMON MILKWORT *Polygala vulgaris*. Low perennial. Lvs lanceolate, pointed, broadest at or below the middle, the lower shorter and broader than the upper. Fls in stalked spikes, usually blue, also often mauve, pink, white or white tipped mauve or blue, 4-7 mm; May-Sept. Fr shorter than inner sepals. Grassy places. T. **4a*, 4b, 4c*, 4d*,** p. 302.

5 SHRUBBY MILKWORT *Polygala chamaebuxus*. More or less prostrate ever-green undershrub. Lvs oval to lanceolate, rather leathery. Fls usually yellow tipped purple, also red, pink or white; Apr-Sept. Woods, grassy places, especially on mountains. F, G, south-eastern.

Daphne Family   Thymelaeaceae

Hairless or almost so. Lvs untoothed, shortly stalked, usually alternate, no stipules. Fls in clusters at base of lvs, petalless with four petal-like sepals joined at base.

6* MEZEREON *Daphne mezereum*. Deciduous shrub to 2 m. Lvs lanceolate, pale green. Fls strongly scented, *pinkish-purple;* Feb-Apr, before the lvs. Fr a *red* berry. Woods on lime. T.

7* SPURGE LAUREL *Daphne laureola*. Evergreen shrub to 1 m. Lvs lanceolate, dark green, rather leathery. Fls slightly scented, *yellow-green;* Jan-Apr. Fr a *black* berry. Open woods, often on lime. B, F, G.

8 ANNUAL THYMELAEA *Thymelaea passerina*. Medium annual. Lvs almost linear, pointed. Fls 1-3 together, greenish, tiny; July-Oct. Fr a downy nut. Dry bare uncultivated places. F, G.

Rue Family   Rutaceae

9 BURNING BUSH *Dictamnus albus*. Medium/tall bushy hairy perennial, pungent from aromatic oil, which may burst into flame if a match is held near it in still summer weather. Lvs pinnate. Fls 5-petalled, white or pink with purple streaks and prominent purple stamens, in stalked spikes. Dry places. G.

9 Burning Bush

139

Currant Family   Grossulariaceae

Small deciduous shrubs, to 1·5 m. Lvs alternate, palmately lobed. Fls 5-petalled, green. Fr a berry.

1* RED CURRANT *Ribes rubrum* agg. Stems erect. Lvs hairless or downy beneath, *not* aromatic. Fls purple-edged, in drooping stalked spikes; Apr-May. Fr red, acid. Damp woods, fens, by fresh water; widely cultivated. T. **1a* Black Currant** *R. nigrum* □ has aromatic lvs and black fr. **1b* Mountain Currant** *R. alpinum* □ is bushier, with smaller, more deeply lobed lvs, fls in erect clusters, male and female on different bushes, each with a conspicuous green bract, and tasteless fr. Upland woods, often on lime; rarely cultivated. **1c,** p.301.

2* GOOSEBERRY *Ribes uva-crispa.* Stems spreading, *spiny.* Lvs smaller than 1. Fls purple-edged, petals turned back, 1-3 together; Mar-May. Fr egg-shaped, green, often tinged red or yellow, usually hairy. Woods, scrub, hedges; widely cultivated. T.

Spindle-Tree Family   Celastraceae

3* SPINDLE-TREE *Euonymus europaeus.* Deciduous shrub, rarely a small tree, to 6 m, with 4-sided green twigs. Lvs elliptical, opposite,scarcely toothed. Fls small, greenish-white, 4-petalled, in small clusters; May-June. Fr a bright *coral-pink* berry. Woods, scrub, hedges, often on lime. T. **3a** *E. latifolius* has grey-brown, more rounded twigs, larger, more obviously toothed lvs, and browner fls, usually 5-petalled. (F), G.

Box Family   Buxaceae

4* BOX *Buxus sempervirens.* Evergreen shrub or small tree, to 5 m, twigs green. Lvs oval, shiny, *leathery,* opposite. Fls petalless, greenish-yellow, in small clusters at base of upper lvs; Mar-May. Fr dry. Dry hillsides, widely planted. B (rare), F, G.

Buckthorn Family   Rhamnaceae

5* BUCKTHORN *Rhamnus catharticus.* Deciduous shrub or small tree to 6 m, often thorny. Lvs broad elliptical, finely toothed, *opposite,* with *conspicuous* veins. Fls small, green, 4-petalled, in small clusters at base of upper lvs; male and female separate; May-June. Fr a black berry. Scrub, hedges, fens, usually on lime. T. **5a,** p. 302.

6* ALDER BUCKTHORN *Frangula alnus.* Deciduous shrub or small tree to 5 m, thornless. Lvs broad elliptical, untoothed, *alternate.* Fls small, green, usually 5-petalled, in small clusters at base of upper lvs; May-June. Fr a black berry, red at first. Damp heaths and woods, fens; usually not on lime. T.

Oleaster Family   Elaeagnaceae

7* SEA BUCKTHORN *Hippophae rhamnoides.* Deciduous shrub or small tree to 11 m, *thorny,* suckering freely and forming thickets. Brown twigs and untoothed narrow lanceolate alternate lvs both covered with *silvery scales,* the older lvs brownish. Fls tiny, green, petalless, male and female on different plants; Apr-May, before lvs. Fr a bright orange berry. Coastal dunes, sea cliffs, mountain river beds. T.

Tamarisk Family   Tamaricaceae

8* TAMARISK *Tamarix gallica.* Evergreen shrub to 3 m, hairless; twigs reddish. Lvs minute, *scale-like,* overlapping, alternate, often greyish, producing a feathery foliage. Fls pink, 5-petalled, in stalked spikes; July-Sept. By salt and fresh water, often planted. (B), F. **8a** *Myricaria germanica,* p. 301.

Bladder-Nut Family   Staphyleaceae

9* BLADDER-NUT *Staphylea pinnata.* Deciduous shrub to 5 m. Lvs pinnate, opposite. Fls white, 5-petalled, in hanging stalked clusters; May-June. Fr *inflated.* Woods on lime; often planted. (B, F), G.

Mallow Family Malvaceae

Downy or softly hairy. Lvs rounded, palmately lobed, toothed, stalked, alternate, with stipules. Fls in whorls at base of lvs, open, with five notched petals; calyx in two rings, the outer smaller; stamens numerous, prominently bunched together. Fr a flat disc.

1* MUSK MALLOW *Malva moschata.* Medium/tall perennial, stem hairs often purple-based. Lvs deeply and *narrowly* cut. Fls 40-60 mm, solitary, *rose-pink;* hairs on stalks unbranched, outer sepal-ring 3-lobed, hairless or almost so; July-Aug. Fr hairy. Grassy places, scrub. T. **1a** *M. alcea* is rather larger, with branched hairs on fl stalks and outer calyx downy. F, G, S.

2* DWARF MALLOW *Malva neglecta.* More or less prostrate annual. Fls 3-6 together, pale lilac or whitish, 10-15 mm; petals twice as long as sepals; June-Sept. Waste places. T. **2a* Chinese Mallow** *M. verticillata* is a short/medium biennial, with smaller fls in denser whorls, the calyx enlarging in fr. (B, F, G), from E Asia.

3* LEAST MALLOW *Malva parviflora.* Short/medium annual, sometimes hairless. Fls 2-4 together, pale lilac-blue, 4-5 mm; petals slightly longer than short-haired or hairless sepals, which greatly enlarge in fr; June-Sept. Fr segments net-veined, winged on angles. Waste places. (B), F. **3a* Small Mallow** *M. pusilla* is usually prostrate, with up to ten pinkish fls in each cluster, sepals long-haired, not enlarging, and angles of fr segments sharp but not winged. T, northern, but (B).

4* COMMON MALLOW *Malva sylvestris.* Variable, often sprawling, medium/tall annual/perennial. Lvs often with a small *dark spot.* Fls 2 or more together, pale to dark pink-purple, 25-40 mm; outer sepal-ring not joined at base; June-Oct. Fr segments with sharp angles. Waste places, waysides. T. **4a** *Lavatera thuringiaca* has larger solitary fls and outer sepal-ring joined at base. G, (S).

5* TREE MALLOW *Lavatera arborea.* Tall biennial to 3 m, softly downy, *woody* at base. Fls pink-purple, veined purple; outer sepal-ring 3-lobed, cup-shaped; Apr-Sept. Rocks and bare ground by sea. B, F.

6* SMALLER TREE MALLOW *Lavatera cretica.* Medium/tall annual/biennial, differing from 4 especially in its less deeply lobed lvs, paler and pinker fls, outer sepal lobes broader and joined at base, and fr segments with rounded angles; Apr-July. Bare places. B (rare), F.

7* MARSH MALLOW *Althaea officinalis.* Tall perennial, covered with velvety down, whose roots are used to make the sweetmeat. Fls pink, 25-30 mm; outer sepal-ring 6-9-lobed; Aug-Sept. Damp saline places, especially near sea. B, F, G. **7a* Hollyhock** *Alcea rosea* ☐ is very much larger, with fls up to 100 mm, often white or mauve. Waste places. (B, F, G), of unknown origin.

8* ROUGH MALLOW *Althaea hirsuta.* Short/medium, sometimes semi-prostrate annual. Lvs becoming more deeply cut up the stems. Fls pale pink, becoming bluer; outer sepal-ring with 6-9 narrow lobes; May-July. Fr rough, hairless. Cf. 1. Grassy and bare places, weed of cultivation; likes lime. (B), F, G.

5* Tree Mallow

7a* Hollyhock

4* Common Mallow

St John's Wort Family   Guttiferae

Hairless, except 4, 9. Lvs opposite, untoothed, with translucent veins. Fls in branched, usually lfy, clusters (except 2), open, yellow, with five petals and sepals and many stamens. Fr a dry capsule, except 1.

1* TUTSAN *Hypericum androsaemum.* Medium/tall half-evergreen undershrub; stems 2-winged, often reddish. Lvs broad oval, faintly aromatic when crushed. Fls 20 mm; sepals oval, unequal; June-Aug. Fr a fleshy *berry,* green, then red, finally purplish-black. Woods, shady places, cliff ledges. B, F. **1a* Stinking Tutsan** *H. hircinum* has stems usually 4-angled, lvs narrower and foetid when crushed, larger (30 mm) fls with narrower sepals, and dry fr. (B, F), from the Mediterranean. **1b* Tall Tutsan** *H. inodorum,* intermediate between 1 and 1a, is taller than both, to 2 m, with no smell when lvs crushed, and fr fleshy and red at first, then dry. (B, F), from Madeira.

2* ROSE OF SHARON *Hypericum calycinum.* Creeping medium evergreen undershrub. Lvs elliptical. Fls solitary, *70-80 mm,* anthers red; June-Oct. Banks, shrubberies. (B, F), from S E Europe.

3* PERFORATE ST JOHN'S WORT *Hypericum perforatum* agg. Medium perennial, with two raised lines down stem. Lvs oval to linear, with *translucent dots.* Fls 20 mm, petals black-dotted, especially on margins; pointed sepals sometimes also black-dotted; July-Sept. Grassy and bushy places. T. **3a* Imperforate St John's Wort** *H. maculatum* has square but not winged stems, normally no translucent dots on lvs, black dots usually on surface rather than margins of petals, and blunt sepals. Cf. 5. Hybridises with 3. Shadier places.

4* HAIRY ST JOHN'S WORT *Hypericum hirsutum.* Medium/tall *downy* perennial. Lvs elliptical, with translucent dots. Fls pale yellow, sometimes red-veined, 15-20 mm; sepals edged with black dots; July-Sept. Woods, scrub, hedge-banks. T. **4a* Pale St John's Wort** *H. montanum* is stiffer, less branched and less downy, with no translucent dots in the lvs but a row of black dots on the margin beneath, and fewer paler fls. Usually on lime.

5* SQUARE-STALKED ST JOHN'S WORT *Hypericum tetrapterum.* Medium/tall perennial; stems *4-winged.* Lvs with translucent dots, almost clasping the stem. Fls pale yellow, 10 mm; sepals pointed, usually not black-dotted; July-Sept. Cf. 3a. Damp places, often by streams. T, southern. **5a* Wavy St John's Wort** *H. undulatum* has slenderer and less conspicuously winged stems, with wavy-edged lvs, larger, usually red-tinged fls, and black-dotted sepals. B, F.

6* SLENDER ST JOHN'S WORT *Hypericum pulchrum.* Slender short/medium perennial. Lvs with translucent dots and often inrolled margins. Fls rich yellow, tinged *red* beneath, 15 mm; petals and sepals with black dots, petals also red-dotted; July-Sept. Heaths, scrub, open woods; not on lime. T, western.

7* TRAILING ST JOHN'S WORT *Hypericum humifusum.* Slender, usually prostrate perennial. Lvs with translucent dots, often also with black dots on margins beneath. Fls 10 mm, sepals sometimes black-dotted at margins; July-Oct. Open woods, heathy places; not on lime. T, southern. **7a* Flax-leaved St John's Wort** *H. linarifolium* □ is short/medium, often erect, and has narrower lvs, usually with neither translucent nor black dots, the margins often inrolled and larger fls, often red beneath. Hybridises with 7. Rocky places. B(rare), F.

8* IRISH ST JOHN'S WORT *Hypericum canadense.* Low/short annual/perennial; stems with four raised lines. Lvs narrow, 1-3-veined. Fls with no black dots, 6 mm; July-Sept. Damp places. B(Ireland), G. **8a** *H. mutilum* is taller with broader 3-5-veined lvs. F, G. **8b** *H. majus* is taller with broader 5-7-veined lvs and larger fls. F, G.

9* MARSH ST JOHN'S WORT *Hypericum elodes.* Creeping low/short *greyish* perennial, conspicuously hairy. Lvs rounded. Fls less open than other St John's Worts, 15 mm; sepals red-dotted at margins; June-Sept. Shallow fresh water, damp mud. B, F, G, western.

7a* Flax-leaved St John's Wort

Violet Family   Violaceae

VIOLETS AND PANSIES. *Viola*. Lvs alternate, toothed, with stipules. Fls solitary, open, with five unequal petals, the lowest spurred; sepals with short appendages. Fr an ovoid capsule. Hybrids frequent. Violets (1-7) are perennials, with more or less heart-shaped lvs and inconspicuous stipules. Pansies (8-9) may be annual and have more deeply toothed, more or less broad lanceolate lvs, with conspicuous lflike stipules, and flatter fls.

1* SWEET VIOLET *Viola odorata*. Low creeping perennial, downy, with rooting runners. Lvs in a tuft, rounded, enlarging in summer. Fls 15 mm, *fragrant,* blue-violet or white, rarely lilac, pink or yellow; sepals *blunt;* Mar-May, Aug-Sept. Woods, scrub, hedgebanks. T. **1a* Hairy Violet** *V. hirta* has no runners, narrower lvs with longer hairs on the stalks, and unscented fls less often white. Grassland, open scrub, often on lime. **1b, 1c, 1d,** p.302.

2* COMMON DOG VIOLET *Viola riviniana*. Low perennial, almost hairless. Lvs in a tuft, heart-shaped, stipules usually toothed. Fls 15-25 mm, blue-violet, *unscented;* spur stout, curved, pale, often creamy, notched at tip; sepals *pointed,* appendages enlarging in fr; Mar-May, July-Sept. Woods, grassy places, mountains. T. **2a* Teesdale Violet** *V. rupestris* is smaller and minutely downy, with rounder lvs, fls 10-15 mm with spur pale violet, and fr downy. Hybridises with 2. Open ground, on lime. T, rare in B. **2b* Early Dog Violet** *V. reichenbachiana* □ has narrower lvs, paler 12-18 mm fls with narrower petals and an unnotched violet spur and sepal appendages less prominent in fr. Shady places. Southern. 2a-2c rarely fl in autumn. **2c,** p.302.

3* HEATH DOG VIOLET *Viola canina*. Low/short perennial, hairless or slightly downy. Lvs not in a tuft, broad to narrow. Fls 15-25 mm, bluer than 2, sometimes white, with spur yellowish or whitish. Sepals *pointed,* appendages not enlarging in fr; Apr-June. Heaths, open woods, fens. T. **3a* Pale Dog Violet** *V. lactea* □ is erecter, with narrower lvs, not heart-shaped, and pale bluish-white fls with a short greenish spur; May-June. Heaths. B, F.

4 MEADOW VIOLET *Viola pumila*. Low/short perennial, hairless. Lvs *narrow.* Fls 15 mm, roundish, *pale blue,* spur greenish, short; May-June. Grassland. F, G, S, eastern. **4a* Fen Violet** *V. persicifolia* has heart-shaped and smaller whitish fls, sometimes tinged blue, with violet veins. Fens, marshes. T, rare in B. **4b,** p.302.

5 NORTHERN VIOLET *Viola selkirkii*. Low perennial, almost hairless. Lvs tufted, heart-shaped. Fls 15 mm, pale violet, long-spurred, sepals pointed; Apr-May. *Coniferous* woods, damp places. S.

6* MARSH VIOLET *Viola palustris*. Low perennial, usually hairless. Lvs kidney-shaped. Fls 10-15 mm, pale lilac veined purple, short spurred, bracts at or below middle of stalks; Apr-July. Bogs, marshes. T. **6a, 6b,** p.302.

7 YELLOW WOOD VIOLET *Viola biflora*. Low creeping perennial, sparsely hairy. Lvs in a tuft, kidney-shaped. Fls 15 mm, bright yellow; June-Aug. Damp, shady places, mainly in mountains. F, G, S.

8* WILD PANSY *Viola tricolor*. Variable low/short annual/biennial, hairless or downy. Lvs oval to broad lanceolate; stipules lfy, pinnately lobed, the end lobe longer. Fls 10-25 mm, violet, yellow or both, petals longer than sepals, spur longer than appendages; Apr-Nov. Grassy and bare places, dunes. T. **8a* Mountain Pansy** *V. lutea* is a creeping perennial with longer fl stalks and 15-30 mm fls, the end segment of the stipules not larger than the others; May-Aug. Mountain grassland. **8b** *V. hispida* has hairy stems. F, north-western, rare.

9* FIELD PANSY *Viola arvensis*. Variable low/short annual, shortly hairy. Lvs broad to narrow; stipules lfy, semi-pinnate, end segment lanceolate. Fls 10-15 mm, cream-coloured, sometimes tinged yellow or violet, petals shorter than (sometimes equalling) sepals, spur equalling appendages; Apr-Nov. Weed of cultivation. T. **9a* Dwarf Pansy** *V. kitaibeliana* □ is smaller, and downy, with more rounded lower lvs and stipule lobes, and 4-8 mm fls never violet-tinged; Apr-July. Dry bare ground. B(rare), F, G, southern.

147

Rock-Rose Family   Cistaceae

Downy. Lvs untoothed. Fls 5-petalled, open, sepals unequal, stamens numerous. Fr egg-shaped.

1* COMMON ROCK-ROSE *Helianthemum nummularium.* Variable, more or less prostrate undershrub. Lvs lanceolate to roundish, 1-veined, usually downy *white* beneath, opposite, with lflike stipules. Fls 20-25 mm, yellow, rarely white, cream, orange or pink; May-Sept. Grassy and rocky places on lime. T, southern. **1a* White Rock-rose** *H. apenninum* □ has narrow lvs with markedly inrolled margins and smaller stipules, and fls white with yellow centre. T, rare in B. **1b** *H. oelandicum* has narrow lvs, not white beneath. Mainly in mountains. F, G, S, eastern. **1c* Hoary Rock-rose** *H. canum* □ is smaller, with narrower lvs, sometimes downy grey above, and 10-15 mm fls; May-June.

2* SPOTTED ROCK-ROSE *Tuberaria guttata.* Low annual. Lvs elliptical, 3-veined, opposite. Fls 10-20 mm, yellow, often with a *red spot,* petals dropping by mid-morning, sepals black-dotted; May-Aug. Dry, rather bare places, often on lime. B (rare), F, G.

3 COMMON FUMANA *Fumana procumbens.* Spreading undershrub to 40 cm. Lvs alternate, linear, no stipules. Fls yellow, usually with a deeper yellow spot; May. Dry, bare, often rocky places. F, G, S, southern.

Sea-Heath Family   Frankeniaceae

4* SEA-HEATH *Frankenia laevis.* Prostrate mat-forming heather-like undershrub, variably downy. Lvs opposite, margins inrolled, sometimes with a white crust. Fls 5-petalled, open, pink, purplish or whitish; July-Aug. Fr a capsule. Sands, shingle and salt-marshes by sea. B, F.

Gourd Family   Cucurbitaceae

5* WHITE BRYONY *Bryonia cretica.* Climbing perennial to 4 m, hairy. Lvs *ivy-shaped,* not glossy, with unbranched tendrils opposite the stalks. Fls 10-18 mm, 5-petalled, greenish-white with darker veins; stamens and styles on different plants, stigma downy; May-Sept. Fr a *red* berry. Hedges, scrub. B, F, G. **5a** *B. alba* has stamens and styles in different fls on same plant, stigma hairless and berry black. F, (F, S). **5b* Prickly Cucumber** *Echinocystis lobata* has branched tendrils, more sharply palmate lvs, male fls smaller, in stalked spikes, female fls solitary, on same plant, and fr prickly. (G) from N America.

Willowherb Family   Onagraceae

6* ENCHANTER'S NIGHTSHADE *Circaea lutetiana.* Short/medium patch-forming downy perennial. Lvs pointed oval, sometimes slightly heart-shaped, opposite. Fls 4-8 mm, in lfless spikes which elongate before fls fall; 2-petalled, white; June-Sept. Fr covered with hooked bristles. Woods, shady places, sometimes a garden weed. T. **6a* Alpine Enchanter's Nightshade** *C. alpina* is smaller and much less downy, with markedly heart-shaped, more deeply toothed lvs and fl spike not elongating until after fls fall. Mountain woods. Northern. **6 × 6a* Upland Enchanter's Nightshade** *C.* × *intermedia* is sterile and frequently found in the absence of the parents.

7* LARGE-FLOWERED EVENING PRIMROSE *Oenothera erythrosepala.* Medium/tall hairy biennial, with conspicuous red spots on stem. Lvs alternate, lanceolate, with finely toothed crinkled margins. Fls 80-100 mm, 4-petalled, open, pale yellow, with red or red-striped sepal-tube; June-Sept. Fr an elongated capsule. Bare and waste places, dunes. (B, F, G), of garden origin. **7a* Common Evening Primrose** *O. biennis* agg. has no red spots and smaller fls (50-60 mm). **7b* Small-flowered Evening Primrose** *O. parviflora* agg. may have faint red spots and smaller fls still (20-30 mm). 7a and 7b both have uncrinkled lvs and green sepals. (T), from N America.

8* FRAGRANT EVENING PRIMROSE *Oenothera stricta.* Differs from 7 in having no red spots and smaller (40-90 mm) fls, fragrant at night and soon turning red; June-Sept. (B, F, G), from S America.

Willowherb Family (contd.)

WILLOWHERBS *Epilobium*. More or less hairy perennials. Lvs lanceolate (except 6), toothed. Fls in stalked spikes (except 6), with four notched petals and eight stamens; June-Aug. Fr a long narrow 4-sided pod, splitting when ripe to reveal the silky plume of hairs attached to each seed. Hybrids are very frequent.

1* ROSEBAY WILLOWHERB *Epilobium angustifolium*. Tall, patch-forming, almost hairless. Lvs alternate. Fls 20-30 mm, bright pink-purple, petals slightly *unequal*. Open woods, heaths, mountains, waste ground. T. **1a River Beauty** *E. latifolium* has shorter lvs and fewer fls in the spike. Iceland. **1b** *E. dodonaei* has much narrower lvs. Bare dry places. F, G.

2* GREAT WILLOWHERB *Epilobium hirsutum*. Tall, softly hairy. Lvs mostly opposite, the upper half-clasping stem. Fls 15-25 mm, purplish-pink, stigma 4lobed. Damp places. T. **2a* Hoary Willowherb** *E. parviflorum* □ is smaller, with woolly hairs on lower stem, upper lvs alternate and not clasping stem, and much smaller (6-9 mm) fls.

3* BROAD-LEAVED WILLOWHERB *Epilobium montanum* agg. Short/medium perennial, stem rounded. Lvs broad lanceolate, opposite, very shortly stalked. Fls 6-9 mm, pinkish-purple, petals well notched, stigma 4-lobed. Shady and waste places. T. **3a** *E. collinum* has smaller fls with blunt not pointed buds. F, G, S. **3b* Spear-leaved Willowherb** *E. lanceolatum* has longer-stalked and mostly alternate lvs, stem slightly 4-angled and fls white becoming pink. B, F, G. **3c* Square-stemmed Willowherb** *E. tetragonum* □ has unstalked strap-shaped lvs and less notched petals. **3d* Short-fruited Willowherb** *E. obscurum* has unstalked lvs and prefers damper places. **3e* Pale Willowherb** *E. roseum* □ has mostly alternate long-stalked lvs and fls white becoming pink. **3f* American Willowherb** *E. adenocaulon* □ has mostly alternate lvs and smaller, paler fls. (T), from N America. **3g** *E. glandulosum* is like 3f with larger purplish-pink fls. (S) from N America. **3h** *E. alpestre* has lvs whorled. Mountains. G. All have somewhat narrower lvs. 3d-3h have four raised lines on stem and club-shaped stigma.

4* MARSH WILLOWHERB *Epilobium palustre*. Slender short/medium perennial, with thread-like runners. Lvs unstalked, narrow, not or scarcely toothed. Fls 4-6 mm, pink to white, stigma club-shaped. Wet places. T. **4a** *E. davuricum* has no runners, narrower lvs and white fls. S. **4b** *E. nutans* has lvs broader and fls pale violet. Mountain moors. G, southern.

5* ALPINE WILLOWHERB *Epilobium anagallidifolium*. Low creeping perennial, almost hairless, with runners above ground. Lvs little toothed. Fls 4-5 mm, reddish-pink, sepals red. Pods red when ripe. Wet places on mountains. T. **5a** *E. hornemanni* has runners below ground and larger, pale violet fls. S. **5b** *E. lactiflorum* is like 5a but fls white. **5c* Chickweed Willowherb** *E. alsinifolium* is larger with more pointed and toothed lvs, and larger, more purplish fls.

6* NEW ZEALAND WILLOWHERB *Epilobium brunnescens*. Prostrate creeping perennial, almost hairless. Lvs stalked, rounded, weakly toothed, often bronzy. Fls 4 mm, long-stalked, very pale pink, petals deeply notched. Bare damp places. (B, F), from New Zealand. **6a/6b** *E. komarovianum* and *E. pedunculare*, with elliptical lvs and sharply toothed lvs respectively, are two closely related and locally established New Zealand plants which may spread.

Loosestrife Family Lythraceae

7* PURPLE LOOSESTRIFE *Lythrum salicaria*. Tall perennial, variably downy; stems with four raised lines. Lvs opposite or in whorls of 3, untoothed, unstalked, lanceolate. Fls 10-15 mm, in whorled spikes, *6-petalled,* bright red-purple; June-Aug. Fr egg-shaped. Damp places, especially by rivers. T.

8* GRASS POLY *Lythrum hyssopifolia*. Low/short hairless annual. Lvs alternate, linear to oblong. Fls tiny, 5-6-petalled, pale pink, unstalked at base of lvs; July-Sept. Bare damp ground, especially in winter-wet hollows. T, southern.

1

2

2a

3f

3e

3

3c

4

5

7

8

6

Key to Umbellifers

(Pages 154-166)

The great majority of umbellifers, members of the Carrot Family, Umbelliferae are easily distinguished by the characteristic umbrella-shape of their flower-heads, which are usually white, sometimes tinged pink, but yellow in a few species. A small number, however, notably sanicle and Sea Holly (p.154) and the *Bupleurum* species (p.164) are quite untypical, while some other plants, such as Danewort and Yarrow (p.238) have superficially umbellifer-like flower-heads. These are all shown in the Main Key on p.13. The fruits of umbellifers are important in identification.

Plant

aromatic: Sweet Cicely, *Chaerophyllum bulbosum* 156; Spignel 158; Hemlock Water Dropwort, Cambridge Milk-parsley 160; Wild Celery, Corn Parsley 162; Scots Lovage 164; Wild Parsnip, Alexanders, Lovage 166

foetid: Coriander 156; Hemlock 160; Stone Parsley 162

growing in water: Marsh Pennywort 154; Cowbane, Greater Water-parsnip 160; Lesser Water-parsnip, Fool's Watercress, Water Dropworts 162

Stems

purplish: Rough and Bur Chervils 156; Angelica, Hemlock 160; Stone Parsley 162; Milk Parsley, Cnidium, Scots Lovage 164; Thorow-wax 166

ridged: Burnet Saxifrage, Moon Carrot, Fool's Parsley 158; Hemlock Water Dropwort, Greater Water-parsnip, Pleurospermum, Cambridge Milk-parsley 160

angled: Honewort 158; Biennial Alexanders 166

rooting: Fool's Watercress 162

Leaves

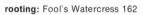

rounded: Marsh.Pennywort, Sea Holly 154; Thorow-wax 166

linear/lanceolate: Hare's-ears 164 and 166

palmate: Sanicle, Astrantia 154

1-pinnate: Bur Parsleys, Coriander 156; Burnet Saxifrage, Moon Carrot 158; Hogweed, Greater Water-parsnip 160; p.162; Hartwort 164; Wild Parsnip, Rock Samphire 166

1–3-trefoil: Ground Elder, Bladderseed 158; Pleurospermum, Longleaf 160; Hog's Fennel, Masterwort, Scots Lovage 164; Alexanders 166

thread-like: Whorled Caraway, *Ammi* 156; Spignel 158; Fine-leaved Water Dropwort 162; Fennel 166

spiny: Sea Holly 154

sheathing: Hogweed, Angelica 160; Masterwort, Scots Lovage 164.

Flowers

yellow: Spignel 158; Hog's Fennel, Slender Hare's-ear 164; all on p.166

blue: Sea Holly 154

enclosed by bracts: Thorow-wax, Small Hare's-ear 166

in simple umbels: all on p.154; Shepherd's Needle 158

with unequal petals: Upright Hedge Parsley, Wild Carrot 156

with lower bracts always present: all on p.154; Upright Hedge and Greater Bur Parsleys, Whorled Caraway, Wild Carrot 156; Great Pignut, Moon Carrot, Spignel, Bladderseed 158; Hemlock, Hemlock Water Dropwort, Greater Water-parsnip, Pleurospermum, Longleaf 160; Lesser Water-parsnip, Parsley Water Dropwort, Corn and Stone Parsleys 162; *Peucedanum alsaticum, P. oreoselinum,* Milk Parsley, Sermountain, Scots Lovage 164; Rock Samphire, Alexanders, Lovage, Small Hare'sear 166

with petal-like bracts: Astrantia 154

with divided bracts: Wild Carrot 156; Lesser Water-parsnip 162; Milk Parsley 164

with spiny bracts: Sea Holly 154

Fruits

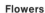

bristly/spiny: Sanicle 154; Bur Chervil, Bur Parsleys, Wild Carrot 156; Hartwort 164

aromatic: Coriander, Caraway 156

Dogwood Family *Cornaceae*

Lvs opposite, untoothed, pointed oval. Fls 4-petalled, in an umbel. Fr a berry.

1* DOGWOOD *Cornus sanguinea.* Deciduous shrub to 4 m, with *red* twigs, conspicuous in winter. Lvs with 3-5 pairs of prominent veins, reddening in autumn. Fls *white;* May-July. Fr black. Scrub, hedgerows, on lime. T. **1a*** *C. alba* with larger 5-7-veined lvs, yellowish-white fls and white fr, is widely planted and locally naturalised. (B, S), from N Russia. **1b* Red Osier** *C. sericea* differs from 1a in its numerous rooting runners and more sharply pointed lvs. (B), from N America.

2* CORNELIAN CHERRY *Cornus mas.* Deciduous shrub or small tree to 8 m, with greenish-yellow twigs. Umbels unstalked with four yellow-green sepal-like bracts, fls *yellow;* Feb-Mar, *before lvs.* Fr red. Woods, scrub, widely planted. (B), F, G.

3* DWARF CORNEL *Cornus suecica.* Short creeping perennial. Lvs unstalked, rounded to elliptical. Umbels with four conspicuous *white petal-like* bracts, fls purplish-black; June-Aug, shy flowerer. Fr red. Heaths, moors, mountains; not on lime. B, G, S.

Ivy Family Araliaceae

4* IVY *Hedera helix.* Evergreen woody climber, carpeting the ground or ascending by means of tiny roots. Lvs glossy, often bronzy, 3-5-lobed (ivy-shaped) on non-flg shoots, pointed oval on flg shoots. Fls in an *umbel,* green with yellow anthers; Sept-Nov. Woods, hedgerows, rocks, walls. T.

Carrot Family Umbelliferae

pp. 154-66. Lvs alternate, no stipules (except Marsh Pennywort, p. 154). Fls small, 5-petalled, in umbels, usually in turn arranged in an umbel. The whole flhead thus looks like a flat-topped umbrella, whose spokes (rays) are the stalks of the secondary umbels. Fls of white-fld species often tinged pink. The strap-like upper and lower bracts at the base of the primary and secondary umbels respectively, and the dry, usually ridged fr, are both important in identification.

Umbellifers on this page are untypical, all having their fls in simple umbels.

5* MARSH PENNYWORT *Hydrocotyle vulgaris.* Prostrate creeping perennial. Lvs almost *circular,* shallowly lobed, erect on long stalks. Fls tiny, pinkish-green, in one or more whorls from base of lvs, often hard to detect among vegetation; June-Aug. Fr rounded. Damp, often grassy places, shallow fresh and brackish water. T.

6* SANICLE *Sanicula europaea.* Short/medium hairless perennial. Lvs long-stalked, shiny, *palmately* 3-5-lobed, toothed. Fls pale pink or greenish-white, in clusters of tight umbels; May-July. Fr roundish with hooked spines. Woods, often on lime. T.

7* ASTRANTIA *Astrantia major.* Medium/tall perennial. Lvs long-stalked, palmately lobed, toothed. Fls white, scented, in umbels with pinkish *petal-like* bracts; June-Sept. Woods, mountain meadows. (B, rare), F, G.

8* SEA HOLLY *Eryngium maritimum.* Short/medium hairless perennial. Lvs leathery, rounded, spiny, blue-green with whitish veins and edges. Fls powder-blue, in tight umbels with spiny lflike bracts; June-Sept. Fr egg-shaped. Sand and shingle by sea. T. **8a* Field Eryngo** *E. campestre* □ is much branched and greener, with smaller heads of usually greenish-white fls. Bare, dry places. B (rare), F, G. **8b** *E. planum* is smaller with shorter bracts. Dry places inland. G (rare).

1

2

3

4

6

7

8

8a

Carrot Family   Umbelliferae (see p. 154)

1* COW PARSLEY *Anthriscus sylvestris* agg. Medium/tall perennial, slightly hairy. Lvs 2-3-pinnate. Fls white, no lower bracts; Apr-June. Fr oblong, beaked, black, *smooth*. Hedge-banks, shady places. T. **1a* Garden Chervil** *A. cerefolium* is a shorter annual, with umbels opposite lvs and narrower long-beaked fr. Escape from cultivation. (B,F,G), from E Europe. **1b* Sweet Cicely** *Myrrhis odorata* □ is aromatic when crushed, with lvs flecked whitish and much longer ribbed fr. Waysides. (T), from S Europe.

2* ROUGH CHERVIL *Chaerophyllum temulentum*. Medium/tall hairy biennial; stem *purple* or purple-spotted. Lvs 2-3-pinnate, turning purple. Fls white, no lower bracts; May-July. Fr long, ridged, narrowed to tip. Hedge-banks, shady places. T, southern. **2a** *C. hirsutum* agg. has green stems and hairy petals. F,G. **2b** *C. bulbosum* has a tuberous rootstock, and lower stem and upper bracts both hairless. F,G,S. **2c* Golden Chervil** *C. aureum* is taller and less hairy, with yellow-green aromatic lvs. (B),F,G. **2d** *C. aromaticum* has undivided lf lobes. G(rare).

3* BUR CHERVIL *Anthriscus caucalis*. Medium annual, slightly hairy; stem often purplish below. Lvs 2-3-pinnate, finely cut. Fls white, umbels *opposite* lvs, no lower bracts; May-June. Fr with hooked bristles. Bare, often sandy places. T, southern.

4* UPRIGHT HEDGE PARSLEY *Torilis japonica*. Medium/tall hairy annual; stems *stiff*. Lvs 1-3-pinnate, narrowly triangular. Fls white or pink, petals slightly unequal, lower bracts 4-6; July-Sept. Fr egg-shaped, with purple *hooked* bristles. Hedge-banks, shady places. T. **4a* Spreading Bur Parsley** *T. arvensis* is shorter, with 0-1 lower bracts and fr bristles not hooked. A weed of cultivation. B,F,G.

5* KNOTTED BUR PARSLEY *Torilis nodosa*. More or less prostrate annual, roughly hairy. Lvs 1-2-pinnate. Fls white or pink, umbels few-fld, *opposite* lvs; no lower bracts, upper bracts longer than fls; May-July. Fr with warts and *straight* bristles. Dry, sparsely grassy places. B,F,G.

6* CORIANDER *Coriandrum sativum*. Short hairless, *foetid* annual. Lvs 1-2-pinnate, the lower with broader segments. Fls white or pink, outer petals longer, no lower bracts; June-Aug. Fr globular, ridged, red-brown, *aromatic* when ripe. Bare places. (B,F,G), from W Asia. **6a** *Bifora radians* is not foetid and has lf-lobes linear, fls larger and fr 2-lobed. (F,G).

7* GREATER BUR PARSLEY *Turgenia latifolia*. Medium annual, roughly hairy. Lvs *1-pinnate*. Fls pink, red-purple or white, with lower bracts 2-3, upper bracts broad with pale edges; May-Aug. Fr ridged, with 2-3 rows of *hooked* bristles on each ridge. A weed of cultivation. (B,rare),F,G. **7a* Small Bur Parsley** *Caucalis platycarpos* is shorter and pale green, with 2-3-pinnate lvs, outer petals much longer, and fr with bristles in single rows. **7b** *Orlaya grandiflora* has 2-3-pinnate lvs, outer petals much longer and fr with bristles in 1-2 rows. Grassy places. F,G.

8* CARAWAY *Carum carvi*. Medium hairless biennial. Lvs 2-3-pinnate. Fls white, usually no bracts; June-July. Fr oblong, ridged, *aromatic,* the familiar caraway seed. Grassy places. T, but (B).

9* WHORLED CARAWAY *Carum verticillatum*. Medium/tall hairless perennial. Lvs oblong, with numerous *whorled* threadlike segments, recalling Yarrow (p.238). Fls white, with both upper and lower bracts; July-Aug. Fr egg-shaped, ridged. Damp grassland, marshes. B,F,G, western.

10* WILD CARROT *Daucus carota*. Medium hairy biennial. Lvs 3-pinnate. Fls white, the centre one of the umbel often red, petals often unequal, *lower bracts conspicuously pinnate or 3-forked;* June-Sept. Fr flattened, with often hooked bristles; fr umbels become hollowly concave. Grassy places; near the sea often fleshy and more luxuriant. T. **10a*** *Ammi visnaga* and **10b*** *A. majus* have tangled linear lf-lobes, resembling Hog's Fennel (p.164). Casual (T). 10a has upright thick rays.

Carrot Family (contd.)

1* PIGNUT *Conopodium majus.* Short/medium perennial, almost hairless; root tuberous, edible. Lvs 2-3-pinnate, the upper with almost *thread-like* lobes. Fls white, lower bracts 0-2; May-July. Fr oblong, obscurely ridged, with erect styles. Open woods, shady grassland. B,F,S. **1a* Great Pignut** *Bunium bulbocastanum* is taller and stouter, with 5-10 lower bracts and fr prominently ridged, the styles turned down. Especially on lime. B(rare),F,G.

2* BURNET SAXIFRAGE *Pimpinella saxifraga.* Variable medium perennial, downy; stems slightly ridged. Lvs 1-2-pinnate, the lower usually *1-pinnate* with broader lflets, the upper usually 2-pinnate with narrow lflets. Fls white, no bracts; May-Sept. Fr egg-shaped. Dry grassland, especially on lime. T. **2a* Greater Burnet Saxifrage** *P. major* □ is taller and hairless, with much larger lvs all 1-pinnate, fls often pink and fr prominently ridged. Damper, shadier grassland. **2b** *Ptychotis saxifraga* has lower lf lobes deeply cut, upper thread-like; and small upper bracts. F.

3* SHEPHERD'S NEEDLE *Scandix pecten-veneris.* Short annual, almost hairless. Lvs 2-3-pinnate, lobes linear. Fls white, umbels often simple but sometimes with 2-3 rays, *opposite* lvs; no lower bracts, upper bracts broad, deeply toothed; May-Aug. Fr conspicuously *long* (15-80 mm). Weed of cultivation. T, southern.

4* GROUND ELDER *Aegopodium podagraria.* Creeping, *patch-forming* short/medium perennial. Lvs 1-2-trefoil. Fls white, usually no bracts; June-Aug. Fr egg-shaped, ridged. Shady and waste places, and a tenacious garden weed. T.

5* MOON CARROT *Seseli libanotis.* Variable medium/tall perennial, downy or hairless; stems ridged. Lvs 1-3-pinnate, lobes narrow. Fls white or pink, umbels many-rayed, bracts numerous; July-Sept. Fr egg-shaped. Dry grassland, especially on lime. T, but rare in B. **5a** *S. montanum* is shorter and always hairless, with fewer rays and no lower bracts. F,G, southern. **5b** *S. annuum* is shorter with lvs oval in outline and 0-1 lower bracts. F,G, southern. **5c** *S. hippomarathrum* is always hairless, with lvs 2-3-pinnate and no lower bracts. G(rare), southern.

6* FOOL'S PARSLEY *Aethusa cynapium.* Low/medium annual, hairless; stems ribbed. Lvs 2-3-pinnate, dark green. Fls white, with long upper bracts making umbels look *bearded;* June-Oct. Fr egg-shaped, ridged. Weed of cultivation. T. Subsp. *cynapioides* is taller; woods, eastern.

7* SPIGNEL *Meum athamanticum.* Short/medium perennial, hairless, *aromatic.* Lvs 3-4-pinnate, lobes thread-like. Fls white, often tinged yellow or purple, lower bracts 0-2; June-Aug. Mountain grassland. B,F,G. **7a** *Athamanta cretensis* is downy and not aromatic, with lvs 3-5-pinnate, lflets rather broader, fls always white and 0-5 lower bracts, sometimes pinnately cut. Mountain rocks. G, southern.

8* BLADDERSEED *Physospermum cornubiense.* Medium/tall perennial, almost hairless. Lvs 2-trefoil, *dark green.* Fls white, bracts lanceolate; July-Aug. Fr globular, ridged. Woods. B(rare),F.

9* HONEWORT *Trinia glauca.* Low *greyish* hairless perennial; stems angled, much branched. Lvs 2-3-pinnate, lobes narrow. Fls white, stamens and styles on different plants, no bracts; May-June. Fr egg-shaped, ridged. Grassland, especially on lime. B(rare),F,G.

2a* Greater Burnet Saxifrage

159

Carrot Family (contd.)

1* HOGWEED *Heracleum sphondylium.* Very variable tall *stout* biennial/perennial, to 3 m, hairless or roughly hairy. Lvs 1-3-pinnate, with very broad toothed lflets and stalks expanded to a sheath. Fls white or pink, umbels up to 20cm across, petals of outer fls *larger,* usually no lower bracts. Fr flattened, elliptical to rounded. Grassy places, open woods. T. **1a* Giant Hogweed** *H. mantegazzianum* □ is much stouter and taller, to 5 m, with stems often blotched purple, umbels up to 50 cm across, and fr narrower; June-July. Especially on damp ground. (T), from S W Asia.

2* ANGELICA *Angelica sylvestris.* Tall perennial, almost hairless; stems often purplish. Lvs 2-3-pinnate, with broad, toothed lflets and *inflated* sheathing stalks. Fls white or pink, sepals untoothed, often no lower bracts, upper bracts thread-like; July-Sept. Fr egg-shaped, flattened, winged. Damp grassy places and woods, cliffs. T. **2a* Garden Angelica** *A. archangelica* has stems usually green, lf-stalks aromatic (the source of the sweetmeat), fls greenish or yellowish, and fr with corky wings. Often by the sea. G, S, (B, F). **2b** *A. palustris* has white sepal-teeth. Wetter places. G(rare). **2c** *Conioselinum tataricum* is always hairless and has fls always white, with no sepals. Stony places. S, Arctic.

3* HEMLOCK *Conium maculatum.* Tall hairless, foetid, very poisonous biennial; stems *purple-spotted.* Lvs 2-4-pinnate, finely cut. Fls white, upper bracts only on *outer edge* of umbel; June-Aug. Fr globular, with wavy ridges. Damp, sparsely grassy places. T.

4* HEMLOCK WATER DROPWORT *Oenanthe crocata.* Tall hairless parsley-scented, very poisonous perennial; stems grooved. Lvs 3-4-pinnate, with broad *wedge-shaped* toothed lflets. Fls white, sometimes no lower bracts; June-Aug. Fr cylindrical, long-styled. Damp grassy places. B, F.

5* GREATER WATER-PARSNIP *Sium latifolium.* Tall hairless perennial; stems ridged. Lvs *1-pinnate* with about five pairs of lflets; also submerged 2-3-pinnate lvs. Fls white, lower bracts large, *lfy;* June-Sept. Fr egg-shaped, ridged, short-styled. In and by fresh water, fens, marshes. T.

6 PLEUROSPERMUM *Pleurospermum austriacum.* Tall biennial/perennial, slightly hairy; stems ridged. Lvs 2-3-trefoil, *triangular,* lobes wedge-shaped, toothed. Fls white, upper bracts unequal, down-turned; June-Sept. Fr egg-shaped, ridged. Mountain grassland. F, G, S, eastern.

7* COWBANE *Cicuta virosa.* Tall, hairless, very poisonous perennial; stems sometimes half-floating. Lvs 2-3-pinnate, lobes toothed. Fls white, no lower bracts; July-Aug. Fr globular, ridged. In and by fresh water. T.

8* LONGLEAF *Falcaria vulgaris.* Medium/tall patch-forming annual/perennial; stems much branched. Lvs 1-2-trefoil with long *strap-like,* finely toothed lflets. Fls white, umbels rather open, bracts linear; July-Sept. Fr oblong, ridged. Grassy and waste places. T but (B, S).

9* CAMBRIDGE MILK-PARSLEY *Selinum carvifolia.* Tall perennial, almost hairless, parsley-scented; stems ridged. Lvs 2-3-pinnate, with winged ridges. Damp grassland and open woods. T, eastern, but rare in B. **9a** *S. pyrenaeum* has unwinged, almost lfless stems and few-rayed umbels with yellowish fls. F, mountains, rare.

1a* Giant and Common Hogweeds

Carrot Family *(contd.)*

1* LESSER WATER-PARSNIP *Berula erecta.* Medium, often sprawling hairless perennial. Lvs 1-pinnate with 7-10 pairs of toothed lflets; also 3-4-pinnate submerged lvs with linear lobes. Fls white, umbels *opposite* lvs, bracts usually lflike, often *3-cleft* or more or less pinnate; July-Sept. Fr globular, almost divided in two, faintly ridged, long-styled. Wet places. T.

2* FOOL'S WATERCRESS *Apium nodiflorum.* More or less prostrate perennial; stems rooting at lower lf-junctions. Lvs 1-pinnate, lflets oval to lanceolate, toothed. Fls white, umbels not or very shortly stalked, *opposite* lvs; *no* lower bracts; June-Sept. Fr egg-shaped, ridged. Wet places. B,F,G. **2a* Creeping Marshwort** *A. repens* is smaller, with stem rooting at every lf-junction, broader lflets and stalked umbels. T, rare in B. **2b* Lesser Marshwort** *A. inundatum* is smaller and usually largely submerged, with linear or thread-like lflets, fewer-rayed and longer-stalked umbels, and narrower fr. Hybridises with 2.

3* WILD CELERY *Apium graveolens.* Medium/tall yellow-green biennial, hairless, *celery-scented.* Lvs 1-2-pinnate, shiny, with large toothed lflets. Fls white. umbels shortly or not stalked, often opposite lvs; no bracts; June-Aug. Fr egg-shaped. Damp grassy places near sea, saline meadows inland. T.

4* TUBULAR WATER DROPWORT *Oenanthe fistulosa.* Medium greyish hairless perennial, little branched. Lvs 1-2-pinnate, the upper with linear lobes and *inflated* stalks. Fls white, umbels few-rayed, becoming globular in fr; usually no lower bracts; June-Sept. Fr cylindrical, long-styled. Wet places, shallow fresh water. T, southern.

5* PARSLEY WATER DROPWORT *Oenanthe lachenalii.* Medium/tall perennial. Lvs 2-pinnate, the upper with narrower pointed lobes and sometimes 1-pinnate. Fls white, umbel rays not thickened in fr, sometimes no lower bracts; June-Sept. Fr egg-shaped, ridged, short-styled. Damp grassland, fens. T, western. **5a* Corky-fruited Water Dropwort** *Oe. pimpinelloides* has rays thickened in fr and fr cylindrical, shorter-styled. B,F,G. **5b* Narrow-leaved Water Dropwort** *Oe. silaifolia* has lower lvs 2-4-pinnate with narrower lobes, rays thickened in fr; cylindrical, long-styled fr. B,F,G. **5c** *Oe. peucedanifolia* is like 5b but short-styled fr. F,G.

6* FINE-LEAVED WATER DROPWORT *Oenanthe aquatica.* Medium/tall, pale green hairless perennial; much branched. Lvs 3-4-pinnate, upper with pointed lobes, lower with lobes linear to thread-like. Fls white, umbels *opposite* lvs as well as terminal, rays not thickened in fr; usually no lower bracts; June-Sept. Fr oblong, often curved, short-styled. Slow or still fresh water. T. **6a* River Water Dropwort** *Oe. fluviatilis* usually floats on the water and has blunter-lobed 2-pinnate upper lvs and still shorter-styled fr; a shy flowerer. B,F,G, western.

7* CORN PARSLEY *Petroselinum segetum.* Medium greyish hairless biennial, *parsley-scented.* Lvs 1-pinnate, lflets broad, toothed. Fls white, umbels few-rayed and few-fld, bracts thread-like; Aug-Oct. Fr egg-shaped. Dry, sparsely grassy places. B,F,G, western. **7a* Wild Parsley** *P. crispum* □ has shiny 3-pinnate lvs, much crisped in cultivated forms, and more numerous, yellowish fls; June-Sept. (T).

8* STONE PARSLEY *Sison amomum.* Medium/tall perennial, hairless, *foetid,* pale green turning purple; much branched. Lvs and fls like 7. Fr globular. Hedge-banks, grassy places. B,F.

7a* Wild Parsley

Carrot Family (contd.)

1* HOG'S FENNEL *Peucedanum officinale.* Tall dark green hairless perennial. Lvs 2-6-trefoil, the repeatedly divided linear lobes making a *tangled mass.* Fls *yellow,* umbels many-rayed, often no lower bracts; July-Sept. Fr egg-shaped, flattened, ridged. Grassy places. B(rare), F, G. **1a** *P. carvifolia* □ has 1-2-pinnate lvs with broader lobes, no bracts and fls sometimes greenish. F, G. **1b** *P. alsaticum* has 2-4-pinnate lvs and always some lower bracts. F, G. **1c** *P. gallicum* is like 1a but has lower lvs 3-pinnate and fls white or pinkish. F. **1d** *P. lancifolium* has shorter pinnate lvs, not tangled; bracts 4-7. F(rare).

2* MILK PARSLEY *Peucedanum palustre.* Tall perennial, almost hairless; stems often purplish. Lvs 2-4-pinnate, lobes blunt. Fls white, bracts unequal, sometimes *forked,* down-turned; July-Sept. Fr egg-shaped, flattened, ridged. Marshes, damp grassland. T. **2a** *P. cervaria* has very sharply toothed lvs, lower bracts often pinnately cut, and fr sometimes rounded. Grassland, woods. F, G.

3* MASTERWORT *Peucedanum ostruthium.* Medium/tall perennial, almost hairless. Lvs 2-trefoil, with broad, toothed lflets and *inflated* sheathing stalks. Fls white or pink, no lower bracts; June-July. Fr rounded, flattened, ridged. Grassy places in hills and mountains. F, G,(B, S). **3a** *P. oreoselinum* is often tinged red and has solid stems, 2-3-pinnate, rather leathery lvs, numerous lower bracts, all bracts down-turned, and more egg-shaped fr; Aug-Sept. F, G, S.

4* HARTWORT *Tordylium maximum.* Medium/tall, coarsely hairy annual/biennial; stems with *down-turned* hairs. Lvs 1-pinnate, root lvs with broader lflets. Fls white, outer petals *larger;* June-Aug. Fr rounded, flattened, ridged, bristly. Grassy and waste places. (B, rare), F, G.

5 SERMOUNTAIN *Lasertrilobum latifolium.* Tall greyish perennial, almost hairless. Lvs 2-pinnate, stalks of upper lvs *inflated,* lobes heart-shaped. Fls white, lower bracts numerous, *down-turned;* June-Aug. Fr egg-shaped with wavy ridges. Mountain woods and rocks. F, G, S. **5a** *L. prutenicum* is a downy green biennial with lf-lobes narrow, lf-stalks not inflated, fls sometimes tinged yellow, upper bracts more numerous, and fr ridges straight. Grassy places, woods. F, G. **5b** *Laser siler* has lvs 2-4 pinnate and lflets lanceolate, untoothed. G, south-eastern. **5c** *L. trilobum* is hairless and green, with 2-3-trefoil lvs, the stalks not inflated, few bracts and broader umbels. On lime. F(rare), G.

6* SLENDER HARE'S-EAR *Bupleurum tenuissimum.* A most atypical umbellifer, hard to detect among long grass. Medium greyish hairless annual. Lvs undivided, more or less *linear.* Fls *yellow,* very small, in head-like umbels with 1-3 rays; no lower bracts; Aug-Sept. Fr globular. Grassy places by the sea, saline meadows inland. T, southern. **6a** *B. gerardi* is green, with lvs partly clasping stem and 5-7 rays. F(rare).

7 CNIDIUM *Cnidium dubium.* Medium/tall perennial, almost hairless. Lvs 2-3-pinnate, lobes with somewhat *recurved* edges and pointed whitish tip; stalks often purplish, sheathing the stem. Fls white, umbels with numerous narrowly winged rays, few or no lower bracts; July-Oct. Fr globular, ridged. Damp woods, disturbed ground, often near sea. G, S.

8* SCOTS LOVAGE *Ligusticum scoticum.* Medium/tall hairless perennial, *celery-scented;* stems often purple. Lvs 2-trefoil, glossy, with broad, toothed lobes and *inflated* stalks sheathing the stem. Fls white, bracts narrow; June-July. Fr egg-shaped, ridged. Cliffs, rocks and shingle by sea. B, G, S. **8a** *L. mutellinum* has pinnate lvs with narrower lobes. G, rare, southern.

1a *Peucedanum carvifolia*

164

Carrot Family (contd.)

1* WILD PARSNIP *Pastinaca sativa.* Medium/tall biennial, roughly hairy, pungent. Lvs 1-pinnate with broad, toothed lflets. Fls yellow, few or no bracts; June-Sept. Fr egg-shaped, winged. Grassy and bare places. T.

2* ALEXANDERS *Smyrnium olusatrum.* Tall hairless pungent biennial. Lvs 3-trefoil, dark green, *glossy,* with broad, toothed lflets. Fls yellow, few or no bracts; Apr-June. Fr globular, black. Hedge-banks and waste ground near sea. (B), F, G. **2a* Biennial Alexanders** *S. perfoliatum* has yellow-green lvs, the upper clasping the angled stems, and no bracts. Woods, rocky places. (B, rare), F, G, southern.

3* ROCK SAMPHIRE *Crithmum maritimum.* Short/medium greyish hairless perennial. Lvs 1-2-pinnate, with narrow *fleshy* lobes. Fls yellow, bracts eventually down-turned; July-Oct. Fr egg-shaped. Cliffs, rocks and sands by sea. B, F, G, southern.

4* FENNEL *Foeniculum vulgare.* Tall greyish strong-smelling biennial/perennial. Lvs 3-4-pinnate, feathery, with long *thread-like* lobes. Fls yellow, no bracts; July-Sept. Fr egg-shaped. (B), F, G. **4a* Dill** *Anethum graveolens* is smaller and slenderer, with flattened, winged, caraway-tasting fr. Waste ground. (B, F, G), from S Asia.

5* PEPPER SAXIFRAGE *Silaum silaus.* Medium/tall hairless perennial. Lvs 2-4-pinnate, lobes linear, pointed. Fls yellow, few or no lower bracts; June-Sept. Fr egg-shaped, short-styled. Grassland. T, southern.

6 LOVAGE *Levisticum officinale.* Tall pungent perennial. Lvs 2-3-pinnate, *glossy,* with broad, toothed lobes. Fls greenish-yellow, lower bracts numerous, upper bracts joined at base; June-Aug. Fr egg-shaped, yellow brown. Grassy places in mountains. (F, G, S), from S W Asia.

7* THOROW-WAX *Bupleurum rotundifolium.* Short/medium, greyish or purplish annual, looking quite unlike most umbellifers. Lvs undivided, roundish, the upper narrower and *joining* round the stem, the lower short- or unstalked. Fls yellow, umbels with 5-10 rays, secondary umbels *cupped* by enlarged upper bracts, no lower bracts; June-Aug. Fr oblong, smooth, blackish. Weed of cultivation, bare ground. (B), F, G. **7a* Narrow Thorow-wax** *B. lancifolium* □ with narrower lvs, 2-3 rays and warty fr is now commoner than 7 in some areas, especially where bird-food is thrown down. (B), F. **7b** *B. longifolium* is much taller with narrower lvs, the upper clasping, the lower long-stalked, and fr strongly ridged. Grassy and rocky places. F, G.

8* SMALL HARE'S-EAR *Bupleurum baldense.* Low hairless greyish annual; another very untypical umbellifer, often hard to detect in short turf. Lvs undivided, *narrower.* Fls yellow, in tiny 3-9-rayed head-like umbels, *enfolded* by the enlarged greyish, yellowish or brownish *bracts;* June-July. Fr egg-shaped, smooth. Dry bare and grassy places; likes lime. B (rare), F.

9* SICKLE HARE'S-EAR *Bupleurum falcatum.* Medium/tall hairless perennial. Lvs *undivided,* variable, often *curved.* Fls yellow, umbels with thread-like rays, lower bracts unequal; July-Oct. Fr egg-shaped. Grassy and bare places. (B, rare), F, G.

7a* Narrow Thorow-wax

Wintergreen Family   Pyrolaceae

Evergreen creeping hairless perennials with 5-petalled fls and fr a dry capsule.

1* ONE-FLOWERED WINTERGREEN *Moneses uniflora.* Low perennial. Lvs rounded, toothed, pale green, opposite, tapering to the stalk. Fls white, *solitary*, open, 15 mm; style long, straight; May-July. Woods, especially coniferous, in hill and mountain districts. T, but rare in B.

2* TOOTHED WINTERGREEN *Orthilia secunda.* Low perennial. Lvs in rosettes, pointed oval, toothed, rather pale green. Fls greenish-white, globular, 5 mm, in a nodding *one-sided* spike; style straight, protruding conspicuously from fl; June-Aug. Woods and rock ledges in hill and mountain districts. T.

3* COMMON WINTERGREEN *Pyrola minor.* Low/short perennial. Lvs in a rosette, rounded, toothed. Fls white or pale pink, globular, 6 mm, in a stalked spike; style *straight*, shorter than stamens, *not* protruding from fl; June-Aug. Woods, scrub, marshes, moors, mountains. T. **3a* Intermediate Wintergreen** *P. media* ☐ has fls 10 mm, in longer spikes, with style protruding from fl and slightly curved.

4* ROUND-LEAVED WINTERGREEN *Pyrola rotundifolia.* Short perennial. Lvs like 3 but more rounded. Fls pure white, larger (12 mm) and more *open* than 3, with a much longer and *S-shaped* style; June-Sept. Woods, short turf, rock ledges, dune slacks. T. **4a** *P. norvegica* ☐ is shorter, with even more rounded lvs. Bare places in mountains. S.

5 YELLOW WINTERGREEN *Pyrola chlorantha.* Short perennial. Lvs in a rosette, roundish, toothed. Fls greenish- or yellowish-white, in a stalked spike, with style protruding, curved; June-July. Woods, rocks. F, G, S, eastern, but only on mountains in the south.

6 UMBELLATE WINTERGREEN *Chimaphila umbellata.* Short perennial. Lvs in *whorls,* narrowly oval, toothed, dark green, leathery. Fls pink, open, in *umbel-like* head, style not protruding; June-July. Woods, rocks. F, G, S.

Diapensia Family   Diapensiaceae

7* DIAPENSIA *Diapensia lapponica.* Low evergreen hairless undershrub, forming *cushions.* Lvs oval, untoothed, shiny, leathery. Fls white, 5-petalled, stigma 3-lobed, sepals reddening in fr; May-June. Fr a dry capsule. Bare mountains and tundra. B(rare), S.

Birdsnest Family   Monotropaceae

8* YELLOW BIRDSNEST *Monotropa hypopitys* agg. Low perennial with no green colouring matter but stems and scale-like lvs *yellow,* turning brown; a saprophyte, feeding on decaying organic matter. Fls yellow, tubular, 4-5-petalled, in nodding spikes, erect in fr; June-Aug. Fr egg-shaped. Cf. Birdsnest Orchid (p. 286). Woods, in Britain especially of beech and pine, and dunes. T.

1* One-flowered Wintergreen

2* Toothed Wintergreen

3* Common Wintergreen

3a* Intermediate Wintergreen

4* Round-leaved Wintergreen

4a *Pyrola norvegica*

5 Yellow Wintergreen

6 Umbellate Wintergreen

1

2

3a

3

4

4a

5

6

7

8

Heath Family Ericaceae

pp. 170-72. Undershrubs or small shrubs with lvs undivided, usually alternate, untoothed and evergreen, and with margins inrolled downwards in many species, including all *Ericas*. Fls globular, bell-shaped or flask-shaped (except Cranberry, p.172), the petals usually joined with 4-5 lobes. Fr usually a dry capsule, but sometimes a berry. The commonly naturalised rhododendron *Rhododendron ponticum* with conspicuous heads of purple fls in June, belongs here, also the less frequent yellow-flowered *R. luteum*.

1* HEATHER *Calluna vulgaris*. Short/medium carpeting undershrub, sometimes downy. Lvs in opposite rows, linear. Fls *pale* purple, in lfy stalked spikes; July-Sept. Heaths, moors, bogs, open woods, fixed dunes, often dominating extensive tracts. T.

2* BELL HEATHER *Erica cinerea*. Short hairless undershrub. Lvs in whorls of three, linear, dark green, often bronzy. Fls *red-purple,* in stalked spikes; May-Sept. Drier heaths and moors, western.

3* CROSS-LEAVED HEATH *Erica tetralix*. Short greyish downy undershrub. Lvs in whorls of four, linear. Fls *pink,* globular, 6-7 mm, in compact heads; June-Oct. Fr downy. Wet heaths and moors, bogs. T, eastern. **3a* Dorset Heath** *E. ciliaris* □ may be taller and more straggly, with lvs in whorls of three and broader; fls larger (8-10 mm), more elongated and in spikes; and fr hairless; Apr-Oct; hybridises with 3; B(rare), F. **3b* Mackay's Heath** *E. mackaiana* has shorter, broader, darker green lvs, hairless above, and deeper pink fls; July-Sept; hybridises with 3. Ireland (rare). **3c** *E. carnea* has bright red fls in one-sided spikes. Mountain woods. G, southern.

4* CORNISH HEATH *Erica vagans*. Short/medium hairless undershrub. Lvs in whorls of 4-5, linear. Fls pink, lilac or white, in lfy stalked spikes, *chocolate-brown* anthers protruding; July-Sept. Drier heaths. B(rare), F.

5* IRISH HEATH *Erica erigena*. Medium/tall hairless shrub *to 2 m*. Lvs in whorls of four, linear. Fls pale pink-purple, in lfy stalked spikes, *reddish* anthers protruding; March-May and occasionally in winter. Wet moors, bogs. Ireland (rare), F.

6* ST DABEOC'S HEATH *Daboecia cantabrica*. Short hairy undershrub. Lvs narrowly elliptical, *whitish* beneath, the margins inrolled downwards. Fls pink-purple, in stalked spikes, calyx-teeth 4; May-Oct. Heaths, moors, open woods. B(Ireland), F.

7* MOUNTAIN HEATH *Phyllodoce caerulea*. Low/short undershrub. Lvs linear, rough-edged, flat, *green* beneath. Fls purple, in head-like clusters of 1-6, calyx-teeth 5; June-July. Moors. B(rare), S.

8* BOG ROSEMARY *Andromeda polifolia*. Low/short hairless undershrub. Lvs narrowly elliptical, greyish and shiny above, whitish beneath. Fls pink or white, in a terminal cluster; May-June. Bogs, wet heaths. T.

9 ARCTIC RHODODENDRON *Rhododendron lapponicum*. Low shrub. Lvs oval, leathery, dark green, reddish hairy beneath. Fls tubular, violet; May-June. Mountain heaths. S.

R. luteum

R. ponticum

9 Arctic Rhododendron

171

Heath Family (contd.)

1* BILBERRY Vaccinium myrtillus. Short/medium erect deciduous hairless undershrub; stems angled, green. Lvs oval, bright green, slightly toothed. Fls pink or greenish-pink, solitary or paired; Apr-July. Fr an edible black berry with characteristic purplish bloom. Hybridises with 2. Heaths, moors, open woods, not on lime. T. **1a* Northern Bilberry** V. uliginosum □ has rounded brownish stems, bluish-green untoothed lvs, clusters of 1-4 pale pink fls, and bluer berries. Mainly higher moors.

2* COWBERRY Vaccinium vitis-idaea. More or less prostrate creeping evergreen undershrub; twigs rounded, slightly downy when young. Lvs oval, broadest in the middle, untoothed, rather leathery, glossy, margins inrolled downwards. Fls white or pink, open-mouthed, in clusters; May-July. Fr an edible red berry. Hybridises with 1. Moors, heaths, mountains, open woods. T.

3* CRANBERRY Vaccinium oxycoccos. Low slender creeping evergreen undershrub. Lvs dark green, oval, whitish beneath. Fls pink, with four spreading or down-turned petals and prominent stamens, slightly downy, 1-4 in stalked spikes; June-Aug. Fr an edible round or pear-shaped white berry, spotted red or brown. Bogs. T. **3a* Small Cranberry** V. microcarpum has more triangular lvs, fls pinker, fls and stalks hairless, fr sometimes lemon-shaped. B,G,S.

4* ALPINE BEARBERRY Arctostaphylos uva-ursi. Prostrate mat-forming evergreen undershrub to 2 m, almost hairless. Lvs dark green, untoothed, broadest at tip, rather leathery, margins flat. Fls pink, in clusters; Apr-July. Fr a shiny edible red berry. Moors, mountains, rocks, open woods, scrub. T. **4a* Black Bearberry** Arctous alpina has brighter green, finely toothed, deciduous lvs; white fls and fr black when ripe. B,S.

5* LABRADOR TEA Ledum palustre. Evergreen bush to 1 m, with rust-coloured down on stems and under the oblong leathery dark green lvs. Fls creamy white, in umbels, with five free petals; May-July. Fr dry. Bogs. (B),G,S.

6* WILD AZALEA Loiseleuria procumbens. Prostrate mat-forming undershrub, hairless. Lvs opposite, oblong. Fls small, pink, in umbels; May-July. Mountain tops, Arctic heaths. T, northern.

7 CASSIOPE Cassiope hypnoides. Prostrate mat-forming undershrub. Lvs scale-like, unstalked, overlapping up the stems. Fls bell-shaped, with white and pink petals and deeper pink sepals, solitary, long-stalked, drooping; June-Aug. Mountain tops, Arctic heaths. S. **7a** C. tetragona is larger and more erect, with fl-stalks downy. On lime. Northern.

8 LEATHERLEAF Chamaedaphne calyculata. Short shrub. Lvs leathery, pointed oval. Fls white, rim yellow, bell-shaped, in a stalked spike; June. Damp places. S, northern.

9* STRAWBERRY TREE Arbutus unedo. Small evergreen tree or shrub, to 12 m, with roughish red-brown bark. Lvs elliptical, leathery, dark green, shiny, slightly toothed, alternate. Fls creamy-white, bell-shaped, in drooping clusters; Aug-Dec. Fr □ a warty berry, 2 cm across, reddening the following autumn, and supposedly strawberry-like. Rocks, scrub. Ireland (rare), F, western.

Crowberry Family Empetraceae

10* CROWBERRY Empetrum nigrum agg. Prostrate mat-forming or erecter tufted evergreen undershrub, resembling the Heath Family. Lvs oblong, dark green, shiny, margins inrolled downwards, alternate. Fls pink, 6-petalled, very small, at base of lvs; stamens and styles usually on separate plants; Apr-June. Fr a berry, green, then pink, then purple, finally black. Moors, bogs. T.

9* Strawberry Tree

Primrose Family Primulaceae

pp. 174-76. Lvs undivided. Fls 5-petalled (except Chickweed Wintergreen, p. 176). Fr a capsule. *Primula* (1-5) has elongated oval lvs, all from the roots, and open fls in an umbel (except 1) on lfless stalks.

1* PRIMROSE *Primula vulgaris.* Low hairy perennial. Lvs *tapering* to stalk. Fls pale yellow, 20-30 mm, *solitary* on long hairy stalks; rarely the common stalk rises just above ground to make a long-stalked umbel; Mar-May and sporadically in autumn and winter. Hybridises with 2 and 3; see 2a. Woods, scrub, grassy banks, sea cliffs, mountains, often in quantity. T, southern.

2* OXLIP *Primula elatior.* Low hairy perennial. Lvs *abruptly* narrowed at base. Fls pale yellow, 15-20 mm, not fragrant, 1-20 in a nodding one-sided cluster; Apr-May. Hybridises with 1 and 3. Woods, scrub, grassland, usually in quantity. T, eastern. **2a* False Oxlip** *P. veris* × *vulgaris* ☐ is the not uncommon hybrid between 1 and 3, with leaves more gradually tapered to base, fls deeper yellow and umbels not one-sided; grows singly, usually near one or both parents.

3* COWSLIP *Primula veris.* Low/short hairy perennial. Lvs *abruptly* narrowed at base. Fls deeper yellow than 1 and 2, with orange spots in centre, 10-15 mm, *fragrant,* 1-30 in an often nodding one-sided cluster; Apr-May. Hybridises with 1 and 2; see 2a. Grassland, open scrub and woods, often in quantity, on lime. T. **3a Auricula** *P. auricula* ☐ has smaller, fleshier lvs and fls twice as large with a white throat and no orange spots. Mountains, G, southern.

4* BIRDSEYE PRIMROSE *Primula farinosa.* Low perennial, mealy white on young stems and beneath lvs. Lvs broadest near *tip, toothed,* in a basal rosette. Fls lilac-pink with a yellow eye; sepal-teeth pointed; May-July. Grassy places in hills and mountains. T.

5* SCOTTISH PRIMROSE *Primula scotica.* Low biennial, mealy white on stems and lvs. Lvs broadest in the *middle, untoothed,* in a basal rosette. Fls purple with a yellow eye, sepal-teeth blunt; May-June and again in July-Aug. Short coastal turf and dunes. N Scotland. **5a** *P. scandinavica* is very similar; meadows, by streams, on lime. S. **5b** *P. stricta* is taller and has no meal on stems and little on lvs; fls lilac or violet. Damp mountain meadows. S. **5c** *P. sibirica* is like 5b with lvs rounded not tapering at base. S, northern.

6* YELLOW LOOSESTRIFE *Lysimachia vulgaris.* Medium/tall downy perennial. Lvs broad lanceolate, often black-dotted, very short-stalked, in whorls of 2-4. Fls yellow, *15-20 mm,* in lfy branched clusters, sepal-teeth with reddish margins; July-Aug. Wet places and by fresh water. T. **6a* Dotted Loosestrife** *L. punctata* has longer-stalked, never black-dotted lvs with hairy margins, larger (35 mm) fls and narrower all-green sepal-teeth. (T), from S E Europe.

7* TUFTED LOOSESTRIFE *Lysimachia thyrsiflora.* Medium perennial, almost hairless. Lvs lanceolate, unstalked, opposite. Fls yellow, 5 mm, with prominent stamens, in *globular clusters* at base of middle lvs, top of stem having lvs only; June-July, a shy flowerer. Fens, marshes, by fresh water. T.

8* CREEPING JENNY *Lysimachia nummularia.* Prostrate creeping hairless perennial. Lvs rounded, opposite. Fls yellow, *bell-like,* 15-25 mm, with broad sepal-teeth; June-Aug. Fr uncommon in B. Damp shady and grassy places. T. **8a* Yellow Pimpernel** *L. nemorum* is smaller, with narrower pointed lvs and longer-stalked open 12 mm fls with narrow sepal-teeth, and fr frequent; May-Aug.

9 ALPINE SNOWBELL *Soldanella alpina.* Low perennial, slightly hairy. Lvs thick, kidney-shaped, long-stalked. Fls blue or white, bell-like, 2-3 together on lfless stems, the petals deeply cut to form a fringe. Mountains. G, southern.

3a Auricula   9 Alpine Snowbell

175

Primrose Family (contd.)

1* WATER VIOLET *Hottonia palustris.* Pale green *aquatic* perennial, almost hairless, with short erect lfless stems above surface. Lvs *pinnate,* submerged. Fls pale lilac with yellow eye, in a whorled cluster; May-July. Still fresh water. T, southern.

2 CYCLAMEN *Cyclamen purpurascens.* Low perennial, almost hairless. Lvs long-stalked, all arising direct from a corm, rounded, heart-shaped, not angled or lobed, sometimes finely toothed, pale-blotched above, reddish-purple beneath. Fls carmine-pink, petals *turned back,* throat rounded; June-Oct, with the lvs. Woods, scrub. F, G, eastern. **2a* Sowbread** *C. hederifolium* □ has sharply angled or lobed lvs, and longer-stalked paler pink or white fls with 5-sided throat; Aug-Nov, before the lvs. (B), F.

3* CHICKWEED WINTERGREEN *Trientalis europaea.* Low/short hairless perennial. Lvs lanceolate, in a *single whorl* near the top of the slender stem. Fls white, star-like, 5-9-petalled, usually solitary; May-Aug. Conifer woods, moors, heaths. T.

4* BROOKWEED *Samolus valerandi.* Low/short hairless perennial, often unbranched. Lvs oval, in a basal rosette, and alternate up the stems. Fls white, in a stalked spike, petals *joined half-way;* June-Aug. Fr globular. Damp places, especially near sea. T, southern.

5* SCARLET PIMPERNEL *Anagallis arvensis.* Prostrate annual, almost hairless; stems square. Lvs pointed oval, black-dotted beneath, unstalked, paired or whorled. Fls usually pale scarlet, but sometimes pink, lilac or blue, *star-like,* solitary at base of lvs, opening in sunshine; petals blunt, often finely toothed at tip, with hairs along margins; May-Oct. Weed of cultivation, dunes. T. **5a* Blue Pimpernel** *A. foemina* □ has fls always blue with narrower pointed hairless petals.

6* BOG PIMPERNEL *Anagallis tenella.* Delicate, often mat-forming hairless perennial. Lvs oval or rounded, short-stalked, opposite. Fls pink, *bell-like,* opening in sunshine; May-Sept. Boggy, peaty and marshy places, fens. B, F, G, western.

7* CHAFFWEED *Anagallis minima.* One of the tiniest European plants, rarely over 5 cm high, often *under 2 cm;* a hairless annual. Lvs oval, with a thin *black line* round the underside, mostly alternate. Fls pink, hidden at base of lvs; June-Aug. Fr pink. Bare damp sandy places on heaths and in open woods. T.

8 NORTHERN ANDROSACE *Androsace septentrionalis.* Low/short downy annual/biennial. Lvs lanceolate, toothed, all in a basal rosette. Fls white or pink, 5 mm, long-stalked, in an *umbel;* May-July. Dry grassy places, mountains. F, G, S, northern. **8a** *A. elongata* □ has lvs almost or quite hairless, and smaller (2-3 mm) white fls with a yellow throat; May. G. **8b** *A. maxima* □ has broader lvs, smaller (3 mm) white fls, and umbels with conspicuous hairy bracts; Apr-May. Lowland grassland. F, G, S, eastern. **8c** *A. lactea* □ has lvs unstalked, untoothed, and fls white, 10 mm. G, southern.

9* SEA MILKWORT *Glaux maritima.* More or less prostrate creeping hairless perennial. Lvs elliptical, almost unstalked, opposite. Fls pale pink, petal-less, unstalked, solitary at base of lvs; May-Sept. *Saltmarshes,* on saline soils inland. T.

8b *A. maxima*

8a *Androsace elongata*

8c *A. lactea*

1

2

2a

3

4

5

5a

6

7

8

9

Sea-Lavender Family   Plumbaginaceae

Perennials. Lvs undivided. Fls 5-petalled, with *papery* bracts. Often in saline habitats. *Limoniums* are hairless and have lvs all basal and fls in one-sided spikes in a branched cluster.

1* THRIFT *Armeria maritima*. Low/short downy cushion-forming perennial. Lvs linear, 1-veined, all basal. Fls dark to pale pink, in *roundish* heads, with a brown bract; sepal-teeth with short bristles; Apr-Aug. Coastal cliffs, rocks and saltmarshes, also inland on mountains; ssp. *elongata* in sandy grassland. T. **1a,** p. 303.

2* COMMON SEA-LAVENDER *Limonium vulgare*. Low/short carpeting perennial; stems rounded. Lvs elliptic to broad lanceolate, long-stalked, with *pinnate* veins. Fls lilac-lavender, tightly packed in flat-topped clusters, at the end of the stalks, which usually start branching well above the middle; July-Sept. Coastal saltmarshes. T, western. **2a* Lax-flowered Sea-lavender** *L. humile* has stems branched below the middle and often angled, lvs narrower and fls widely scattered along whole length of branches.

3* MATTED SEA-LAVENDER *Limonium bellidifolium*. Low perennial, with stems branched almost from the base into numerous *zigzag* branches, rough to the touch. Lvs oblong. Fls pink-lavender, with whitish bracts, only on the uppermost branches; June-Aug. Coastal saltmarshes. B (rare), F.

4* ROCK SEA-LAVENDER *Limonium binervosum* agg. Variable low/short perennial; stems branched from near base. Lvs oval to lanceolate, not pinnately veined, with a winged 3-veined stalk. Fls lilac-lavender, on all but the lowest branches, spikes curved; July-Sept. Coastal cliffs and rocks, less often shingle and sand. B, F. **4a** p. 303.

Olive Family   Oleaceae

5* WILD PRIVET *Ligustrum vulgare*. Half-evergreen shrub to 4 m. Lvs lanceolate, untoothed, opposite, rather leathery, often bronzing. Fls white, strong-smelling, in short spikes, with four joined petals; May-June. Fr a shiny *black* berry. Scrub, open woods. T. **5a* Garden Privet** *L. ovalifolium* has much broader lvs and is commonly planted. (B, F, G), from China.

Bogbean Family   Menyanthaceae

6* BOGBEAN *Menyanthes trifoliata*. Short hairless aquatic or creeping perennial. Lvs *trefoil*, held conspicuously above surface. Fls pink and white, in spikes, petal-tube with five lobes, *fringed* with long white hairs; Apr-June. Fr globular. Shallow fresh water, marshes, fens, bogs. T.

7* FRINGED WATER-LILY *Nymphoides peltata*. Hairless aquatic perennial. Lvs rounded, shallowly toothed, floating, purple beneath and sometimes purple-spotted above; much smaller than true water-lilies (p. 66). Fls yellow, with five *fringed* petal-lobes; June-Sept. Fr egg-shaped. Still and slow-moving fresh water. B, F, G.

Periwinkle Family   Apocynaceae

8* LESSER PERIWINKLE *Vinca minor*. More or less prostrate evergreen undershrub, almost hairless; stems rooting. Lvs elliptic, opposite, short-stalked, rather leathery. Fls *blue-violet*, 25-30 mm, solitary, with five joined petals; Feb-May. Fr elongated, forked. Woods, hedge-banks, rocks. (B), F, G. **8a* Greater Periwinkle** *V. major* □ is quite hairless and has stems rooting only at tip, much broader longer-stalked lvs, and larger (40-50 mm) fls with fringed sepals.

Milkweed Family   Asclepiadaceae

9 VINCETOXICUM *Vincetoxicum hirundinaria*. Variable medium/tall perennial, slightly hairy. Lvs *heart-shaped*, sharply pointed, opposite. Fls greenish- or yellowish-white, in clusters at base of upper lvs, with five joined petals; June-Sept. Fr elongated, forked. Woods, rocks, bare ground; on lime. F, G, S.

Gentian Family Gentianaceae

pp. 180-82. Hairless. Lvs undivided, untoothed, opposite, usually unstalked; no stipules. Fls in a branched cluster, with joined petals, often opening only in sunshine. Fr a dry capsule.

1* COMMON CENTAURY *Centaurium erythraea.* Variable low/short annual. Lvs elliptical, 3-7-veined, mostly in a basal *rosette* but a few narrower stem lvs. Fls pink, 5-petalled, almost *unstalked* in tightly packed stalked clusters, but occasionally solitary; June-Sept. Grassy places, dunes. T. **1a* Seaside Centaury** *C. littorale*□ is often shorter, with much narrower, rather leathery, 3-veined lvs, and larger paler fls. Dunes, sandy ground near sea. Western. **1b** *C. glomeratum* is shorter; fls with shorter tube. S.

2* LESSER CENTAURY *Centaurium pulchellum.* Slender low/short annual, varying from unbranched to well branched. Lvs elliptical, with *no* basal rosette. Fls reddish-pink, with five (rarely four) rather narrow petals, *stalked,* in loose, widely spreading clusters; June-Sept. Damp grassy places, drier salt-marshes. T. **2a* Slender Centaury** *C. tenuiflorum*□ has broader lvs and erecter tighter flat-topped clusters of fls. B(rare), F.

3* PERENNIAL CENTAURY *Centaurium scilloides.* Low/short perennial; stems numerous, semi-prostrate, some *not* flg. Lvs roundish and stalked on the non-flg stems. Fls pink, larger than 1, in heads of 1-6, 5-petalled; July-Aug. Grassy sea cliffs. B(rare), F.

4* YELLOW-WORT *Blackstonia perfoliata.* Short/medium greyish annual. Lvs of basal rosette oblong, stem lvs more or less triangular and *joined* round stem. Fls yellow, 6-8-petalled, petals longer than sepals, in a loose branched cluster; June-Oct. Grassy places, dunes; on lime. B, F, G. **4a** *B. serotina*□ has lvs not completely joined, fls paler and sepals equalling petals. G.

5* YELLOW CENTAURY *Cicendia filiformis.* Low annual, sometimes only 2 cm high, often little branched. Lvs linear. Fls yellow, solitary, 4-petalled; June-Oct. Fr egg-shaped. Bare damp places, often near sea. B, F, G, western.

6 GUERNSEY CENTAURY *Exaculum pusillum.* A minute low annual, sometimes only 1 cm high, well branched, very hard to detect in short turf. Lvs linear, greyish. Fls pink, pale yellow or white, 4-petalled, minute; sepal-teeth long, narrow; June-Oct. Bare damp places. F.

7 GREAT YELLOW GENTIAN *Gentiana lutea.* Medium/*tall* perennial. Lvs pointed oval, up to 30 cm, bluish-green, the upper clasping the stem, the lower stalked. Fls yellow, 5-9-petalled, in tight whorls at base of upper lvs; June-Aug. Grassy places in mountains. F, G.

8 NORTHERN GENTIAN *Gentianella aurea.* Short biennial, branched from the base. Lvs oval, the upper unstalked. Fls in a lfy spike, the lvs almost concealing the dull blue-violet fls, dull yellow at the base; sepals 4-5; July-Aug. Near coast. S, northern.

1a* Seaside Centaury   **2a*** Slender Centaury   **4a** *Blackstonia serotina*

Gentian Family (contd.)

1* SPRING GENTIAN *Gentiana verna*. Low tufted perennial. Lvs oval to elliptical, mostly in a basal rosette. Fls bright blue, *solitary*, petal-tube 5-lobed, 15-25 mm long; Apr-June. Short, often stony turf in hills and mountains. B, F, G. **1a* Alpine Gentian** *G. nivalis* ☐ is a slenderer annual, often unbranched, with fls half the size (10-15 mm). Also on Arctic heaths. T, but rare in B. **1b Bladder Gentian** *G. utriculosa* is an annual with fls intermediate between 1 and 1a and a slightly inflated calyx. Damp places. G, southern. **1c** *G. detonsa* may be taller and more branched, with no marked rosette and larger fls. Often near the sea. S, northern.

2* MARSH GENTIAN *Gentiana pneumonanthe*. Low/short perennial. Lvs linear. Fls bright blue, striped *green* outside, trumpet-shaped, 25-40 mm long, often solitary, petal-tube 5-lobed; July-Sept. Wet heaths, bogs. T. **2a Willow Gentian** *G. asclepiadea* ☐ is much taller, with arching stems, numerous lanceolate lvs, and even larger fls with pale but not green stripes outside. F, G.

3 CROSS GENTIAN *Gentiana cruciata*. Medium perennial. Lvs oval to broad lanceolate, rather leathery, the upper clasping the stem, the lower stalked. Fls *dull* blue, oblong, in tight clusters up the stem, petal-tube 4-lobed; June-Sept. Dry grassy places, open woods. F, G, eastern.

4 PURPLE GENTIAN *Gentiana purpurea*. Short/medium perennial. Lvs elliptical, greyish, the lower stalked. Fls bright *red-purple,* yellow at base and with green stripes and spots, bell-shaped, petal-tube 5-8-lobed; sepal-tube split to base on one side; July-Aug. Grassy and rocky places in mountains. F, G, S. **4a Spotted Gentian** *G. punctata* ☐ is taller and has fls yellow with purple spots, and sepal-tube not split. G.

5* FIELD GENTIAN *Gentianella campestris* agg. Short annual/biennial, variably branched. Lvs oval and blunt at base, narrower and pointed on stem. Fls bluish-mauve, bell-shaped, 15-30 mm long, in small clusters up the stem, *petal-tube 4-lobed;* sepal-tube with two outer lobes much larger and *overlapping* the inner ones; June-Oct. Grassy places, dunes. T.

6* AUTUMN GENTIAN *Gentianella amarella*. Low/short biennial. Lvs elliptical to lanceolate. Fls dull purple, sometimes whitish, bell-shaped, 12-22 mm long, in small clusters up the stem, petal-tube lobes four or five; sepal-teeth equal; July-Sept. Hybridises with 6a. Grassy places, dunes; often on lime. T. **6a* Chiltern Gentian** *G. germanica* ☐ is larger with broader lvs and longer (25-35 mm) fls, wider at the mouth. Always on lime. B (rare), F, G. **6b* Early Gentian** *G. anglica* differs chiefly in flg May-June. Always on lime. **6c* Dune Gentian** *G. uliginosa* is annual and much smaller (rarely over 7 cm), and has unequal sepal-teeth; Aug-Nov. T, rare in B. **6d** *G. aspera* is hairier than 6a. G, southern, rare.

7 SLENDER GENTIAN *Gentianella tenella*. Slender low annual, often unbranched. Lvs oblong, the lower *broader* at the tip. Fls dull blue-violet, 8-12 mm long, *long-stalked,* with usually four petal-lobes; July-Aug. Grassy places in mountains. F, G, S.

8 FRINGED GENTIAN *Gentianella ciliata*. Slender low/short perennial. Lvs linear, not in a rosette. Fls blue, in small clusters at base of upper lvs, the four spreading oval petal-lobes 40 mm across and *fringed* with hairs; Aug-Oct. Dry grassy, rocky places. F.

9 MARSH FELWORT *Swertia perennis*. Medium perennial, unbranched. Lvs lanceolate, yellowish-green, the upper clasping the stem, the lower long-stalked. Fls purple, 5-petalled, *star-like,* 25-35 mm across; July-Sept. Wet grassy places in mountains. F, G. **9a** *Lomatogonium rotatum* ☐ is shorter with much broader petals; fls long-stalked, blue. S, northern.

9a *Lomatogonium rotatum*

4a Spotted Gentian

2a Willow Gentian

Bindweed Family   Convolvulaceae

Often climbers, twining anticlockwise. Lvs alternate, no stipules. Fr a capsule. *Convolvulus* and *Calystegia* have fls open trumpet-shaped.

1* HEDGE BINDWEED *Calystegia sepium.* Creeping or climbing perennial to 3 m, more or less hairless. Lvs arrow-shaped. Fls white, rarely pink, unscented, *30-35 mm* across; two large but not overlapping sepal-like bracts enclosing the five narrower sepals; June-Sept. Fr a capsule. Bushy and waste places, woods, fens. T. **1a* Great Bindweed** *C. sylvatica* □ has larger (60-75 mm) fls, occasionally pink or pink-striped, and larger, almost inflated overlapping bracts, hiding the sepals. (T) from S Europe, often commoner in and around human settlements. **1b* Hairy Bindweed** *C. pulchra* differs from 1a chiefly in being sparsely hairy and having fls always pink; widely naturalised, probably of garden origin.

2* SEA BINDWEED *Calystegia soldanella.* Prostrate creeping hairless perennial. Lvs *kidney-shaped,* fleshy. Fls pink with white stripes, 40-50 mm across, bracts shorter than sepals; June-Sept. Coastal sand, less often on shingle. B, F, G, western.

3* FIELD BINDWEED *Convolvulus arvensis.* Creeping or climbing perennial to 2 m, slightly downy when young. Lvs arrow-shaped. Fls pink and/or white, *15-30 mm* across, faintly scented; June-Sept. Weed of cultivation, waste places. T.

4* COMMON DODDER *Cuscuta epithymum.* Slender *climbing* annual, parasitic on Gorse (p. 118), Heather (p. 170) and many smaller plants; stems red. Lvs reduced to *scales.* Fls pale pink, scented, bell-shaped with five pointed petal-lobes, stamens and styles slightly protruding, sepals pointed, in tight heads; June-Oct. Heaths, grassy places. T. **4a* Greater Dodder** *C. europaea* has stems sometimes yellowish-green, larger heads of larger fls with petal-lobes and sepals blunt and stamens and styles not protruding; parasitic on Nettles (p. 38) and other large coarse plants. **4b,** p. 303.

Phlox Family   Polemoniaceae

5* JACOB'S LADDER *Polemonium caeruleum.* Medium/tall perennial, slightly downy. Lvs *pinnate,* alternate. Fls in a cluster, purplish-blue, open, with five petal-lobes; June-Aug. Fr a capsule. Grassy places and open woods in hill and mountain districts, also as a garden escape. B (rare), F, G. **5a,** p. 303.

Bedstraw Family   Rubiaceae

pp. 184-86. Stems *square,* weak, often scrambling over other vegetation. Lvs usually elliptical or lanceolate, unstalked, in *whorls,* with often lflike stipules. Fls in clusters or loose heads, small, open, with usually four petal-lobes. Only Field Madder has sepals. Fr a 2-lobed nutlet, a berry in Wild Madder (p. 186).

6* FIELD MADDER *Sherardia arvensis.* More or less prostrate annual, hairy. Lvs in whorls of 4-6, elliptical. Fls *pale purple,* in heads surrounded by lflike bracts; May-Sept. Fr globular, surrounded by enlarged sepal-teeth. Bare and cultivated ground. T.

7* SQUINANCYWORT *Asperula cynanchica.* Slender, more or less prostrate hairless perennial. Lvs *linear,* sometimes unequal, in whorls of four. Fls pale pink outside, white inside, in terminal clusters, petal-lobes 4, bracts lanceolate; June-Sept. Dry grassland, dunes; on lime. B, F, G. **7a** p. 303.

8* WOODRUFF *Galium odoratum.* Short *carpeting* perennial, almost hairless; unbranched. Lvs elliptical, in whorls of 6-9, edged with tiny forward-pointing prickles. Fls *white,* in loose heads; Apr-June. Fr covered with hooked bristles. Woods. T. **8a,** p. 303.

9* BLUE WOODRUFF *Asperula arvensis.* Slender short annual, hairless. Lvs linear, blunt, in whorls of 6-9. Fls bright *blue;* in heads; Apr-June. Weed of cultivation. (B, rare), G.

Bedstraw Family *(contd.)*

1* WILD MADDER *Rubia peregrina.* Scrambling evergreen perennial to 2 m; stems rough on angles with down-turned prickles. Lvs lanceolate, dark green, shiny, *prickly,* rather leathery, in whorls of 4-6. Fls yellowish-green, petal-lobes 5, in small stalked clusters at base of lvs; June-Aug. Fr a black *berry.* Woods, scrub, rocks; in Britain near the sea. B,F.

2* CROSSWORT *Cruciata laevipes.* Short perennial, *softly* hairy. Lvs elliptical, yellowish-green, in whorls of four. Fls *pale* yellow, scented, with bracts on stalks, in whorls at base of lvs; Apr-June. Fr blackish. Grassy places, open woods; on lime. B,F,G. **2a** *C. glabra* is slenderer and hairless, except for young lvs, and has no bracts on fl stalks. F, G.

3* HEDGE BEDSTRAW *Galium mollugo.* Variable medium/tall, often scrambling perennial, sometimes downy; stems *smooth,* square. Lvs elliptical, *1-veined,* ending in a point, in whorls of 6-8. Fls white, petal lobes pointed, in loose branched clusters; June-Sept. Fr black. Hybridises with 7. Grassy places, hedge-banks. T. **3a** *G. sylvaticum* has rounded stems, lvs greyish beneath and broader and blunter, and smaller fls. Eastern.

4* NORTHERN BEDSTRAW *Galium boreale.* Short, rather stiff perennial, sometimes downy. Lvs dark green, rough-edged, *3-veined,* in whorls of four. Fls white, in clusters; June-Aug. Fr covered with hooked bristles. Grassy, bushy and rocky places. T. **4a** *G. rotundifolium* has weak stems, smaller, more rounded lvs, and fr sometimes greenish. Mountains. F,G,S.

5* HEATH BEDSTRAW *Galium saxatile.* More or less prostrate hairless perennial. Lvs elliptical, sharply pointed, in whorls of 4-6, edged with forwardly directed prickles. Fls white, with a sickly fragrance, in short-stalked clusters along the stems; June-Aug. Fr with very pointed warts. Dry grassy and heathy places; not on lime. T. **5a Slender Bedstraw** *G. pumilum* is slenderer, the lvs with a few backwardly directed prickles, longer fl stalks, and fr with rounded warts. On lime. F,G. **5b* Limestone Bedstraw** *G. sterneri* differs from 5a in having many prickles on lf-margins and fr with shortly pointed warts. G, S. **5c** *G. timeroyi* is like 5a, but with lvs not broadest at tip and fl spikes twisted. F.

6* MARSH BEDSTRAW *Galium palustre.* Variable, rather straggling, short/medium hairless perennial; stems *rough* at angles. Lvs elliptical, blunt, in whorls of 4-5. Fls white, anthers red, in loose stalked clusters; June-Aug. Fr black. Wet places. T. **6a* Slender Marsh Bedstraw** *G. debile* is smaller and slenderer, with shorter, narrower, more pointed rough lvs, and pinkish fls. B(rare),F. **6b* Fen Bedstraw** *G. uliginosum* has rougher stems with down-turned prickles on angles, and narrower lvs, ending in a minute point, 6-8 in a whorl, and yellow anthers. **6c** *G. trifidum* is much shorter and has only three petal-lobes. S. **6d** *G. triflorum* is like 6b with much broader lvs. Rocky woods. S.

7* LADY'S BEDSTRAW *Galium verum.* Low/short sprawling perennial, almost hairless. Lvs linear, dark green, shiny, with margins inrolled beneath, in whorls of 8-12. Fls *bright* yellow, in lfy clusters; June-Sept. Fr black. Hybridises with 3 and *Galium glaucum* (p.184). Dry grassy places. T.

8* COMMON CLEAVERS *Galium aparine.* Straggling medium/tall annual, *clinging* to animal fur and human clothing by numerous tiny down-turned prickles on stems, lvs and fr. Lvs lanceolate, in whorls of 6-8. Fls *dull* white, in small stalked clusters at base of and longer than lvs; May-Sept. Fr green or purplish, bristles with white swollen bases, on straight spreading stalks. Hedge-banks, fens, disturbed and waste ground, shingle. T. **8a* Corn Cleavers** *G. tricornutum* has creamier fls on stalks shorter than the narrower lvs, and fr rough but with no prickles, on down-curved stalks; cornfields. **8b* False Cleavers** *G. spurium* has smaller green fls and smaller blackish fr whose bristles have unswollen bases; July. A weed of cultivation. T, but (B, rare). **8c* Wall Bedstraw** *G. parisiense* has lvs downturned, tiny fls green inside, reddish outside, and blackish, sometimes smooth fr; June-July. Sandy places, old walls. B,F,G.

Borage Family Boraginaceae

pp. 188-92. Hairy (except Oyster Plant, p. 192), often roughly hairy. Lvs undivided, alternate. Fls often pink in bud, later turning blue, five joined petals, usually in *one-sided* stalked spikes, which at first are tightly coiled. Fr four nutlets.

1* COMMON COMFREY *Symphytum officinale.* Tall stout perennial, softly hairy; well branched. Lvs broad lanceolate, *running down* on to winged stems, root lvs largest, to 25 cm. Fls creamy white or mauve, bell-like, in forked clusters; sepal-teeth pointed, at least equalling tube; May-June. Fr black, shiny. By fresh water, fens. T. **1a* Russian Comfrey** *S.* × *uplandicum* □, a hybrid between 1 and 1b, is stiffly hairy and has stem wings narrower and not reaching next lf-junction, and blue or purple-blue fls. Dry waysides, waste places, a widespread escape from cultivation. (T). **1b* Rough Comfrey** *S. asperum* is roughly hairy and has stems not winged, stalked heart-shaped lvs, larger blue fls, red in bud, and blunter sepal-teeth; a much less frequent escape. (B, F, G), from SW Asia. **1c* Soft Comfrey** *S. orientale* is smaller, paler green and earlier flg (Apr-May), with rounder upper lvs, whiter fls and short blunt sepal-teeth. Hedge-banks, a garden escape. (B), from SW Asia. **1d* Tuberous Comfrey** *S. tuberosum* is smaller and sometimes unbranched, with middle lvs the longest, fls yellowish-white only, and pointed sepal-teeth three times as long as tube. Damp, shady places. B, F, G. **1e*,** p. 303.

2* HOUNDSTONGUE *Cynoglossum officinale.* Medium/tall greyish biennial, softly downy, *smelling of mice.* Lvs lanceolate. Fls *maroon,* open, in clusters; May-Aug. Fr flattened, covered with short hooked spines and so adhering to fur and clothing, with a thickened flange. Dry grassy places, dunes. T. **2a* Green Houndstongue** *C. germanicum* is not downy, does not smell, and has shiny green lvs, almost hairless above, fls half the size, and no thickened flange on fr. Woods, shady places. B (rare), F, G.

3* AMSINCKIA *Amsinckia intermedia.* Short/medium annual, rather bristly. Lvs lanceolate. Fls *orange-yellow,* on upper side of coiled spikes; Apr-Aug. Fr wrinkled. Bare or disturbed ground. (B, G), from America. **3a Heliotrope** *Heliotropium europaeum* □ is greyish and foetid, with fls white or lilac, unstalked and no bracts. (F, G, S), from Mediterranean. **3b Lesser Honeywort** *Cerinthe minor* □ is like 3a but almost hairless, with lvs often white-spotted and fls yellow, in lfy spikes, not coiled. On lime. G. southeastern.

4* YELLOW ALKANET *Anchusa ochroleuca.* Medium perennial, roughly hairy. Lvs lanceolate. Fls creamy yellow, open, in clusters; *white-edged* blunt sepals cut to less than half-way; July-Aug. Waste places. (B, F, G), from Caucasus.

5 NONEA *Nonea pulla.* Short/medium greyish perennial. Lvs oblong to lanceolate, hairy but not rough, the upper clasping the stem. Fls dark brown, in a loose lfy one-sided spike, sepals enlarging in fr; Apr-Aug. Fr egg-shaped, beaked. Dry grassy, stony places. (F, S), G. **5a** *N. rosea* is a short annual, with oblong blunt lvs and pink fls with a yellow throat. (G, S), from Caucasus.

6* CORN GROMWELL *Lithospermum arvense.* Short/medium annual; *little* branched. Lvs strap-shaped to lanceolate, without prominent side veins, the lower shortly stalked. Fls creamy white, in clusters; Apr-Sept. Fr grey-brown, warty. A weed of cultivation. T.

7* COMMON GROMWELL *Lithospermum officinale.* Medium/tall perennial, *well* branched. Lvs lanceolate, unstalked, with prominent side veins. Fls creamy white, in lfy clusters; May-July. Fr white, shiny, smooth. Woods, scrub. T.

8* PURPLE GROMWELL *Lithospermum purpuro-caeruleum.* Short/medium perennial, downy, unbranched. Lvs narrow lanceolate, dark green. Fls reddish-purple turning *blue,* in lfy terminal clusters; Apr-June. Fr white, shiny. Woods, scrub, on lime. B (rare), F, G.

3b *Cerinthe minor*

3a Heliotrope

Borage Family (contd.)

1* LUNGWORT *Pulmonaria officinalis.* Low./short perennial, downy. Lvs lanceolate, *pale spotted*, abruptly narrowed at base, those from roots greatly enlarging after flg, to 40-60 cm. Fls pink and blue, bell-shaped, sepals with short broad teeth, in lfy clusters; Mar-May. Fr egg-shaped, pointed. Shady places. T, but (B). **1a* Narrow-leaved Lungwort** *P. longifolia* □ has narrower lvs tapering to base, smaller fls with longer narrower sepal-teeth, and flattened fr. B(rare),F,G, western. **1b, 1c,** p.303.

2* BLUE-EYED MARY *Omphalodes verna.* Low/short creeping perennial, slightly downy. Lvs *pointed oval,* often heart-shaped, long-stalked. Fls bright blue, 10 mm across, in a loose cluster; Mar-May. Fr with a hairy border. Woods. (B,F,G). **2a** *O. scorpioides* has yellow-centred fls, solitary at base of lvs. G.

FORGETMENOTS *Myosotis.* Lvs oblong, usually unstalked. Fls usually pink in bud, later blue, open. Fr shiny, on usually elongated spikes.

3* WOOD FORGETMENOT *Myosotis sylvatica.* Short perennial, softly hairy. Fls *sky-blue*, 6-10 mm, flat, sepal-tube with spreading hooked hairs; Apr-July. Fr dark brown, on spreading stalks longer than sepal-tube. Woods, grassland, mountains. T. **3a*,** p.303.

4* FIELD FORGETMENOT *Myosotis arvensis.* Low/short, softly hairy annual/biennial. Lower lvs stalked, in a rosette. Fls *grey-blue*, 3-5 mm across, cup-shaped, sepal-tube with numerous hooked hairs; Apr-Oct. Fr dark brown, shiny, on stalks longer than sepal-tube. Dry bare or disturbed ground, shady places, dunes. T. **4a** *M. stricta* has stems with smaller fls almost from base, lvs often with hooked hairs and fr stalks shorter than sepal-tube. Sandy places. F,G,S, southern. **4b* Early Forgetmenot** *M. ramosissima* is still smaller, often only 2-3 cm high, with smaller bluer fls and sepal-tube longer-stalked with teeth spreading in fr. Dry open places. **4c,** p.303.

5* CHANGING FORGETMENOT *Myosotis discolor.* Slender low/short annual. Fls creamy or pale *yellow* at first, turning blue, smaller, almost unstalked; May-June. Fr dark brown, shiny, on stalks shorter than sepal-tube, which has *incurved teeth*. Bare places, usually on light soils. T.

6* WATER FORGETMENOT *Myosotis scorpioides.* Low/short creeping perennial; hairs closely *pressed* to stems and sepal-tube. Fls sky-blue, occasionally pink or white, flat, petals notched, 4-10 mm across, sepal-teeth short, cut up to ⅓; June-Sept. Fr black, on stalks 1-2 times as long as sepal-tube. *Wet* places. T. **6a* Creeping Forgetmenot** *M. secunda* □ has spreading hairs on lower stem, fls 4-6 mm, sepal-tube narrowly cut to at least half-way, and fr dark brown on stalks 3-5 times as long as sepal-tube, down-turned. B,G,S. **6b* Pale Forgetmenot** *M. stolonifera* has much shorter bluish-green lvs, fls pale blue and 5 mm, petals not always notched, sepal-teeth cut up to ⅔, and fr olive-brown on sometimes down-turned stalks. B. **6c* Tufted Forgetmenot** *M. caespitosa* agg.□ may be hairless and has fls 2-4 mm, petals not notched, sepal-teeth cut up to ¾, and fr dark brown on stalks 2-3 times as long as sepal-tube. **6d Jersey Forgetmenot** *M. sicula* □ is smaller with fls pale blue and 2-3 mm, petals not notched, blunt sepal-teeth half as long as sepal-tube, and brown fr on stalks 1-3 times as long as sepal-tube. F. **6e** *M. baltica* is like 6c but has longer fl stalks lengthening after flg. S.

7* MADWORT *Asperugo procumbens.* Prostrate annual, very bristly; stems *angled*. Fls blue, 3 mm across, 1-3 together on short down-turned stalks much shorter than the elliptical lvs; May-Nov. Fr *surrounded* by enlarged sepals. Bare and disturbed ground. T,but(B,rare).

8 BUR FORGETMENOT *Lappula myosotis.* Short greyish annual/biennial, roughly hairy; well branched. Lvs lanceolate, unstalked. Fls blue, 2-4 mm, in a loose lfy spike; June-July. Fr with *hooked* spines, surrounded by *star-like* sepal-teeth. Dry bare places, dunes. F,G,S. **8a,** p.303.

1a* Narrow-leaved Lungwort

Borage Family (contd.)

1* VIPER'S BUGLOSS Echium vulgare. Short/medium perennial, roughly hairy. Lvs lanceolate, with no prominent side veins, the upper unstalked. Fls pink turning vivid blue, 15-20 mm, in a branched spike, with all stamens protruding; May-Sept. Fr hidden by sepal-teeth which give fr spike a mossy appearance. Dry bare and waste places. T. **1a* Purple Viper's Bugloss** E. lycopsis □ is more softly hairy, the lvs with prominent side veins, the upper heart-shaped and clasping the stem, and larger (25-30 mm) fls, red-purple turning blue-purple, and only two stamens protruding. Dry sandy places. B (rare), F. **1b** Onosma arenarium has a symmetrical, scarcely toothed petal tube. G, rare.

2* GREEN ALKANET Pentaglottis sempervirens. Medium perennial, roughly hairy. Lvs pointed oval, the lower stalked. Fls bright blue, white-eyed, 10 mm, in long-stalked lfy clusters, sepal-teeth blunt; Apr-July. Woods, hedge-banks. T, western, but (B).

3* BORAGE Borago officinalis. Medium annual, roughly hairy, with cucumber-scented juice. Lvs pointed oval with wavy margins, the lower stalked. Fls bright blue, 20-25 mm, in loose lfy clusters, with narrow petal and sepal lobes and a prominent column of purple stamens; May-Sept. Fr rough. Waysides, waste places. (B, F, G), from S Europe. **3a* Abraham, Isaac and Jacob** Trachystemon orientalis□ is shorter, with large heart-shaped lvs mainly from the roots, and smaller (15 mm) purplish fls in almost lfless clusters, and petal-lobes down-turned, making stamen column more conspicuous.

4* ALKANET Anchusa officinalis. Medium biennial/perennial, softly hairy; well branched. Lvs lanceolate, the lower narrowed to a stalk. Fls blue-purple, 10 mm, in an elongating coiled cluster; June-Aug. Fr conical, unstalked. Grassy places. T, but (B). **4a* Large Blue Alkanet** A. azurea □ is taller and roughly hairy, its much larger (15-25 mm) fls with a tuft of white hairs in the middle (F, G), from Mediterranean.

5* BUGLOSS Lycopsis arvensis. Short annual, roughly hairy. Lvs lanceolate, wavy-edged. Fls blue, 5-6 mm, in elongating, often forked clusters, the petal-tube bent; Apr-Sept. Fr unstalked. Bare, often sandy or cultivated places. T.

6* OYSTER PLANT Mertensia maritima. Prostrate mat-forming greyish hairless perennial. Lvs oval, fleshy, oyster-tasting. Fls pink turning blue-purple, in lfy clusters; June-Aug. Fr smooth, fleshy. Coastal shingle. B, S, northern.

Verbena Family   Verbenaceae

7* VERVAIN Verbena officinalis. Short/medium perennial, roughly hairy; stems stiff, square. Lvs pinnately lobed, toothed, opposite, the upper unstalked. Fls lilac, more or less 2-lipped, petals 5-lobed, in long slender lfless spikes; June-Oct. Dry bare or sparsely grassy places. T, southern.

3a* Abraham, Isaac and Jacob

1a* Purple Viper's Bugloss

4a* Large Blue Alkanet

1
2
3
4
5
6
7

Key to Two-lipped Flowers
(Mint and Figwort Families)
(Pages 196-216)

This key does not include all two-lipped flowers, but only those belonging to the Mint and Figwort Families – the Labiatae (pp. 196-204) and Scrophulariaceae (pp. 208-16). Labiates are distinctive for their square stems, and many of them are aromatic, but members of the Figwort Family lack any common distinctive features. Other two-lipped flowers include the Broomrapes (p. 218) and Orchids (pp. 274-86). The aberrant labiates with only one lip, the Bugles and Germanders (p. 196) are excluded. For all these groups, see the Main Key (p. 13).

Plant

white or grey: Cut-leaved Self-Heal, White Horehound, Catmint 198; Downy Woundwort 202; Lesser Calamint 204; Hoary Mullein 208

aromatic: Ground Ivy, White Horehound, Catmints, Winter Savory, Hyssop 198; Red and White Dead-nettles 200; Meadow Clary, Balm, Yellow Woundwort 202; most on p. 204.

foetid: Bastard Balm, Black Horehound 198; Motherwort, Yellow Archangel 200; Hedge Woundwort 202; Common Figwort 208

purplish: Ground Ivy 198; Claries and Wild Sage 202; Red Bartsia 212; Marsh Lousewort 216

woody at base: Ground Ivy 198; Yellow Archangel 200

with winged stems: Water Figwort 208

Leaves

untoothed: Spear-leaved Skullcap 196; Self-heal, *Dracocephalum ruyschiana*, Winter Savory, Hyssop 198; Marjoram 204; most on p. 210; Cow-wheats 216

narrow: *Dracocephalum ruyschiana, D. thymifolium*, Winter Savory, Hyssop 198; Red Hemp-nettle, Yellow Archangel 200; Marsh and Annual Woundworts 202; Spear Mint, Gipsywort 204; most on p. 210; Large Yellow Foxglove, Gratiola, Red Bartsia, Yellow Odontites 212; Yellow Rattle, Cow-wheats 216

pinnate: Cut-leaved Self-heal 198; Wild Clary 202; French Figwort 208; Louseworts and Moor-king 216

heart- or kidney-shaped: Ground Ivy, Catmint 198; Dead-nettles, *Galeopsis pubescens* 200; Wild Sage, Balm, Hedge and Field Woundworts 202; Balm-leaved and Yellow Figworts 208; Cornish Moneywort 216

triangular: Fluellens, Ivy-leaved Toadflax 210

alternate: most on p. 210; Foxgloves and Gratiola 212; Eyebright, Louseworts, Moor-king and Cornish Moneywort 216

Flowers

white: Self-heals, White Horehound, Catmint, Bastard Balm, Winter Savory 198; White Dead-nettle, Common Hemp-nettle 200; Balm 202; Gipsywort 204; *Lindernia pyxidaria* 210; Foxglove, Gratiola 212; Eyebright 216

blue/violet: Skullcap 196; Self-heal, Ground Ivy, Hairless Catmint, Hyssop 198; Claries and Wild Sage 202; Basil Thyme 204; Pale, Purple and Ivy-leaved Toadflaxes 210

yellow: Common and Downy Hemp-nettles, Yellow Archangel 200; Yellow Figwort 208; Snapdragon, Common and Sand Toadflaxes 210; Monkey Flower, Large Yellow Foxglove, Yellow Odontites, Yellow Bartsia 212; Yellow Rattle, Moor-king, Leafy Lousewort, Common Cow-wheat 216

pink/purple: Lesser Skullcap 196; Self-heal, Ground Ivy, Bastard Balm, Black Horehound, Winter Savory 198; Spotted, Red and Henbit Dead-nettles, False Motherwort, Common and Red Hemp-nettles 200; Whorled Clary, Woundworts, Betony 202; all on p.204; Purple Mullein, French Figwort 208; Snapdragons, Purple and Small Toadflaxes 210; Foxglove, Gratiola, Fairy Foxglove, Alpine and Red Bartsias 212; Lousewort and Marsh Lousewort 216

yellow and purple: Motherwort, Large-flowered Hemp-nettle 200; Yellow Woundwort 202; Small, Pale and Ivy-leaved Toadflaxes, Fluellens 210; Yellow Rattle, Common Cow-wheat, Cornish Moneywort 216

red-brown: Figworts 208

in whorls: Ground Ivy, White Horehound, Catmint, Bastard Balm, Black Horehound, Hyssop 198; most on pp.202-4

in spikes: Skullcaps 196; Hairless Catmint 198; Betony 202; Spear Mint 204; Snapdragon, Toadflaxes 210; most on pp.212-14

in heads: Self-heals 198; Water Mint, Wild Thyme 204

in clusters: Winter Savory 198; Common Calamint, Marjoram 204; Figworts 208

solitary: Lesser Snapdragon, Small and Ivy-leaved Toadflaxes, Fluellens 210; Gratiola 212

in pairs: Common Cow-wheat 216

with mouth closed and/or spurred: all on p.210

bell-shaped: Mints and Gipsywort 204

tubular: Foxgloves and Gratiola 212

with calyx inflated: Yellow Rattle, Louseworts 216

Labiate Family   Labiatae

Often *aromatic*. Stems *square*. Lvs usually undivided, in *opposite* pairs. Fls with both petals and sepals, normally *2-lipped* and open-mouthed (except p.204, nos. 1-5), though *Ajuga* and *Teucrium* both have upper lip absent or very short, but lower lip 3-lobed and 5-lobed respectively. Fr four nutlets.

1* BUGLE *Ajuga reptans.* Low/short creeping perennial; runners *rooting*, stems hairy on two sides. Lvs oblong, scarcely toothed, hairless, often bronzy, the lower long-stalked. Fls powder-blue, rarely pink or white, in a lfy, often purplish, spike; Apr-June. Damp woods, grassland. T. **1a Blue Bugle** *A. genevensis* has no runners, stems often hairy all round, root lvs larger, fls bright blue, and stamens protruding. Dry grassy, stony places. F,G,(S). **1b* Pyramidal Bugle** *A. pyramidalis* has no runners, stems hairy all round, root lvs hairy, pyramidal bluish spikes of smaller blue-violet fls, and stamens protruding; a shy flowerer; grassy, rocky places in mountains and near sea.

2* GROUND-PINE *Ajuga chamaepitys.* Low greyish annual, faintly *aromatic* of pine resin. Looks somewhat like a bushy pine seedling with its narrowly *linear* 3-lobed lvs. Fls yellow, red-dotted, 1-2 together up the lfy stems; May-Sept. Bare, stony, sparsely grassy places. B,F,G.

3* SKULLCAP *Scutellaria galericulata.* Short creeping perennial, often downy. Lvs lanceolate, bluntly toothed, short-stalked. Fls bright *blue*, in *pairs* up the lfy stem; petal-tube slightly curved; June-Sept. Hybridises with 4. Wet grassy places, by fresh water. T. **3a* Spear-leaved Skullcap** *S. hastifolia* □ is slenderer, with untoothed arrow-shaped lvs, purple beneath, and larger fls in a denser, less lfy spike, with petal-tube strongly curved. Rare in B. **3b*** *S. columnae* has bracts much smaller than lvs and larger fls in a one-sided spike. (B,rare),F.

4* LESSER SKULLCAP *Scutellaria minor.* Low/short creeping perennial, sometimes downy. Lvs lanceolate to oval, scarcely toothed. Fls pale *pinkish-purple*, spotted darker, in *pairs* up the lfy stems; petal-tube almost straight; July-Oct. Damp *heathy* places. T, southern.

5* WOOD SAGE *Teucrium scorodonia.* Short/medium perennial, downy. Lvs heart-shaped, bluntly toothed, *wrinkled*, sage-like. Fls greenish-yellow, in pairs in lfless stalked spikes, with prominent maroon stamens; July-Sept. Open woods, scrub, heaths, fixed dunes, screes; not on lime. T, western.

6 MOUNTAIN GERMANDER *Teucrium montanum.* Low/short spreading undershrub. Lvs linear/lanceolate, *white* beneath, margins inrolled. Fls yellowish-white, in rounded clusters; May-Aug. Rocky places in mountains. F,G.

7* WALL GERMANDER *Teucrium chamaedrys.* Short tufted perennial, slightly hairy. Lvs pointed oval, *shiny*, dark green, rather leathery, with rounded teeth. Fls dark pinkish-purple, in short lfy whorled spikes; May-Sept. Dry bare places, including walls. (B,rare),F,G.

8* WATER GERMANDER *Teucrium scordium.* Low/short sprawling perennial, softly hairy. Lvs oblong, toothed, unstalked. Fls pinkish-purple, in whorls up lfy stems; June-Aug. Damp places, by *fresh water*, dune slacks. T, rare in B.

9* CUT-LEAVED GERMANDER *Teucrium botrys.* Low/short downy annual/biennial. Lvs oval, *deeply cut*, almost pinnate. Fls pink or pinkish-purple, in whorls along lfy stems; June-Oct. Bare, stony ground; on lime. B(rare),F,G.

3a* Spear-leaved Skullcap

Labiate Family (contd.)

1* SELF-HEAL *Prunella vulgaris.* Low creeping perennial, downy, not aromatic. Lvs pointed oval, *scarcely toothed,* the lower stalked. Fls *violet,* rarely pink or white, 10-15 mm long, in *oblong* or square purplish heads; June-Nov. Flhds with lvs at base. Hybridises with 1a and 2. Grassy places, bare ground. T. **1a Large Self-heal** *P. grandiflora* ☐ is larger, with fls 20-25 mm long. Flhds with no lvs at base. F, G, S.

2* CUT-LEAVED SELF-HEAL *Prunella laciniata.* Low creeping perennial, downy, not aromatic; stems with white hairs. Lvs pointed oval, the upper *pinnate* or pinnately lobed. Fls creamy *white,* occasionally tinged violet (indicating hybridisation with 1), in oblong or square heads; June-Oct. Dry grassy and bare places. (B), F, G.

3* GROUND IVY *Glechoma hederacea.* Low creeping, often purplish perennial, softly hairy, aromatic; runners long, rooting. Lvs *kidney-shaped,* blunt-toothed, long-stalked. Fls blue-violet, rarely pink in loose whorls at base of lvs; Mar-June. Woods, hedge-banks, sparsely grassy and bare places, sometimes carpeting. T. **3a** *Dracocephalum ruyschiana* ☐ has narrow lanceolate untoothed unstalked lvs and dense heads of larger fls; July-Aug. G, S. **3b** *D. thymifolium* ☐ is like 3a but with toothed lvs and looser heads of smaller bright blue fls. G.

4* WHITE HOREHOUND *Marrubium vulgare.* Medium perennial, *white* with down, *thyme-scented.* Lvs roundish, bluntly toothed, wrinkled, stalked. Fls white, in dense whorls up lfy stems; sepal-teeth 10, hooked; June-Sept. Dry bare and waste places. T, rare in B.

5* CATMINT *Nepeta cataria.* Short/medium perennial, *grey* with down, *mint-scented.* Lvs heart-shaped, more markedly toothed than 4, stalked. Fls white, red-dotted, in whorls up lfy stems; sepal-teeth 5, straight; June-Sept. Hedge-banks, waysides, rocks. T.

6 HAIRLESS CATMINT *Nepeta nuda.* Medium perennial, almost *hairless.* Lvs oblong, toothed. Fls violet, in branched spikes; June-Sept. Open woods, scrub, grassy and rocky places. F, G.

7* BASTARD BALM *Melittis melissophyllum.* Medium perennial, hairy, *strong smelling.* Lvs pointed oval, toothed, stalked. Fls pink, or white with pink spots, *large* (35-45 mm long), in few-fld whorls at base of lvs; May-July. Woods, hedge-banks, shady rocks. B, F, G.

8* BLACK HOREHOUND *Ballota nigra.* Medium, rather straggly perennial, hairy, *strong smelling.* Lvs pointed oval, toothed, stalked. Fls pinkish-purple, in whorls at base of upper lvs, sepal-teeth *curved* back in fr; June-Sept. Waste places, waysides. T. **8a** *B. alba* has sepals scarcely toothed, tube expanded at tip. S.

9* WINTER SAVORY *Satureja montana.* Short half-evergreen undershrub, almost hairless, pleasantly aromatic; a culinary herb. Lvs narrow lanceolate, *pointed,* untoothed, rather leathery, hairy on margins, *shiny-dotted,* unstalked and joined at base. Fls pale pink or white, in clusters; July-Oct. Old walls. (B, rare), from S Europe; cultivated T.

10* HYSSOP *Hyssopus officinalis.* Medium perennial, almost hairless, pleasantly aromatic; a culinary herb. Lvs linear oblong, *blunt.* Fls violet-blue, in few-fld whorls making a long narrow spike; July-Sept. Old walls, dry banks. (B, F, G, rare), from S Europe.

3a *Dracocephalum ruyschiana*

3b *D. thymifolium*

199

Labiate Family (contd.)

1* WHITE DEAD-NETTLE *Lamium album.* Short creeping perennial, hairy, faintly aromatic; unbranched. Lvs heart-shaped, toothed, stalked, resembling benign Nettle (p. 38) lvs. Fls *white* with greenish streaks on lower lip, 20-25 mm long, in whorls at base of lvs; Mar-Nov. Waysides, waste places. T. **1a* Spotted Dead-nettle** *L. maculatum* ☐ is strong smelling and has lvs often pale-spotted and fls usually pink-purple. T, but (B).

2* RED DEAD-NETTLE *Lamium purpureum.* Low/short, often purplish annual, downy, aromatic. Lvs heart-shaped, bluntly toothed, wrinkled, *all stalked.* Fls pale to dark pinkish-purple, 10-15 mm long; all year. A weed of cultivation. T. **2a* Cut-leaved Dead-nettle** *L. hybridum* ☐ has more deeply toothed lvs and smaller fls.

3* HENBIT DEAD-NETTLE *Lamium amplexicaule* agg. Low/short annual, downy, little branched. Lvs rounded, bluntly toothed, the topmost usually *unstalked* and often half-clasping the stem. Fls pink-purple, 15 mm long, often also smaller and unopened; in whorls at base of upper lvs; Mar-Oct. A weed of cultivation. T.

4* MOTHERWORT *Leonurus cardiaca.* Medium/tall perennial, slightly downy, strong smelling; unbranched. Lvs *palmately* 3-7-lobed, whitely downy beneath, stalked. Fls pinkish-purple or white, the lower lip spotted purple, furry outside, 12 mm long, in whorls up the lfy stems; July-Sept. Waysides, waste places. T, but (B). **4a False Motherwort** *L. marrubiastrum* is biennial, with lower lvs undivided, the upper only toothed, and pale pink, much less hairy fls. F,G,S, south-eastern.

5* COMMON HEMP-NETTLE *Galeopsis tetrahit* agg. Short/medium annual, roughly hairy; well branched, stems *swollen* at lf-junctions. Lvs broad lanceolate, toothed, stalked. Fls pinkish-purple, less often yellow or white, 10-20 mm long, in whorls at base of upper lvs; July-Sept. Probably originated as 5a × 7 hybrid. Open woods, heaths, fens, also a weed of cultivation. T. **5a** *G. pubescens* has lvs square or heart-shaped at base and fls bright purple. G.

6* RED HEMP-NETTLE *Galeopsis angustifolia* agg. Short/medium annual, softly hairy; well branched, stems *scarcely* swollen at lf-junctions. Lvs narrow to broad lanceolate, sparsely toothed, stalked. Fls *deep* pink, white-flecked on lip, 15-25 mm long, in whorls at base of upper lvs; July-Sept. A weed of cultivation, bare stony places. T, southern.

7* LARGE-FLOWERED HEMP-NETTLE *Galeopsis speciosa.* Medium/tall annual, roughly hairy; well branched, stems *swollen* at lf-junctions. Lvs broad lanceolate, toothed, stalked. Fls pale *yellow,* lower lip *purple,* 20-45 mm long, in whorls at base of upper lvs; July-Sept. A weed of cultivation, waste places. T.

8* DOWNY HEMP-NETTLE *Galeopsis segetum.* Short/medium annual, *softly* hairy; well branched, stems *not* swollen at lf-junctions. Lvs lanceolate, toothed, stalked. Fls pale *yellow,* 20-30 mm long, in whorls at base of upper lvs; July-Sept. A weed of cultivation. (B, rare), F, G. Hybridises with 6.

9* YELLOW ARCHANGEL *Lamiastrum galeobdolon.* Short/medium creeping perennial, hairy, strong smelling; *runners* long, lfy, not rooting. Lvs narrow to broad lanceolate, dark green, stalked. Fls *butter* yellow, usually streaked red-brown on lower lip, in whorls at base of upper lvs; May-June. Woods. T.

Nettle (p.38)

1* White Dead-nettle

Labiate Family (contd.)

1* MEADOW CLARY *Salvia pratensis.* Medium perennial, hairy, slightly aromatic. Lvs broad lanceolate, the few on the stem narrower and unstalked, bluntly toothed, wrinkled. Fls *brilliant blue,* in whorled lfless spikes; June-July. Grassland on lime. T. **1a* Whorled Clary** *S. verticillata* is strong smelling, with upper lvs and stems often purplish, all lvs stalked and often with a purplish sheen, smaller pinkish-purple fls, and purple sepals; dry bare and waste places; May-Aug. (T), from S Europe. **1b Jupiter's Distaff** *Salvia glutinosa* □ has yellow fls. Mountain woods. G, southern.

2* WILD SAGE *Salvia sylvestris.* Short/tall hairy perennial. Lvs heart-shaped, much longer than broad, pointed, the upper and middle ones unstalked. Fls blue, in purplish lfy whorled and branched spikes; June-Aug. Bare and waste places. (B, rare, G), from S Europe.

3* WILD CLARY *Salvia horminoides.* Variable medium perennial, hairy; little branched, purplish on upper lvs and stems. Lvs oval, *jaggedly* toothed, sometimes almost pinnately lobed, the upper unstalked. Fls blue-violet, white-spotted on lower lip, 5-15 mm long, in whorled spikes; often not opening and so appearing much smaller than bracts; June-Sept. Dry grassy places. B, F. **3a** *S. verbenaca* □ has longer root lvs and fls paler, with no white spots, and always open, in less purplish spikes. F.

4* BALM *Melissa officinalis.* Medium hairy perennial, *lemon-scented* when crushed, a culinary and medicinal herb. Lvs pointed oval, the lowest heart-shaped, toothed, stalked, yellow-green. Fls *white,* in lfy whorled spikes; July-Sept. An escape from cultivation. (B, F, G), from S Europe.

5* HEDGE WOUNDWORT *Stachys sylvatica.* Medium creeping perennial, roughly hairy, *strong smelling.* Lvs heart-shaped, toothed, stalked. Fls dark *beetroot-purple* with whitish blotches, in lfless whorled spikes; June-Oct. Hybridises with 5a. Hedge-banks, shady places. T. **5a* Marsh Woundwort** *S. palustris* □ is only faintly aromatic and has narrower short- or unstalked lvs, and pale pinkish-purple fls. Damp places, by fresh water. **5b* Limestone Woundwort** *S. alpina* □ is soft to the touch and not creeping, and has lfy spikes of slightly larger pinkish fls. On lime. B (rare), F, G.

6* DOWNY WOUNDWORT *Stachys germanica.* Medium/tall perennial, thickly covered with a *felt of white hairs.* Lvs pointed oval, toothed, the upper short- or unstalked. Fls pale pinkish-purple, in a lfy whorled spike; July-Sept. Bare ground; on lime. B (rare), F, G.

7* YELLOW WOUNDWORT *Stachys recta.* Variable short/medium hairy perennial, pleasantly aromatic. Lvs toothed, oval to oblong, stalked, the upper much narrower and unstalked. Fls *pale yellow* with purple streaks, in lfless whorled spikes, the stamens protruding; June-Sept. Dry rocky or waste places. (B, rare), F, G. **7a* Annual Woundwort** *S. annua* is a shorter annual, with all lvs lanceolate and stalked, and smaller fls; T, but (B, rare). **7b** *Sideritis montana* has stamens enclosed within the petal-tube. (F), G, (S).

8* FIELD WOUNDWORT *Stachys arvensis.* Low/short, often sprawling hairy annual. Lvs heart-shaped, bluntly toothed, the upper unstalked. Fls dull purple, *scarcely* exceeding sepals; Apr-Oct. A weed of cultivation; not on lime. T.

9* BETONY *Betonica officinalis.* Variable short/medium hairy perennial, usually unbranched. Lvs oblong, slightly toothed, all but the topmost stalked. Fls reddish-purple, in tight *oblong* spikes; June-Oct. Grassy and heathy places. T.

1b Jupiter's Distaff

Labiate Family (contd.)

MINTS *Mentha.* Aromatic perennials, each with a distinctive scent. Lvs stalked, except 3. Fls small, usually lilac or pale purple, bell-shaped, with four more or less equal petal-lobes, in *tight whorls* at base of lvs; sepal-tube 5-toothed, stamens four, protruding except in the numerous hybrids.

1* WATER MINT *Mentha aquatica.* Variable short/medium, often purplish hairy perennial, pleasantly aromatic. Lvs pointed oval, toothed. Fls lilac or pinkish-lilac, with a *round* terminal head; sepal-tube hairy, long-toothed; July-Sept. Hybridises with 2 (q.v.) and with 3, from which the resultant hybrid is the often escaping culinary herb Peppermint *M.* × *piperita,* with narrower, more sharply toothed lvs, reddish-lilac fls, fl stalks and sepal-tube usually hairless and stamens not protruding. Wet places. T.

2* CORN MINT *Mentha arvensis.* Low/medium hairy perennial, rather acridly aromatic. Lvs pointed oval, toothed. Fls lilac, with *no* terminal head; sepal-tube hairy, short-toothed; July-Sept. Hybridises with 1, the resulting *M.* × *verticillata* having no terminal head but longer sepal-teeth and stamens not protruding. Damp places, and as a weed of cultivation. T.

3* SPEAR MINT *Mentha spicata.* Short/medium perennial, hairy or hairless; the most commonly grown garden mint. Lvs lanceolate, almost *unstalked,* either green, shiny and hairless or greyish, wrinkled and downy. Fls lilac, with a *pointed* terminal spike; July-Oct. Hybridises with 1 (q.v.) and 3a. Damp and waste places. T, southern. **3a*** and hybrids, p.303.

4* PENNYROYAL *Mentha pulegium.* Prostrate, sometimes erect and short, perennial, downy, strongly aromatic. Lvs oval, scarcely toothed. Fls lilac, with *no* terminal head; sepal-tube hairy, well ribbed, the two lower teeth narrower; July-Oct. Damp places. B, F, G.

5* GIPSYWORT *Lycopus europaeus.* Medium/tall perennial, slightly hairy, *not* aromatic; unbranched. Lvs lanceolate, deeply toothed or lobed, short-stalked. Fls small, white spotted purple, bell-shaped, in tight whorls at base of upper lvs; July-Sept. By fresh water, marshes. T.

6* COMMON CALAMINT *Calamintha sylvatica.* Short/medium hairy perennial, *mint-scented.* Lvs oval, slightly toothed, stalked, dark green. Fls lilac-purple spotted darker, 10-15 mm, in short-stalked opposite clusters at base of lvs; sepal-tube often purple with no hairs protruding from throat after flg; July-Sept. Dry grassy places, on lime. In open woods or scrub fls may be 17-22 mm (**6a*** □). B, F, G. **6b* Lesser Calamint** *C. nepeta* has greyer lvs, paler, sometimes almost white, less spotted fls, and hairs protruding from mouth of sepal-tube after flg.

7* WILD BASIL *Clinopodium vulgare.* Short/medium perennial, softly hairy, faintly aromatic; little branched. Lvs pointed oval, slightly toothed, stalked. Fls pinkish-purple, in whorls at base of upper lvs; bristle-like bracts and purplish sepal-tube covered with *white hairs;* July-Sept. Dry grassy places, scrub, on lime. T.

8* MARJORAM *Origanum vulgare.* Medium downy perennial, pleasantly aromatic; a culinary herb. Lvs oval, often slightly toothed, stalked. Fls purple, darker in bud and with *dark purple* bracts, in loose clusters; July-Sept. Dry grassland, scrub, screes; on lime. T.

9* BASIL THYME *Acinos arvensis.* Low sprawling hairy annual, not aromatic. Lvs oval, slightly toothed, stalked. Fls *violet* with *white* patch on lower lip, in whorls at base of upper lvs; June-Sept. Dry bare and sparsely grassy places; on lime. T.

10* WILD THYME *Thymus serpyllum* agg. Variable prostrate hairy *mat-forming* undershrub, aromatic; with runners. Lvs oval, sometimes quite woolly, short-stalked. Fls pinkish- or reddish-purple, in *rounded* heads, sometimes also in whorls below; June-Sept. Dry grassy and heathy places, dunes. T. **10a* Large Thyme** *T. pulegioides* is larger and taller, with no runners, and more strongly aromatic, and has lvs always almost hairless and fls in elongated whorled spikes. Often on lime.

Nightshade Family   Solanaceae

Often poisonous. Lvs alternate, normally stalked. Fls 5-petalled, joined at base.

1* BITTERSWEET *Solanum dulcamara.* Clambering, sometimes prostrate downy perennial, to 2 m. Lvs pointed oval, often with two lobes at base. Fls in loose clusters, bright *purple,* petals turning down, yellow anthers in a conspicuous *column;* May-Sept. Fr a poisonous egg-shaped berry, green then yellow then *red.* Woods, scrub, hedges, waste and damp places, shingle. T.

2* BLACK NIGHTSHADE *Solanum nigrum.* Variable low/short annual, hairless or downy; stems often blackish. Lvs pointed oval, often toothed or lobed. Fls in loose clusters, *white,* petals turning down, with prominent column of yellow anthers; July-Oct. Fr a poisonous berry, green then *black.* A weed of cultivation, waste places. T. **2a* Green Nightshade** *S. sarrachoides* is hairier, with stems never blackish and sepals enlarging to exceed the always green berry. **2b* Tomato** *Lycopersicum esculentum* □, a straggling plant with 2-pinnate lvs, fls all yellow and large red fr, is a widespread escape from cultivation.

3 HAIRY NIGHTSHADE *Solanum villosum.* Short hairy annual. Lvs pointed oval, more deeply lobed than 2. Fls like 2 but larger. Fr *orange* or yellowish-brown. Bare and waste places. F, G.

4* DEADLY NIGHTSHADE *Atropa bella-donna.* Tall stout perennial, sometimes downy; well branched. Lvs pointed oval, up to 20 cm long. Fls dull purple or greenish, *bell-shaped,* solitary, 25-30 mm; June-Sept. Fr a glossy *black* berry, extremely poisonous. Woods, scrub, rocky places; on lime. B,F,G. N.B. Bittersweet is often miscalled Deadly Nightshade.

5* HENBANE *Hyoscyamus niger.* Medium/tall annual/biennial, *stickily* hairy, unpleasantly foetid, very poisonous. Lvs oblong, sometimes with a few teeth, up to 20 cm long, the upper unstalked, clasping the stem. Fls dull creamy yellow with purple veins, bell-shaped, in a lfy cluster; May-Sept. Fr a capsule. Bare and disturbed ground, especially by sea. T.

6* APPLE OF PERU *Nicandra physaloides.* Medium/tall annual, often hairless, foetid, very poisonous. Lvs pointed oval, toothed or lobed. Fls blue or pale *violet* with white throat, bell-shaped, solitary, 30-40 mm across, opening only for a few hours; June-Oct. Fr a brown berry, encased in the net-veined *bladder-like* swollen sepals. Bare and waste places, waysides. (B,F,G), from Peru.

7* THORN-APPLE *Datura stramonium.* Stout medium/tall hairless annual, foetid and poisonous. Lvs pointed oval, jaggedly toothed. Fls white, rarely purple, *trumpet-shaped,* solitary, 60-80 mm across; July-Oct. Fr egg-shaped, *spiny.* Bare, cultivated and waste ground. (T).

8 SMALL TOBACCO PLANT *Nicotiana rustica.* Medium/tall annual, stickily hairy, strong smelling and poisonous. Lvs *heart-shaped,* shiny. Fls greenish-yellow, bell-shaped, in clusters; June-Aug. Fr a capsule. An escape from cultivation. (F, G, southern), from S America.

2b* Tomato

207

Figwort Family   Scrophulariaceae

pp. 208-16. Lvs with no stipules. Fls either open and more or less flat with four or five joined petals (Mulleins, below, and Speedwells, p. 214), or two-lipped with the lips either open or closed. Fr a capsule.

MULLEINS *Verbascum*. Biennials, usually tall. Lvs in a basal rosette and alternate up stems. Fls flat, 5-petalled, in spikes; anthers orange. Fr egg-shaped.

1* GREAT MULLEIN *Verbascum thapsus*. Tall, stout, to 2 m, covered with *thick* white woolly down; stem round, usually unbranched. Lvs broad lanceolate, bluntly toothed, running down on to the *winged* stems. Fls yellow, almost flat, 15-30 mm, with three whitely hairy and two hairless stamens; June-Aug. Dry grassy and bare places, open scrub. T. **1a*** *V. thapsiforme* ☐ has yellowish-grey down, more pointed and coarsely toothed lvs, and sometimes branched spikes of larger (30-50 mm), completely flat fls. (B), G. **1b* Orange Mullein** *V. phlomoides* has lvs not or hardly running down stem and larger (30-50 mm), sometimes orange-yellow fls. (B), F, G.

2* DARK MULLEIN *Verbascum nigrum*. Medium/tall, hairy, stem ridged, usually unbranched, often purplish. Lvs heart-shaped, toothed, dark green, stalked. Fls yellow with *purple* hairs on all stamens; June-Sept. Dry grassy places. T.

3* WHITE MULLEIN *Verbascum lychnitis*. Medium/tall, downy; stems angled at top, *branched* like a candelabrum. Lvs lanceolate, bluntly toothed, shortly stalked, dark green and almost hairless above, *white* with down beneath. Fls yellow or white, 15-20 mm, with all stamens whitely hairy; June-Sept. Dry bare and sparsely grassy places. T. **3a* Hoary Mullein** *V. pulverulentum* is thickly covered all over with mealy white down, and has rounded stems, broader lvs and always yellow fls. B (rare), F, G.

4* MOTH MULLEIN *Verbascum blattaria*. Medium/tall, hairless below but with sticky hairs above; stems angled, usually unbranched. Lvs lanceolate, *shiny*, toothed, gradually narrowed to base. Fls yellow, 20-30 mm, solitary, *long-stalked*; stamens with *purple* hairs; June-Aug. Bare and waste places. B, F, G. **4a* Twiggy Mullein** *V. virgatum* is more stickily hairy above, with 1-5 short-stalked fls together. B (rare), F. **4b Purple Mullein** *V. phoeniceum* ☐ has purple fls. (F), G.

FIGWORTS *Scrophularia*. Square-stemmed perennials. Lvs opposite, stalked, toothed. Fls two-lipped, open-mouthed, with five small blunt lobes; in a branched terminal cluster.

5* COMMON FIGWORT *Scrophularia nodosa*. Medium/tall, almost hairless, foetid; stems not winged. Lvs *pointed oval*, not lobed. Fls red-brown, sepals *all green;* June-Sept. Woods, shady places. T. **5a* Water Figwort** *S. auriculata* has winged stems and lf-stalks, lvs blunter, often with two lobes at base, and sepals white-edged. Wet places. B, F, G. **5b* Green Figwort** *S. umbrosa* is completely hairless and has lf stalks running down on to even more broadly winged stem than 5a, and olive-brown fls.

6* BALM-LEAVED FIGWORT *Scrophularia scorodonia*. Medium/tall, *downy* all over. Lvs *heart-shaped*, double-toothed, wrinkled. Fls red-brown, sepals *white-edged;* June-Sept. Dry shady places near sea. B, F.

7* FRENCH FIGWORT *Scrophularia canina*. Medium/tall, hairless; well branched. Lvs *pinnately lobed*, the lobes toothed. Fls *purplish-black*, upper lip sometimes white, with purple stamens and orange anthers; June-Aug. Bare and waste places. (B, rare), F, G.

8* YELLOW FIGWORT *Scrophularia vernalis*. Medium/tall biennial/perennial, softly hairy. Lvs heart-shaped, wrinkled, well toothed, yellowish-green. Fls pale *greenish-yellow;* Apr-June. Woods, scrub, bare places, old walls. T, but (B).

209

Figwort Family (contd.)

All plants on this page have 2-lipped fls, with *mouth closed,* except 9, and spurred, except 1 and 2. Some are semi-parasitic on the roots of other plants.

1* SNAPDRAGON *Antirrhinum majus.* Short/medium perennial, downy; woody at base; well branched. Lvs narrow lanceolate, untoothed. Fls very variable, most commonly red-purple, less often pale yellow, with a *pouch* instead of a spur, *30-40 mm;* in stalked spikes; May-Oct. Dry bare places, rocks, walls. (B), F, G.

2* LESSER SNAPDRAGON *Misopates orontium.* Short annual, usually downy. Lvs linear, untoothed. Fls pink-purple, *10-15 mm, pouched,* scarcely stalked at base of upper lvs; July-Oct. A weed of cultivation, bare sandy places. T.

3* COMMON TOADFLAX *Linaria vulgaris.* Short/medium greyish hairless perennial. Lvs linear, untoothed, numerous up stems. Fls *yellow,* with orange spot on lower lip, and long straight spur 15-30 mm, in stalked spikes; June-Oct. Bare and waste places. T. **3a** *L. genistifolia* is more branched with broader lvs and paler fls with a browner orange spot. (F), G. **3b* Prostrate Toadflax** *L. supina* is smaller and more or less prostrate, and has smaller (10-15 mm) fls in a roundish cluster. B, F.

4* SAND TOADFLAX *Linaria arenaria.* Low annual, stickily hairy; well branched. Lvs linear, untoothed. Fls yellow with short spur often *violet,* 4-6 mm, in stalked spikes; May-Sept. Coastal dunes. (B, rare), F. **4a** *L. bipartita* is rather taller with purple upper lip and orange and white lower lip. (G).

5* PALE TOADFLAX *Linaria repens.* Short greyish hairless perennial. Lvs linear, untoothed, numerous up stems. Fls pale *lilac,* veined violet, with orange spot on lower lip and short curved spur, 7-14 mm, in stalked spikes; June-Sept. Dry bare and sparsely grassy places. T. Hybridises with 3. **5a** *L. arvensis* □ has lower lvs whorled in fours and smaller (4-8 mm) blue-lilac fls with a white spot. F, G, decreasing. **5b Alpine Toadflax** *L. alpina* □ has broader whorled lvs and larger (20 mm) violet fls with an orange spot. Mountain screes and river beds. F, G.

6* PURPLE TOADFLAX *Linaria purpurea.* Medium greyish hairless perennial; often unbranched. Lvs linear, untoothed, numerous up stems. Fls bright *violet,* occasionally pink, 8 mm, with long curved spur, in stalked spikes; June-Aug. Walls, bare ground. F, (B, G, S). **6a Jersey Toadflax** *Linaria pelisseriana* □ is shorter than *L. purpurea* and annual, with lower lvs whorled and larger (10-20 mm) bluer fls with a white spot and a straight spur. Bare and disturbed ground. F.

7* IVY-LEAVED TOADFLAX *Cymbalaria muralis. Trailing,* often purplish hairless perennial. Lvs palmately lobed, ivy-like, long-stalked. Fls lilac with yellow spot, 8-10 mm, with a short curved spur; solitary on long stalks at base of lvs; Apr-Nov. *Walls,* rocks. (B, F, G), from S Europe.

8* SHARP-LEAVED FLUELLEN *Kickxia elatine.* More or less prostrate annual, stickily hairy. Lvs all *pointed,* almost triangular, stalked. Fls with upper lip and throat *purple* and *lower lip yellow,* 8-11 mm, spur straight; solitary on almost hairless stalks at base of lvs; July-Oct. A weed of cultivation, bare places. T. **8a* Round-leaved Fluellen** *K. spuria* □ has lvs oval or roundish, fls with upper lip maroon, spur curved, stalks hairy. **8 × 8a** has hairy fl-stalks, pointed lvs. F, G.

9* SMALL TOADFLAX *Chaenorhinum minus.* Slender low downy annual. Lvs linear, untoothed, greyish, scarcely stalked. Fls pale purple, mouth *slightly open,* 6-8 mm, with a short blunt spur, solitary on long stalks at base of lvs; May-Oct. A weed of cultivation, railway tracks, bare places. T, southern. **9a**, p.303.

10 DAISY-LEAVED TOADFLAX *Anarrhinum bellidifolium.* Medium/tall hairless biennial/perennial. Lvs in basal rosette oval to narrowly elliptic, toothed, those up stems narrowly *palmately lobed.* Fls blue or violet, 3-5 mm, in long slender lfless spikes, spur slender, curved; Mar-Aug. Dry, rather bare places, rocks, walls, coniferous woods. F, G, southern.

SNAPDRAGONS, TOADFLAXES

1

2

3

4

5

5a

5b

6

6a

7

8

8a

9

10

211

Figwort Family (contd.)

1* MONKEY FLOWER *Mimulus guttatus*. Short creeping perennial, downy above, hairless below. Lvs oblong to roundish, toothed, opposite, the upper clasping the stem. Fls bright yellow with small red *spots*, 2-lipped, open-mouthed, 25-45 mm, in lfy stalked spikes; June-Sept. Hybridises freely with 1a. Wet places. T, from N America. **1a* Blood-drop Emlets** *M. luteus* □ is smaller and almost hairless, and has large red blotches on fls. Less widely naturalised, from Chile. **1b* Musk** *M. moschatus* □ is smaller and stickily hairy all over, and has much smaller (10-20 mm) all yellow fls. Less widely naturalised.

2* FOXGLOVE *Digitalis purpurea*. Tall perennial to 1½ m, downy; unbranched. Lvs broad lanceolate, wrinkled, *soft* to the touch. Fls *pink-purple*, occasionally white, tubular, 2-lipped, in long tapering spikes; June-Sept. Woods, scrub, heaths, mountains. T.

3 LARGE YELLOW FOXGLOVE *Digitalis grandiflora*. Medium/tall hairy perennial; unbranched. Lvs lanceolate, toothed, hairless above. Fls *yellow* with purple-brown veins, tubular, 2-lipped, hairy outside, 30-40 mm long, in a long slender spike; June-Sept. Open woods, rocks in mountains. F, G. **3a Small Yellow Foxglove** *D. lutea* □ has lvs hairless on both sides, and smaller (15-20 mm) plain yellow fls hairless outside. Woods, stony hillsides, mainly on lime. $3 \times 3a$ and $2 \times 3a$ occasionally occur.

4 GRATIOLA *Gratiola officinalis*. Medium hairless perennial. Lvs lanceolate, opposite, unstalked, toothed near the tip. Fls white or pale pink, tubular, scarcely 2-lipped, long-stalked at base of upper lvs; May-Oct. Wet places. F, G.

5* FAIRY FOXGLOVE *Erinus alpinus*. Low tufted hairy perennial. Lvs oval, broadest at tip, toothed, mostly in rosettes. Fls purple, with five *notched* petal-lobes, in short stalked spikes; May-Oct. Rocks, screes, walls, mainly in mountains. (B), F.

6* RED BARTSIA *Odontites verna*. Low/short semi-parasitic, often purplish downy annual; well branched. Lvs lanceolate, toothed, unstalked, opposite. Fls *pink*, 2-lipped, open-mouthed, lower lip 3-lobed, in lfy one-sided spikes; June-Sept. Bare and disturbed ground. T.

7* YELLOW ODONTITES *Odontites lutea*. Short/medium semi-parasitic annual, almost hairless; *well branched*. Lvs narrow lanceolate, toothed, unstalked, opposite, with margins inrolled. Fls 6 mm, bright yellow, 2-lipped, open-mouthed, lower lip 3-lobed, with stamens protruding, in one-sided spikes; July-Sept. Dry bare or sparsely grassy places, often on lime. (B, rare), F, G. **7a** *O. jaubertiana* is downy, with fls 8 mm, sometimes reddish and stamens not protruding; Aug-Oct. F.

8* ALPINE BARTSIA *Bartsia alpina*. Low semi-parasitic downy perennial; unbranched. Lvs oval, toothed, unstalked, opposite. Fls dark *purple*, 2-lipped, open-mouthed, upper lip hooded, lower 3-lobed, in a short spike with *purple* bracts; July-Aug. Rocks, grassland in hills and mountains. T, but rare in B.

9* YELLOW BARTSIA *Parentucellia viscosa*. Short semi-parasitic annual, *stickily* hairy; usually *unbranched*. Lvs lanceolate, toothed, opposite, unstalked. Fls yellow, 2-lipped, open-mouthed, lower lip 3-lobed, upper hooded, in lfy spikes; June-Sept. Damp grassy and sandy places, dune slacks. B, F.

2* Foxglove

Figwort Family (contd.)

SPEEDWELLS *Veronica*. Lvs opposite, usually toothed. Fls blue, with four joined petals and (except 8) four sepals. Fr flattened, 2-lobed or notched.

1* SPIKED SPEEDWELL *Veronica spicata*. Short/medium downy perennial; unbranched. Lvs oval, stalked, often in a rosette, the upper narrower and unstalked. Fls bright blue, 5 mm, with prominent stamens and blunt sepals, in long dense *Ifless* spikes; July-Oct. Dry grassland, rocks, often on lime. T, but rare in B. **1a***, p.303.

2 LARGE SPEEDWELL *Veronica teucrium*. Short/medium hairy perennial. Lvs oblong, well toothed, unstalked. Fls bright blue, 10 mm, *5-sepalled*, in stalked spikes at base of upper lvs; June-Aug. Grassy places. F,G. **2a, 2b**, p.303.

3* GERMANDER SPEEDWELL *Veronica chamaedrys*. Low/short sprawling hairy perennial, with two *opposite* lines of hairs on stems. Lvs pointed oval, short- or unstalked. Fls bright blue with a *white eye*, 10 mm, in opposite stalked spikes at base of upper lvs; Apr-June. Grassy places. T. **3a* Wood Speedwell** *V. montana* ☐ has stems hairy all round, longer-stalked paler green lvs, purplish beneath and smaller (7 mm) paler blue fls. Damp woods. **3b* Rock Speedwell** *V. fruticans* ☐ is almost hairless and has much less toothed lvs and fls with a red eye; July-Aug. Mountains. **3c**, p.304.

4* COMMON FIELD SPEEDWELL *Veronica persica*. Low sprawling hairy annual. Lvs oval, short-stalked, pale green. Fls sky-blue with darker veins, the lowest petal usually *white*, 8-12 mm, solitary on long stalks at base of upper lvs; all year. Fr with widely diverging lobes. A weed of cultivation. (T), from W Asia. **4a* Green Field Speedwell** *V. agrestis* ☐ is smaller, with paler 4-8 mm fls on shorter stalks, and fr only notched; Mar-Nov. T. **4b* Grey Field Speedwell** *V. polita* ☐ is smaller and greyish, with darker and uniformly blue 4-8 mm fls on shorter stalks and fr only notched; Mar-Nov. **4c** *V. opaca* ☐ is like 4a but has spoon-shaped sepals and hairless fr. F,G,S. **4d* Slender Speedwell** *V. filiformis* ☐ is a mat-forming perennial with rounder lvs, mauver fls on very long slender stalks, and rounded fr; Apr-June. Increasing on lawns and in grassy places. (B,F,G), from W Asia.

5* LILAC IVY-LEAVED SPEEDWELL *Veronica sublobata*. Low sprawling downy annual. Lvs palmately lobed, *ivy-like*. Fls *purple-lilac* to white, rather small, stalked, solitary at base of lvs; Mar-Aug. A weed of cultivation. T. **5a* Blue Ivy-leaved Speedwell** *V. hederifolia* has more sharply toothed lvs and bright blue fls.

6* THYME-LEAVED SPEEDWELL *Veronica serpyllifolia*. Low creeping perennial, almost hairless. Lvs oval, shiny, *scarcely* toothed, short- or unstalked. Fls pale blue or white with purple veins, 5-6 mm (larger on mountains, ssp *humifusa* ☐ **(6a)**), stalked, in *Ify* spikes; Apr-Oct. Bare and sparsely grassy places. T. **6b* Alpine Speedwell** *V. alpina* ☐ is scarcely creeping and has bluish-green lvs and deep blue fls with a white eye; July-Aug. Mountains, Arctic heaths.

7* HEATH SPEEDWELL *Veronica officinalis*. Low creeping hairy perennial. Lvs oval, short-stalked. Fls *lilac* with darker veins, in stalked spikes at base of lvs; May-Aug. Dry grassy and heathy places. T.

8* WALL SPEEDWELL *Veronica arvensis*. Low *erect* annual, hairy. Lvs pointed oval, toothed, short-stalked. Fls blue, tiny, shorter than sepals, in Ify spikes; Mar-Oct. Dry bare places, walls. T. **8a*, 8b*, 8c*, 8d*, 8e*, 8f**, p.304.

9* BROOKLIME *Veronica beccabunga*. Low creeping rooting hairless fleshy perennial. Lvs *oval*, stalked. Fls blue, 7-8 mm, in opposite stalked spikes at base of upper lvs; May-Sept. Wet places. T.

10* WATER SPEEDWELL *Veronica anagallis-aquatica*. Short creeping rooting hairless fleshy perennial. Lvs *lanceolate*, slightly toothed, unstalked. Fls pale blue, 5-6 mm, in opposite stalked spikes at base of upper lvs; June-Aug. Fr rounded. Wet places. T. **10a* Pink Water Speedwell** *V. catenata* ☐ has stems often purplish and fls pink. **10b* Marsh Speedwell** *V. scutellata* is slenderer with lvs often purplish, whiter fls in alternate spikes, and fr deeply notched. **10c**, p.304.

Figwort Family (contd.)

1* EYEBRIGHT *Euphrasia officinalis* agg. Variable low/short, often bronzy green semi-parasitic hairy annual; often well branched. Lvs more or less oval, deeply toothed, the upper sometimes alternate. Fls *white,* often tinged violet or with purple veins or a yellow spot; 2-lipped, open-mouthed, the lower lip 3-lobed; in lfy spikes with broad, toothed bracts; June-Oct. Fr hairy. Grassy places, mountains. T. **1a* Irish Eyebright** *E. salisburgensis* has narrower, less markedly toothed bracts and fr hairless or almost so.

2* YELLOW RATTLE *Rhinanthus minor* agg. Variable low/medium semi-parasitic annual, usually almost hairless. Lvs oblong to linear, toothed, unstalked, opposite. Fls *yellow,* 2-lipped, usually open-mouthed, often with two purple teeth, in loose lfy spikes; May-Sept. Fr *inflated,* the seeds rattling inside when ripe. Grassy places, cornfields. T. **2a** *Pedicularis lapponica* is shorter, with very narrow lvs, longer, paler yellow fls and less inflated sepal-tube. S.

LOUSEWORTS *Pedicularis.* Semi-parasites, especially on grasses. Lvs pinnately cut, toothed, short-stalked, alternate. Fls 2-lipped, open-mouthed, in lfy spikes. Sepal-tube inflated in fr.

3* MARSH LOUSEWORT *Pedicularis palustris.* Short/medium biennial, almost hairless; stem *single,* branched, often purplish. Fls reddish-pink, the lips equal, sepal-tube *downy;* May-Sept. Wet grassy and heathy places. T. **3a** *P. hirsuta* has woolly stems and less deeply toothed lvs. S.

4* LOUSEWORT *Pedicularis sylvatica.* Low perennial, almost hairless; stems *numerous,* unbranched. Fls pink, the upper lip longer, sepal-tube usually *hairless;* Apr-July. Moors, damp heaths, bogs. T.

5 MOOR-KING *Pedicularis sceptrum-carolinum.* Stout medium/tall perennial. Lvs with *oval lobes,* mostly in a basal rosette. Fls yellow, *30-35 mm,* mouth often closed but if not lower lip purplish, in almost lfless spikes. Damp grassy places and scrub. G,S. **5a** *P. oederi* is short, with rounded lflets and fls in a dense head, the upper lip red-brown with two red spots. **5b** *P. flammea* is like 5a, but lip unspotted.

6 LEAFY LOUSEWORT *Pedicularis foliosa.* Stout medium perennial. Lvs 2-pin-nate, with pointed *triangular* lobes. Fls pale yellow, 15-20 mm, in dense lfy spikes; June-Aug. Mountain grassland. F,G.

7* COMMON COW-WHEAT *Melampyrum pratense.* Variable short semi-parasitic annual, slightly downy. Lvs lanceolate, untoothed, unstalked, opposite. Fls yellow, sometimes pinkish-purple, 2-lipped, with mouth *shut,* in pairs facing the same way at base of toothed lfy bracts; May-Sept. Woods, heaths, grassland. T. **7a* Small Cow-wheat** *M. sylvaticum* □ is slenderer with smaller, always deep yellow open-mouthed fls and less toothed bracts; June-Aug. Woods. T, rare in B. **7b** *M. nemorosum* □ has lvs stalked, much broader at base and slightly toothed, rather larger fls with orange lower lip, and upper bracts purple. Heaths, woods. F,G,S. **7c** *M. paludosum* has linear lvs and whitish fls streaked purple. G, southern, rare.

8* CRESTED COW-WHEAT *Melampyrum cristatum.* Short/medium semi-parasi-tic annual, slightly downy. Lvs narrow lanceolate, untoothed, unstalked, opposite. Fls yellow and purple, 2-lipped, mouth almost closed, in a short *squarish* spike with conspicuously *purple* short-toothed bracts; June-Sept. Dry grassy, rocky places, wood margins. T. **8a* Field Cow-wheat** *M. arvense* □ has a longer looser fl spike with bracts magenta, broader and much longer-toothed. T, rare in B.

9* CORNISH MONEYWORT *Sibthorpia europaea.* Prostrate *mat-forming* peren-nial, hairy; stems thread-like. Lvs *kidney-shaped,* palmately lobed with very blunt lobes, long-stalked, alternate. Fls 5-petalled, the two upper pale yellow, the three lower pale pink, tiny, short-stalked at base of lvs; June-Oct. Damp shady places. B,F.

217

Broomrape Family   Orobanchaceae

Parasitic on roots of other plants. *No* green pigment. Lvs replaced by *scales,* more or less pointed oval, on stems. Fls usually coloured as rest of plant, 2-lipped, in spikes. Fr a capsule, usually egg-shaped. Broomrapes *Orobanche* (2-5) are a difficult group, with the host often the best clue to identity. Their erect stems persist rigid and brown with the dead fls through the winter. Grassy places.

1* TOOTHWORT *Lathraea squamaria.* Low/short creamy perennial, slightly downy. Fls pale *pink,* short-stalked, in a drooping one-sided spike; Apr-May. Woods, hedges; hosts: Hazel and various trees. T.   **1a* Purple Toothwort** *L. clandestina* □ is much shorter and hairless, with longer-stalked purple fls. (B), F.

2* GREATER BROOMRAPE *Orobanche rapum-genistae.* Medium/tall perennial, honey-brown tinged purple. Fls 20-25 mm, upper lip hooded *not lobed,* stamens hairless below and stigma lobes yellow; only one bract; June-July. Hosts: shrubby peaflowers, especially Broom and Gorse (p. 118). B, F, G.   **2a* Knapweed Broomrape** *O. elatior* is shorter and sometimes reddish, with fls 15-20 mm, upper lip 2-lobed and stamens hairy below; hosts: Knapweeds (p. 248) and Globe-thistles (p. 244). **2b, 2c,** p. 304.

3* COMMON BROOMRAPE *Orobanche minor.* Variable short/medium annual, purplish, reddish or yellowish. Fls 10-18 mm, shorter than or equalling their single bract, stamens hairless, stigma lobes purple; June-Sept. Hosts: peaflowers (pp. 116-32) and composites (pp. 228-56). B, F, G.   **3a* Thyme Broomrape** *O. alba* is usually much shorter and deep purplish-red, with fls longer than bract and stamens hairy below; on labiates (pp. 194-204), especially Wild Thyme (p. 204). **3b* Ox-tongue Broomrape** *O. picridis* is yellowish-white tinged purple, with stamens very hairy below: on ox-tongues (p. 256) and hawksbeards (pp. 254, 256). B (rare), F, G. **3c* Thistle Broomrape** *O. reticulata* is yellowish tinged purple with sparse hairs on fls and stamens sometimes sparsely hairy; on thistles (p. 246). B (rare), F, G. **3d* Ivy Broomrape** *O. hederae* is yellowish tinged purple, with yellow stigma lobes; on Ivy (p. 154). **3e, 3f, 3g*,** p. 304.

4* PURPLE BROOMRAPE *Orobanche purpurea.* Short/medium annual, bluish-purple; rarely branched. Fls large (18-30 mm), with *three* bracts to each fl, petal lobes pointed; June-July. Hosts: composites (pp. 228-56), especially Yarrow (p. 238). T, southern, rare in B. **4a*, 4b,** p. 304.

5* CLOVE-SCENTED BROOMRAPE *Orobanche caryophyllacea.* Short annual, yellowish tinged purple. Fls *fragrant,* large (20-35 mm), much longer than their single bract, stamens hairy below; June-July. Hosts: bedstraws (p. 186). B (rare), F, G. **5a,** p. 304.

Globularia Family   Globulariaceae

6 GLOBULARIA *Globularia vulgaris.* Low/short perennial. Lvs oval, untoothed, the larger often *notched* or 3-lobed at tip, stalked, in a basal rosette; lanceolate, unstalked up stems. Fls blue, rarely lilac or white, 2-lipped, the upper lip very short, the lower 3-lobed, in a *globular* head; Apr-June. Dry grassy or stony places. F, G, S, southern.

Butterwort Family   Lentibulariaceae

Butterworts *Pinguicula.* Low perennials, stickily hairy. Lvs oblong, untoothed, all in a starfish-like basal rosette, the margins rolling inwards to trap and digest insects. Fls 2-lipped, spurred, solitary on long lfless stalks.

7* COMMON BUTTERWORT *Pinguicula vulgaris.* Lvs yellow-green. Fls *violet* with a *white* throat-patch, spur pointed, 10-15 mm; May-July. Bogs, fens, wet heaths and moors. T. **7a* Large-flowered Butterwort** *P. grandiflora* □ has much larger (25-30 mm) fls with a stouter spur. B, F.

8* PALE BUTTERWORT *Pinguicula lusitanica.* Lvs olive green with red-brown veins. Fls pale *lilac* with *yellow* throat and down-turned cylindrical spur, 6-7 mm; June-Oct. Bogs, wet heaths. B, F, western. **8a, 8b,** p. 304.

Plantain Family Plantaginaceae

Lvs strongly veined or ribbed, all in a basal rosette (except 4). Fls tiny, 4-petalled, in dense spikes or heads, on long *lfless* stalks (except 4), with prominent *stamens* providing most of the colour. Fr a capsule.

1* GREATER PLANTAIN *Plantago major.* Low/short perennial, hairless or downy. Lvs *broad oval.* Fls pale greenish-yellow and pale brown, in long greenish spikes about equalling the unfurrowed stalks; anthers pale purple then yellowish-brown; June-Oct. Waste and well trodden places, paths, lawns. T. **1a* Hoary Plantain** *P. media* □ is always downy and greyish, with longer, often narrower lvs, and shorter long-stalked spikes of fragrant whitish fls with a fuzz of pinkish-lilac stamens; May-Aug. Grassy places; on lime. T.

2* RIBWORT PLANTAIN *Plantago lanceolata.* Low/medium perennial, hairless or downy. Lvs *lanceolate,* slightly toothed. Fls brown, in short blackish spikes on furrowed stalks, anthers pale yellow; Apr-Oct. Grassy and waste places; one of the commonest European plants. T.

3* SEA PLANTAIN *Plantago maritima.* Low/short perennial, usually hairless. Lvs *linear,* fleshy, 3-5-veined, sometimes slightly toothed. Fls brownish-pink with yellow anthers, in greenish spikes on unfurrowed stalks; June-Sept. Coastal saltmarshes, occasionally on mountains inland. T.

4* BRANCHED PLANTAIN *Plantago indica.* Short downy annual; much branched. Lvs linear, sometimes obscurely toothed, not in a rosette, the lower with short lfy shoots at their base. Fls pale brown, in long-stalked *egg-shaped* heads up the lfy stems; May-Aug. Waste and sandy places, dunes. T, but (B).

5* BUCKSHORN PLANTAIN *Plantago coronopus.* Low biennial, usually downy. Lvs variable, usually *pinnately lobed,* less often linear and deeply toothed, 1-veined. Fls yellow-brown with yellow anthers, in short greenish spikes on unfurrowed stalks; May-Oct. Dry bare, often sandy places, most frequent on the coast. T.

Arrow-Grass Family Juncaginaceae

A monocotyledonous family. (See p.10 and p.258.)

6* MARSH ARROW-GRASS *Triglochin palustris.* Short slender hairless perennial. Lvs linear, deeply furrowed to appear almost *cylindrical.* Fls green, purple-edged, 3-petalled, tiny, short-stalked, in an interrupted spike, the style showing as a white tuft; May-Aug. Fr narrow, opening arrow-shaped when ripe. *Marshes,* fens, damp grassland. T. **6a* Rannoch Rush** *Scheuchzeria palustris* □ (Scheuchzeriaceae) has flat lvs with inflated sheathing bases, longer than the small cluster of yellower fls; bog pools. T. Scattered, rare in B.

7* SEA ARROW-GRASS *Triglochin maritima.* Short/medium hairless perennial. Lvs linear, fleshy, *flat,* not veined. Fls green, 3-petalled, tiny, short-stalked, in a long plantain-like spike; May-Sept. Fr egg-shaped. Coastal *saltmarshes.* T.

Moschatel Family Adoxaceae

8* MOSCHATEL *Adoxa moschatellina.* Low carpeting hairless perennial. Lvs 2-trefoil, lobed, long-stalked. Fls greenish, in tight heads of five, at *right angles* to each other; Mar-May. Fr a green berry. Woods, shady places, also on mountains. T.

fruit

6a* Rannoch Rush

Valerian Family   Valerianaceae

Lvs opposite, no stipules. Fls with five joined petals, small, in clusters.

1* COMMON VALERIAN *Valeriana officinalis* agg. Variable short/tall perennial, hairy mainly below; usually unbranched. Lvs *pinnate,* the lower stalked, lflets lanceolate, toothed. Fls pale pink, pouched at base, in more or less rounded clusters; June-Aug. Woods, grassy places, both damp and dry. T.

2* MARSH VALERIAN *Valeriana dioica.* Short perennial, almost hairless, with creeping *runners.* Root lvs *oval,* untoothed, long-stalked; stem lvs more or less pinnate, unstalked. Fls pale pink, pouched at base, in rounded clusters; stamens and styles on different plants; May-June. Marshes, fens. T. **2a** *V. montana* has no runners and stem lvs oval, scarcely toothed; June-Aug. Mountains. F, G. **2b,** p. 304.

3* RED VALERIAN *Centranthus ruber.* Medium/tall greyish tufted hairless perennial. Lvs pointed oval or *lanceolate,* the lower stalked and untoothed. Fls red, pink or sometimes white, fragrant, spurred, in rounded clusters; May-Sept. Cliffs, rocks, quarries, walls, steep banks. (B), F, G.

4* CORNSALAD *Valerianella locusta* agg. Low/short annual, almost hairless, well branched. Lvs oblong, unstalked. Fls pale lilac, *tiny,* in flat-topped umbel-like clusters; Apr-Aug. Fr flattened. Bare and cultivated ground, dunes, walls. T. **4a,** p. 304.

Honeysuckle Family   Caprifoliaceae

Lvs opposite. Fls with five joined petals.

5* DWARF ELDER *Sambucus ebulus.* Stout tall foetid patch-forming perennial to 1 m. Lvs *pinnate,* lflets lanceolate, sharply toothed, with lfy stipules. Fls white, often tinged pink, in a flat-topped umbel-like cluster; anthers violet; July-Aug. Fr a black berry. Waysides, waste places, scrub. T, but (B). **5a* Elder** *S. nigra* is a deciduous shrub to 10 m, with broader lflets, no or tiny stipules, fragrant fls, and yellow anthers; May-July. Woods, hedges, scrub, waste places, often at rabbit warrens. T. **5b* Red-berried Elder** *S. racemosa* differs from 5a in its more pointed yellow-green lvs, greenish-white fls in a dense oval cluster, and red berries; Apr-May. Hill and mountain districts. T, but (B).

6* GUELDER ROSE *Viburnum opulus.* Deciduous shrub or small tree to 4 m, scarcely downy. Lvs palmately 3-5-lobed, *ivy-like.* Fls white, slightly fragrant, in flat umbel-like clusters, outer petals of outer fls *much larger;* May-July. Fr a shiny *red* berry. Damp woods, fens, scrub, hedges. T.

7* WAYFARING TREE *Viburnum lantana.* Downy deciduous shrub to 4 m. Lvs *oval,* minutely toothed, wrinkled. Fls creamy white, in a flat-topped umbel-like cluster, fragrant; Apr-June. Fr a berry, red then *black.* Scrub, open woods, hedges; on lime. T.

8* TWINFLOWER *Linnaea borealis.* Delicate prostrate mat-forming downy evergreen undershrub. Lvs oval, toothed, stalked. Fls pink, *bell-shaped,* drooping, fragrant, in pairs; June-Aug. Fr egg-shaped. Coniferous woods. T, northern.

9* HONEYSUCKLE *Lonicera periclymenum.* Deciduous woody *climber* to 6 m, hairless or downy; twining clockwise. Lvs oval, untoothed, short- or unstalked, appearing in midwinter. Fls creamy becoming orange-buff, often reddish outside, tubular, 2-lipped, very fragrant, in a *head;* June-Oct. Fr a red berry. Woods, scrub, hedges. T. **9a* Perfoliate Honeysuckle** *L. caprifolium* □ is always hairless, with upper lvs and bracts joined at the base, cup-like; May-July. (B), F, G.

10* FLY HONEYSUCKLE *Lonicera xylosteum.* Deciduous shrub to 2 m, downy. Lvs pointed oval, short-stalked, greyish. Fls yellowish, tubular, 2-lipped, *in pairs* at base of upper lvs; May-June. Fr a red berry. Woods, scrub. T, rare in B. **10a, 10b, 10c,** p. 304.

9a* Perfoliate Honeysuckle

Teasel Family   Dipsacaceae

Lvs opposite. Fls small, 4-5-petalled, tightly packed into a composite-like (p. 230) head, but with four stamens projecting from each fl. Each head cupped by sepal-like bracts. Fr small, with one seed.

1* FIELD SCABIOUS *Knautia arvensis.* Medium/tall hairy perennial. Lvs *pinnately lobed,* the upper sometimes undivided. Fls bluish-lilac, in rather *flat* heads, 30-40 mm, outer petals of outer fls much larger; bracts in two rows, shorter than fls; anthers pink; June-Oct. Dry grassy places, cornfields. T. **1a Wood Scabious** *K. sylvatica* □ has brighter green lvs, toothed but not lobed, with outer petals only slightly larger, and bracts nearly as long as fls. Shady places in mountains. F, G.

2* SMALL SCABIOUS *Scabiosa columbaria.* Variable short/medium hairy perennial; little branched. Lvs *pinnate* with narrow lobes, basal lvs sometimes undivided. Fls like 1, but heads smaller (15-25 mm) with *dark bristles* among the fls, and only one row of bracts; June-Oct. Dry grassland; on lime. T, southern. **2a Yellow Scabious** *S. ochroleuca* □ is annual/biennial, with upper lvs more but lower lvs less deeply cut, and fls yellow. G. **2b** *S. canescens* □ has basal lvs always lanceolate and anthers bright purple. F, G, S, southern.

3* DEVILSBIT SCABIOUS *Succisa pratensis.* Medium/tall hairy perennial. Lvs *elliptical,* untoothed, stalked, often blotched purplish, narrower and sometimes toothed up stems. Fls dark blue-purple, occasionally pink, in *rounded* heads, 15-25 mm; June-Oct. Damp grassy places. T.

4* SMALL TEASEL *Dipsacus pilosus.* Medium/tall biennial, hairy above, *prickly* on stems and lvs. Lvs oblong, stalked, the upper narrower, all often with a pair of lflets at base. Fls *white,* in woolly spiny *globular* heads, anthers violet; July-Sept. Woods, scrub, streamsides. T, but (S).

5* TEASEL *Dipsacus fullonum.* Tall hairless perennial, *prickly* on stems and lvs; dead stems and flhds persist through winter. Lvs lanceolate, covered with white *pimples,* in a basal rosette that withers before flg; narrower up stems, often cupped at base. Fls pale *purple,* in a *conical* spiny head; July-Aug. Bare and sparsely grassy places, often damp. B, F, G. **5a** *D. laciniatus* has deeply cut lvs and paler fls. Waste places. G, southern.

Bellflower Family   Campanulaceae: see p. 226

6* ROUND-HEADED RAMPION *Phyteuma tenerum.* Low/short perennial, usually hairless, unbranched. Root lvs variable, toothed, broad lanceolate, rarely heart-shaped; stem lvs lanceolate to linear, unstalked. Fls *dark* blue, in globular heads, petal lobes narrow; bracts short, narrowly triangular; July-Aug. Dry grassland; on lime. B, F, G. **6a** *P. orbiculare* has lower lvs more often heart-shaped, and bracts broader, lanceolate. F, G. **6b** *P. nigrum* has root lvs always heart-shaped, stem lvs broader, sometimes untoothed, and flhds more elongated. F.

7* SPIKED RAMPION *Phyteuma spicatum.* Short/medium hairless perennial. Root lvs heart-shaped, toothed, long-stalked; stem lvs narrower, less stalked. Fls yellowish-white, small, in *elongated* heads, petal lobes narrower; May-July. Woods, grassy places. T, rare in B and S.

8* SHEEPSBIT SCABIOUS *Jasione montana.* Low/short biennial, slightly hairy; no dead lvs at base. Lvs narrow oblong, untoothed but sometimes *wavy,* the upper unstalked. Fls blue, in globular heads, petal lobes narrow; May-Sept. Dry grassy places, heaths, sea cliffs, shingle; not on lime. T. **8a** *J. perennis* is perennial with runners and has longer lvs persisting at base when dead. F, G.

1a Wood Scabious

1

2a

2b

2

3

4

5

6

7

8

Bellflower Family   Campanulaceae

Lvs undivided, alternate, no stipules. Petals joined, with five lobes. Fr a capsule. Bellflowers *Campanula* have more or less bell-shaped fls.

1* HAREBELL *Campanula rotundifolia*. Slender short/medium hairless perennial. Root lvs *roundish*, withering early; stem lvs *linear*, the upper unstalked. Fls blue, 15 mm, on long thin stalks, in loose clusters; petal lobes short; July-Oct. Dry grassland, heaths. T. **1a, 1b, 1c, 1d, 1e,** p.304

2* CLUSTERED BELLFLOWER *Campanula glomerata*. Variable stiff low/medium hairy perennial. Lower lvs oval, sometimes heart-shaped, stalked; upper lvs narrower, unstalked. Fls deep violet, in tight *heads;* sepal lobes *pointed;* June-Oct. Grassland; on lime. T. **2a** *C. cervicaria* has lanceolate lower lvs, smaller pale blue fls and blunt sepal lobes. F,G,S.

3* NETTLE-LEAVED BELLFLOWER *Campanula trachelium*. Medium/tall hairy perennial; stems sharply angled, unbranched. Lvs broad lanceolate, long-pointed, irregularly toothed. Fls violet-blue, 30-40 mm long, 1-3 together in a lfy spike, opening from the top; petal lobes short, triangular; July-Sept. Woods, scrub, hedges. T. **3a* Giant Bellflower** *C. latifolia* is stouter and has bluntly angled stems, more evenly toothed lvs, and larger (40-50 mm), paler fls, with longer, narrower petal lobes, solitary at base of lvs, the lower ones opening first. Mainly in hills and mountains. **3b* Creeping Bellflower** *C. rapunculoides* is smaller, slenderer and creeping, with stems hardly angled, narrower, more evenly toothed lvs, and smaller (20-30 mm) fls with longer petal lobes. Sometimes on bare ground. T, but (B). **3c** *C. bononiensis* is like 3b but has lvs white-woolly beneath and still smaller (10-20 mm) fls in a denser one-sided lfless spike-like cluster. Mountains. F,G.

4* SPREADING BELLFLOWER *Campanula patula*. Medium/tall biennial/perennial, rough to the touch; stems angled. Lvs oblong, the upper narrower and unstalked. Fls violet-blue, 15-25 mm, in a spreading cluster, on stalks with a small bract in the middle; petal-lobes long, sepal-teeth linear; June-July. Woods, grassy places. T. **4a* Rampion Bellflower** *C. rapunculus* has milky juice, and bluer fls with shorter petal lobes and small bracts at base of stalks. T, but (B).

5* PEACH-LEAVED BELLFLOWER *Campanula persicifolia*. Medium/tall *hairless* perennial. Lower lvs oblong, stalked; upper lvs linear/lanceolate, unstalked. Fls violet-blue, *30-40 mm,* in a stalked spike, petal-lobes broadly triangular; May-Aug. Woods, scrub, also naturalised from gardens. T.

6 BEARDED BELLFLOWER *Campanula barbata*. Short perennial, hairy; unbranched. Lvs oblong, wavy, mainly in a basal rosette. Fls pale blue, with long *white hairs* inside the bell, in a one-sided cluster; sepals in two rows, like a Canterbury Bell *(C. medium);* June-Aug. Woods, grassy places, in mountains. F,G,S.

7* IVY-LEAVED BELLFLOWER *Wahlenbergia hederacea*. Low creeping rather delicate hairless perennial. Lvs palmately lobed, *ivy-like,* stalked. Fls pale blue, bell-shaped with short petal lobes, on hairlike stalks at base of lvs; July-Aug. Damp woods, heaths and moors. B,F,G.

8* VENUS'S LOOKING GLASS *Legousia hybrida*. Low/short hairy annual. Lvs oblong, wavy, unstalked. Fls purple, in clusters, petal lobes only 8-12 mm across, *shorter* than sepal teeth, closing in dull weather and then hard to detect; May-Aug. A weed of cultivation, bare sandy or stony places. B,F,G.

9 LARGE VENUS'S LOOKING GLASS *Legousia speculum-veneris*. Short annual. Lvs oblong, wavy, unstalked. Fls violet, opening *wide* to 20 mm, in clusters; May-July. Bare, waste and cultivated ground. F,G.

10* HEATH LOBELIA *Lobelia urens*. Short/medium perennial, more or less hairless; juice *milky,* acrid. Lvs oblong, toothed. Fls purple-blue, 2-*lipped,* upper lip 2-lobed, lower lip 3-lobed, in a stalked spike; Aug-Sept. Damp woods and heaths; not on lime. B (rare), F.

Key to Composites

(Pages 230-256)

Composites are members of the Daisy Family, Compositae; see p. 230, where the nature of their flowers as a tight head of small flowers or florets is explained. There are three main types of composite flower-heads, consisting of disc florets only (thistle-like), ray florets only (dandelion-like) or both (daisy-like).

Plant

aromatic: Stinking Cocklebur 304; Chamomiles, Scented Mayweed 232; Mountain Everlasting 234; Mugworts, Wormwoods, Buttonweed 238; Tansy, Feverfew 240

a shrub: Silver Ragwort 242

spiny: Spiny Cocklebur 230; Star-thistles 248; Prickly, Perennial and Marsh Sow-thistles, Prickly Lettuce 252

Stems

winged: Cotton Thistle 244; Spear, Musk, Slender, Welted and Great Marsh Thistles 246; Yellow Star-thistle 248; Alpine Sow-thistle 250

Leaves

clasping stem: Michaelmas Daisies, Fleabanes 232; Jersey Cudweed, Small Fleabane 234; Common Fleabane, Elecampane, Irish Fleabane, Leopardsbane 236; Garden Marigold 240; Hoary, Marsh and Oxford Ragworts, Marsh Fleawort, Groundsel 242; Carline, Milk and Cabbage Thistles 244; most on pp. 250 and 252; Spotted Catsear 254; Bristly Ox-tongue 256

in rosette: Daisy, Alpine Fleabane 232; Mountain Everlasting, Dwarf Cudweed 234; Arnica, Leopardsbane 236; Carline Thistle 244; Lamb's Succory 252; most on p. 254; Few-leaved and Alpine Hawkweeds, Crepis praemorsa 256

fleshy: Golden Samphire 230; Sea Aster 232

opposite: Hemp Agrimony, Ragweed 230; Shaggy Soldier 232; Sunflowers, Arnica 236; Bur Marigolds 238

white-veined: Milk Thistle 244; Goatsbeard, Salsify 250

Heads with ray-florets only (Dandelion-like):

pink/purple: Salsify, Purple Viper's Grass, Alpine Sow-thistle, Purple Lettuce 250

brown: Podospermum laciniatum 250

yellow: Goatsbeard, Viper's Grass 250; Sow-thistles and Lettuces 252; Dandelions and allies 254 and 256

blue: Chicory, Alpine Sow-thistle, Blue Lettuce 250

Heads with disc florets only (Thistle-like):

pink/purple: Hemp Agrimony 230; Mountain Everlasting 234; Butterbur 238; Purple Coltsfoot 240; Burdocks, Cotton and Milk Thistles, Alpine Sawwort 244; all on p.246 and most on p.248

red: Mountain Everlasting 234; *Artemisia austriaca* 304

yellow: Ragweed 230; Sea Aster, Pineapple Mayweed 232; all on 234; Bur Marigolds, Buttonweed, Cottonweed 238; Tansy 240; Groundsel 242; Yellow Star-thistle 248

blue: Globe Thistle 244; Cornflowers 248

white: Mountain Everlasting 234

brown: Cockleburs 230; Cudweeds 234; Mugworts, Wormwoods, Bur Marigolds 238; Carline and Cabbage Thistles 244

Heads with discs and rays (Daisy-like):

rays pink/purple: Sea Aster, Michaelmas daisies, Fleabanes 232; Yarrow 238; Purple Coltsfoot 240

rays yellow: most on p.230; Small Fleabane 234; all on p.236; Bur Marigolds 238; Coltsfoot, Corn and Garden Marigolds 240; all on p.242

rays white: Daisy, Mayweeds, Shaggy Soldier 232; Yarrow, Sneezewort 238; Ox-eye Daisy, Feverfew 240

Heads solitary

Golden Samphire 230; Daisy, Mayweeds, Pineapple Mayweed, Alpine Fleabane 232; Small, Dwarf and Heath Cudweeds 234; all on p.236; Bur Marigolds 238; most on pp.240, 244 and 248; Goatsbeard, Salsify, Viper's Grasses, Chicory 250; most on p.254; Alpine Hawkweed 256

Heads more than 30 mm

Golden Samphire 230; Mayweeds 232; all on p.236; Ox-eye Daisy, Corn Marigold 240; Fen Ragwort 242; Globe, Cotton, Milk and Cabbage Thistles 244; Spear, Woolly, Dwarf, Musk and Welted Thistles 246; Greater and Black Knapweeds 248; Blue Lettuce 250; Perennial and Marsh Sow-thistles 252; Dandelion, Common and Spotted Catsears, Rough Hawkbit 254; Few-leaved and Alpine Hawkweeds 256

Heads in spikes

Goldenrods, Canadian Fleabane, Ragweed 230; Dwarf and Heath Cudweeds 234; Butterbur, Mugworts, Wormwoods 238; Burdocks 244; Chicory, Alpine Sow-thistle 250; Prickly and Least Lettuces 252.

Daisy Family   Compositae

The Composites are the largest and most versatile family of flowering plants. Fls tiny, closely packed into a compound head, surrounded by sepal-like bracts. Petals are joined in a tube, and are of two kinds, with the tube ending either in five short teeth, *disc florets,* or in a conspicuous flat flap, *ray florets.* Composite flhds are thus of three kinds: *rayless,* with disc florets only, like thistles; *rayed,* with disc florets in the centres and ray florets round the edge, like daisies; and *dandelion-like,* with all ray florets. Fr tiny, often surmounted by a feathery pappus on which it floats away in the wind. Sometimes the pappuses form a rounded "clock".

1* HEMP AGRIMONY *Eupatorium cannabinum.* Medium/tall perennial, downy; stems often reddish, usually branched. Lvs *palmate,* the segments lanceolate, toothed, the upper undivided. Flhds whitish pink to reddish mauve, *rayless,* in dense, rather flat-topped clusters; July-Sept. Damp woods, marshes, fens, by fresh water, and on waste ground. T.

2* GOLDEN-ROD *Solidago virgaurea.* Variable short/medium perennial, hairless or downy; little branched. Lvs bluntly oval or lanceolate, toothed, stalked; narrower, untoothed and unstalked up stems. Flhds bright yellow, *shortly* rayed, in branched spikes; June-Sept. Woods, scrub, heaths, grassy and rocky places. T.

3* CANADIAN GOLDEN-ROD *Solidago canadensis.* Tall downy perennial; unbranched. Lvs lanceolate, toothed, 3-veined. Flhds yellow, *very shortly* rayed, in *one-sided* spikes in a branched cluster; July-Sept. Waste places, streamsides. (T), from N America. **3a*** *S. gigantea* is usually shorter, hairless below and greyish, with more deeply toothed lvs and longer ray florets. (B,F,G), from N America. **3b*** *S. graminifolia* has linear-lanceolate lvs and flhds in flat heads. (B,G).

4* GOLDILOCKS *Crinitaria linosyris.* Short/medium hairless perennial; unbranched. Lvs *linear,* 1-veined, numerous up stems. Flhds yellow, *rayless,* in a flat-topped umbel-like cluster; Sept-Nov. Woods, grassy places, on rocks, often by the sea; on lime. T, southern, rare in B.

5* CANADIAN FLEABANE *Conyza canadensis.* Low/tall annual, slightly hairy; branched above. Lvs narrow lanceolate, sometimes toothed. Flhds with yellow disc florets and *short* whitish rays, in loose branched spikes; June-Oct. Bare and waste places. (T), from N America.

6* PLOUGHMAN'S SPIKENARD *Inula conyza.* Short/tall perennial, downy; stems purplish, branched above. Lvs broad lanceolate, *foxglove-like,* slightly toothed. Flhds dull yellow with no or obscure rays and purplish inner bracts, in a flat-topped umbel-like cluster; July-Sept. Open woods, scrub, dry grassy and rocky places; on lime. B,F,G. **6a* Stink Aster** *I. graveolens* is foetid and stickily hairy, with linear lvs and a long spike of smaller flhds. Damp places. (B, rare), F.

7* GOLDEN SAMPHIRE *Inula crithmoides.* Tufted short/medium hairless perennial. Lvs linear, *fleshy,* sometimes 3-toothed at tip. Flhds yellow, *rayed,* in a loose flat-topped cluster; July-Oct. Coastal cliffs, shingle and saltmarshes. B,F.

8* RAGWEED *Ambrosia artemisiifolia.* Medium/tall annual, hairy; stems branched, often reddish, angled. Lvs oval, deeply pinnately cut, *greyish* beneath, mostly opposite. Flhds greenish-yellow, rayless, those with stamens in spikes, those with styles solitary at base of lvs; June-Sept. Fr prickly. Bare and waste ground. (B,F,G), from N America.

9* SPINY COCKLEBUR *Xanthium spinosum.* Stiff short/medium annual, hairless or not. Lvs diamond-shaped, 3-5-lobed, dark shiny green above, downy white beneath, with 1-2 sharp 3-forked orange *spines* at their base. Flhds greenish, rayless, covered with hooked spines, globular male heads with stamens above egg-shaped female heads with styles; July-Oct. Fr covered with hooked spines. A weed of cultivation, waste places. (B,F,G), from N America. **9a*, 9b***, p. 304.

1

2

3

4

5

6

7

8

9

9

Daisy Family *(contd.)*

1* DAISY *Bellis perennis.* Low hairy perennial. Lvs spoon-shaped, slightly toothed, in a basal rosette. Flhds with disc florets yellow and rays white, often tipped red, *solitary* on lfless stalks; all year. Lawns, short turf. T. **1a* Mexican Fleabane** *Erigeron mucronatus* is taller and branched with lanceolate lvs; Apr-Oct. Walls. (B,F), from Mexico. **1b,** p. 305.

2* SCENTLESS MAYWEED *Tripleurospermum inodorum.* Variable short half-prostrate hairless annual/biennial, *not* aromatic. Lvs 2-3-pinnate with hairlike segments. Fls with yellow disc florets and white rays, 15-40 mm, solitary; sepal-like bracts bordered brown; Apr-Oct. Bare and disturbed ground. T. **2a* Sea Mayweed** *T. maritimum* is stouter, slightly fleshy and earlier flg. By the sea. **2b* Corn Chamomile** *Anthemis arvensis* is erecter and grey with down, and has broader lf segments and bracts all green. **2c* Stinking Chamomile** *A. cotula* is erecter and has a sickly smell, slightly hairy stems, broader lf segments and whitish bracts with a green midrib. **2d* Lawn Chamomile** *Chamaemelum nobile* is perennial, spreading, hairy and pleasantly aromatic, and has lf segments with inrolled margins and bracts white-edged; June-Sept. Grassy and heathy places. B, F, G. **2e* Scented Mayweed** *Matricaria recutita* is erecter and aromatic, with bristle-tipped lf segments, smaller (10-25 mm) fls, rays turned downwards and bracts edged greenish-white.

3* PINEAPPLE MAYWEED *Matricaria matricarioides.* Low/short hairless annual, *pineapple*-scented; well branched. Lvs 2-3-pinnate with thread-like segments. Flhds yellowish-green, *rayless,* solitary; sepal-like bracts pale-edged; May-Nov. Bare and waste, especially well trodden places. (T), from N E Asia.

4* SEA ASTER *Aster tripolium.* Short/medium hairless, *fleshy* perennial; often well branched. Root lvs bluntly oval, untoothed, dark green; stem lvs narrow lanceolate. Fls with yellow disc florets and pale purple or whitish rays, but frequently *rayless,* in loose clusters; sepal-like bracts blunt; July-Oct. Coastal *saltmarshes,* cliffs. T. **4a,** p. 305.

5* MICHAELMAS DAISY *Aster novi-belgii.* Medium/tall perennial, almost hairless; well branched. Lvs lanceolate, untoothed, the upper unstalked, numerous up stems. Fls with yellow disc florets and purple, violet or whitish rays, in a widely branched cluster; outer sepal-like bracts half as long as the inner, *pointed;* Aug-Nov. Damp and waste places, streamsides. (B, F, G), from N America. **5a*** *A. novae-angliae* is roughly and stickily hairy with rays usually reddish. (B, G). **5b*** *A. salignus* has narrower, sometimes toothed lvs, all unstalked and narrowed at base; rays white turning violet, and bracts purplish. Several other introduced species occur. **5c** *A. alpinus* is much shorter, with shorter, broader lvs and larger (to 50 mm) solitary fls; July-Aug. Mountains. G, southern.

6* BLUE FLEABANE *Erigeron acer.* Low/medium roughly hairy annual/biennial; stems often purple. Lvs lanceolate, untoothed, the upper unstalked. Flhds with yellow disc florets and short *erect* dingy purple rays, 10-15 mm, solitary or in loose clusters; sepal-like bracts dull purple; June-Sept. Dry bare and sparsely grassy places, walls, mountains. T. **6a, 6b,** p. 305.

7* ALPINE FLEABANE *Erigeron alpinus.* Low/short perennial, hairy, unbranched. Lvs spoon-shaped in basal rosette, narrower and fewer up stems. Flhds with yellow disc florets and spreading pinkish-purple rays, *20-30 mm,* usually *solitary,* bracts *shortly* hairy, grey-violet; July-Sept. Grassy places and rock-ledges in mountains. T, rare in B. **7a** *E. uniflorus* has lvs almost hairless, flhds smaller (10-15 mm) and bracts with long woolly hairs. G, S. **7b** *E. unalaschenensis* is shorter, with bracts black or dark violet. S.

8* SHAGGY SOLDIER *Galinsoga ciliata.* Low/short annual, whitely hairy. Lvs pointed oval, toothed, dark green, opposite, narrowing up the stem. Flhds with yellow disc florets and 4-5 flat *white* trifid rays, in open clusters; June-Oct. Bare and disturbed ground. (T), from S America. **8a* Gallant Soldier** *G. parviflora* is yellower green and much less hairy, with shorter narrower grooved rays. Hybridises with 8.

Daisy Family (contd.)

1* MOUNTAIN EVERLASTING *Antennaria dioica.* Low/short creeping perennial, with rooting runners. Lvs whitely woolly *beneath,* those in basal rosette broadest near tip. Flhds rayless, pink or red, with sepal-like bracts pink or white, in umbels; stamens and styles on separate plants; June-July. Heaths, moors, mountain grassland. T. **1a** *A. alpina* is shorter, with rosette lvs grey hairy above and smaller flhds. Mountain rocks on lime. S. **1b** *A. carpatica* is like 1a, but lvs longer, linear and 3-veined. **1c** *A. porsildii* has shorter runners, lvs white or grey woolly on both sides, and bracts greenish or brownish. S, Arctic. **1d* Pearly Everlasting** *Anaphalis margaritacea* is a medium perennial, whitely woolly, with yellow flhds appearing white from the bracts, and stamens and styles not always on separate plants. Waste places. (B,F,G), from N America.

2* COMMON CUDWEED *Filago vulgaris.* Low/short annual, covered with silvery, sometimes yellow, hairs; widely branched. Lvs narrow oblong, *wavy-edged,* blunt or pointed, spirally up stems. Flhds rayless, white tipped red, but appearing yellow from tips of sepal-like bracts; in globular clusters overtopping upper lvs; July-Aug. Heaths, sandy places. T. **2a* Red-tipped Cudweed** *F. lutescens* has hairs always yellowish, and bracts red-tipped. **2b* Broad-leaved Cudweed** *F. pyramidata* has hairs always silvery, flhds well overtopped by 3-5 large lvs and bracts with down-curved points. B,F. 2a and 2b both have lvs broader, scarcely wavy and sharply pointed.

3* SMALL CUDWEED *Filago minima.* Slender low/short annual, covered with silvery hairs. Lvs linear, spirally up stems. Flhds rayless, white tinged red, but appearing yellow from tips of sepal-like bracts, which spread *star-like* in fr; in narrow conical clusters, overtopping upper lvs; June-Sept. Heaths, sandy places. T. **3a* Narrow-leaved Cudweed** *F. gallica* has longer, even narrower lvs, the topmost exceeding the flhds. (B, rare), F,G. **3b** *F. arvensis* is like 3a but has bracts tipped with white hairs. F,G,S. **3c** *F. neglecta* is like 3a but has longer broader lvs. F,G.

4* MARSH CUDWEED *Gnaphalium uliginosum.* Low/short silvery grey annual; well branched. Lvs narrow oblong, green above, alternate. Flhds rayless, yellow-brown, in clusters overtopped by upper lvs; July-Oct. Damp bare places. T. **4a* Dwarf Cudweed** *G. supinum* is perennial and tufted with narrower lvs, red fls hidden by brown bracts and flhds in a spike-like cluster not overtopped by upper lvs. Mountains.

5* JERSEY CUDWEED *Gnaphalium luteoalbum.* Short/medium annual, thickly covered with white woolly hairs; little branched. Lvs narrow lanceolate, margins inrolled, the upper *wavy-edged,* hairy on both sides. Flhds rayless, yellow, with red stigmas, egg-shaped, in umbels; July-Sept. Sandy places. T, but rare in B. **5a Cape Cudweed** *G. undulatum* is taller, bushier and foetid, with broader lvs green above and running down on to stem. (F), from S Africa.

6* HEATH CUDWEED *Gnaphalium sylvaticum.* Short/medium perennial, grey with wool; unbranched. Lvs linear, 1-veined, green above, white-felted beneath, alternate. Flhds reddish or yellowish, hidden by brown sepal-like bracts, solitary or in small clusters in a *long lfy spike;* July-Sept. Heaths, open woods, dry grassland. T. **6a* Highland Cudweed** *G. norvegicum* is shorter, with broader 3-veined lvs downy on both sides, and a much shorter fl spike. Mountains. B (rare),F,S.

7 HELICHRYSUM *Helichrysum arenarium.* Short perennial, grey with woolly hairs. Lvs oblong, *flat,* alternate. Flhds *bright yellow,* rayless, in a cluster; July-Sept. Bare ground, often sandy. F,G.

8 MICROPUS *Micropus erectus.* Short annual, covered with woolly hairs. Lvs short, narrow, alternate, numerous up stems, Flhds straw yellow, rayless in a *tight* cluster; June-July. Grassy places. F,G, southern.

9* SMALL FLEABANE *Pulicaria vulgaris.* Low/short downy annual; much branched. Lvs oblong, *wavy-edged,* alternate. Flhds yellow, *shortly* rayed, rays no longer than sepal-like bracts, solitary in loose clusters; Aug-Sept. Damp bare places, especially where water stands in winter. T, southern, rare in B.

Daisy Family (contd.)

1* COMMON FLEABANE *Pulicaria dysenterica.* Medium hairy perennial. Lvs lanceolate, *wavy-edged,* the upper clasping the stem. Flhds daisy-like, yellow, *15-30 mm,* in a loose flat-topped cluster; July-Sept. Damp grassy places. T, southern.

2* IRISH FLEABANE *Inula salicina.* Medium perennial, more or less hairless; stems brittle. Lvs *narrow* elliptical, the upper unstalked and half-clasping the stem. Fls daisy-like, yellow, 25-40 mm, solitary or in small clusters; sepal-like bracts lanceolate; July-Aug. A shy flowerer in Ireland. Marshes, scrub, rocky slopes. Ireland, F,G,S. **2a** *I. britannica* has lvs lanceolate, bracts linear and both hairy. F,G. **2b** *I. ensifolia* has lvs lanceolate or linear, parallel-veined and hairy on margins, and larger (25-55 mm) fls. G. **2c** *I. hirta* has long hairs on both stems and lvs. Grassy places. F,G. **2d** *I. germanica* has lvs hairy below and flhds 10 mm. G, southern. **2e** *I. helvetica* has lvs hairy below. G (Upper Rhine valley).

3* ELECAMPANE *Inula helenium.* Tall hairy perennial. Lvs elliptical, toothed, basal ones up to *40 cm* long, the rest unstalked, alternate, clasping the stem. Flhds daisy-like, rays *narrow,* yellow, 60-80 mm, in small clusters; June-Sept. Grassy places, woods. T, but (B). **3a Large Yellow Ox-eye** *Telekia speciosa* has more deeply toothed, heart-shaped lower lvs, broader orange-yellow rays and brownish-yellow disc florets. Especially by streams. (F,G), from E Europe.

4 YELLOW OX-EYE *Buphthalmum salicifolium.* Medium/tall hairy perennial; little branched. Lvs oblong-lanceolate, feebly toothed, alternate, clasping the stem. Flhds daisy-like, rays *broad,* yellow, 30-60 mm, *solitary;* July-Aug. Woods, rocks, in hills and mountains. F,G.

5* CONE FLOWER *Rudbeckia laciniata.* Tall perennial, almost hairless. Lvs lanceolate, mostly *lobed,* sometimes toothed, alternate. Flhds daisy-like, rays broad and down-turned, yellow, disc florets blackish-brown, making a conspicuous *cone,* 70-120 mm, solitary or in small clusters; June-Oct. Damp woods, waste places. (B,F,G), from N America. **5a* Black-eyed Susan** *R. serotina* is roughly hairy, with lvs not lobed and fls smaller. (T).

6* PERENNIAL SUNFLOWER *Helianthus rigidus.* Tall patch-forming perennial to 2 m, *rough* to the touch. Lvs broad lanceolate, scarcely toothed, opposite, very shortly stalked. Flhds daisy-like, rays long and broad, yellow, disc florets turning dark *purple,* 60-100 mm, long-stalked; Aug-Oct. Waste places. (B,F,G), from N America. **6a* Jerusalem Artichoke** *H. tuberosus* has coarsely toothed lvs with winged stalks, and smaller all-yellow fls rarely seen; Sept-Nov. **6b* Sunflower** *H. annuus* is annual and can reach nearly 3 m, with broader, often heart-shaped lvs and much larger (100-300 mm) fls. From S America. **6c*** *Guizotia abyssinica* is stickily hairy above with unstalked lvs and flhds 20-30 mm. (T, casual), from Africa.

7 ARNICA *Arnica montana.* Short/medium downy *aromatic* perennial. Lvs lanceolate, mainly in a basal *rosette,* a few opposite on stems. Flhds daisy-like, rays long, yellow, 40-80 mm, solitary; June-Aug. Grassy places in *hills and mountains.* F,G,S. **7a** *A. alpina* is shorter, with narrower lvs and smaller (30 mm) flhds. S.

8* YELLOW CHAMOMILE *Anthemis tinctoria.* Short/medium perennial, *grey* with woolly down. Lvs 2-*pinnate,* downy beneath. Flhds daisy-like or sometimes rayless, yellow, 25-40 mm, solitary, long-stalked; July-Aug. Dry bare and waste places. T, but (B). **8a** *A. austriaca* has white fls. G, southern.

9* LEOPARDSBANE *Doronicum pardalianches.* Medium patch-forming hairy perennial. Lvs broad *heart-shaped,* toothed, the upper unstalked. Flhds daisy-like, rays long, yellow, 40-60 mm, in small clusters; May-July. Woods, shady places. (B),F,G. **9a* Green Leopardsbane** *D. plantagineum* has narrower lvs, not heart-shaped, and larger (50-80 mm) solitary flhds.

Daisy Family (contd.)

1* BUTTERBUR *Petasites hybridus.* Low/medium patch-forming perennial, downy. Lvs *very large,* to 1 m across, heart-shaped, toothed, downy grey beneath, long-stalked, all from roots, appearing after fls. Flhds rayless, brush-like, lilac-pink, unscented, in spikes; stalks with strap-like bracts; stamens and styles *on separate plants;* Mar-May. Damp places, roadsides, streamsides. T. **1a* Winter Heliotrope** *P. fragrans* □ is shorter, and its much smaller lvs appear with the fewer fragrant fls; Nov-Mar. (B), F. **1b* White Butterbur** *P. albus* has fragrant white fls and smaller lvs whitely woolly beneath. T, but (B). **1c** *P. frigidus* is like 1b but shorter and with smaller lvs. S. **1d** *P. spurius* has stems pale, lvs triangular and flhd much branched. F, G, southern. **1e***, p.305.

2* YARROW *Achillea millefolium.* Short/medium perennial, downy, *aromatic.* Lvs 2-3-pinnate, *feathery,* dark green. Flhds shortly rayed, rays white or pink, disc florets creamy, in flat umbel-like clusters; June-Nov. Grassy places. T. **2a, 2b, 2c, 2d**, p.305.

3* SNEEZEWORT *Achillea ptarmica.* Medium hairy perennial, *not aromatic.* Lvs *linear,* minutely saw-toothed, dark green, half-clasping stem. Flhds shortly rayed, white, with creamy disc florets, in a loose umbel-like cluster; July-Sept. Damp grassy places, on acid soils. T.

4* MUGWORT *Artemisia vulgaris.* Medium/tall downy perennial, *slightly* aromatic. Lvs 1-2-pinnate, the upper unstalked, dark green and almost hairless above, silvery downy beneath. Flhds rayless, egg-shaped, yellowish- or purplish-brown, numerous in branched spikes; July-Sept. Waste places, roadsides. T. **4a* Chinese Mugwort** *A. verlotorum* is shorter and more aromatic, has darker lflets less downy beneath, and is much later flg (Oct-Dec). (B), from China. **4b* Wormwood** *A. absinthium* is strongly aromatic and has lvs whitely downy above and larger yellower flhds. **4c* Field Wormwood** *A. campestris* agg. is shorter and not aromatic, with almost hairless stems, 2-3-pinnate lvs downy above at first, and yellower fls; Aug-Sept. Bare, often sandy places. T, rare in B. **4d, 4e, 4f, 4g, 4h**, p.305.

5* SEA WORMWOOD *Artemisia maritima.* Short greyish perennial, downy, *pungently* aromatic. Lvs 2-pinnate. Flhds rayless, egg-shaped, yellow or orange-yellow, in branched spikes; Aug-Oct. *Saltmarshes,* occasionally inland. T.

6* TRIFID BUR MARIGOLD *Bidens tripartita.* Short/medium annual, almost hairless. Lvs lanceolate, with two smaller *lobes at base,* toothed, opposite, on winged stalks. Flhds normally unrayed, button-like, yellow, *erect,* often solitary; July-Oct. Fr flattened, adhering to clothing with two barbed bristles. Damp places, by water. T. **6a* Nodding Bur Marigold** *B. cernua* □ is hairier, with unstalked unlobed lvs and larger nodding flhds. **6b* Beggar Ticks** *B. frondosa* □ is usually taller, with lvs pinnate and smaller flhds with outer bracts shorter than inner. (B,G), from America. **6c** *B. connata* □ has unlobed lvs clasping the stem. (F,G, spreading), from N America. **6d** *B. radiata* □ has yellowish lvs with teeth bent inwards. (G).

7* COTTONWEED *Otanthus maritimus.* Short creeping perennial, covered with *thick* white down. Lvs oblong, toothed. Flhds rayless, button-like, yellow, in small clusters; Aug-Nov. Coastal sand and shingle. Ireland (rare), F.

8* BUTTONWEED *Cotula coronopifolia.* Low/short yellow-green hairless fleshy *aromatic* annual. Lvs variable, linear to pinnate, alternate, sheathing the stem. Flhds very shortly rayed, appearing rayless and button-like, yellow, long-stalked, *solitary,* always facing the sun. July-Oct. Sandy places, especially on *coast.* (B,F,S), from S Africa. **8a* Scottish Wormwood** *Artemisia norvegica* is shorter, tufted, not fleshy, and grey with silky hairs, and has 1-2-pinnate lvs and larger rayless fls not always solitary; June-Sept. Mountains. B(rare), S.

6a* Nodding Bur Marigold

6b* Beggar Ticks

1

1a

2

3

4

5

6

7

8

Daisy Family *(contd.)*

1* OX-EYE DAISY *Leucanthemum vulgare.* Medium perennial, slightly hairy; little branched. Lvs dark green, variable, the lower rounded or spoon-shaped, toothed or lobed, the upper narrower, clasping the stem. Flhds daisy-like, rays white, disc florets yellow, 25-50 mm, *solitary;* May-Sept. Grassy places. T.
1a* Shasta Daisy *L. maximum,* with much larger (50-80 mm) flhds, is a frequent patch-forming garden escape; July-Sept. (B, G), from the Pyrenees.

2* FEVERFEW *Tanacetum parthenium.* Short/medium perennial, downy, highly *aromatic.* Lvs yellowish, 1-2-pinnate. Flhds daisy-like, rays white (occasionally absent), disc florets yellow (a "double" garden form often escapes), 10-25 mm, in an umbel-like *cluster;* June-Sept. Walls, waste places. (T), from S E Europe.
2a *T. corymbosum* is not aromatic and has lobed lanceolate lvs and larger (25-40 mm) fls. F, G.

3* TANSY *Tanacetum vulgare* Medium/tall perennial, almost hairless, strongly *aromatic;* unbranched. Lvs pinnate, toothed. Fls rayless, *button-like,* yellow, in large umbel-like clusters; July-Oct. Grassy and waste places. T.

4* CORN MARIGOLD *Chrysanthemum segetum.* Short/medium greyish hairless perennial. Lvs fleshy, oblong, toothed or lobed, the upper clasping the stem. Fls daisy-like, bright *yellow,* 35-55 mm, solitary; June-Oct. A weed of cultivation. T.

5* COLTSFOOT *Tussilago farfara.* Low/short creeping downy perennial; unbranched. Lvs heart-shaped with pointed teeth, from roots only, appearing after fls, much smaller than Butterbur (p. 238). Fls yellow, with narrow rays, solitary on stems covered with purplish scales; *Feb*-Apr. Fr a white "clock". Bare and waste ground. T.

6* PURPLE COLTSFOOT *Homogyne alpina.* Low/short creeping hairy perennial; unbranched. Lvs kidney-shaped, dark green, shiny, purplish beneath, all from roots, much smaller than 5. Fls purple, *rayless,* solitary, on stems with narrow scales; June-Sept. Fr a white "clock". Damp grassy places in mountains. B (rare), S.

7* GARDEN MARIGOLD *Calendula officinalis.* Short/medium perennial, roughly hairy. Lvs oblong, often broadest at tip. Fls daisy-like, *orange,* 40-50 mm, solitary; May-Oct. Waste ground, a frequent garden escape. (B, F, G), from Mediterranean.
7a* Marigold *C. arvensis* has lvs narrower and toothed and flhds· 10-20 mm. Arable weed, especially in vineyards. F, G.

5* Coltsfoot

1

2

3

4

5

6

7

Daisy Family (contd.)

RAGWORTS and GROUNDSELS Senecio. Flhds normally rayed, daisy-like, all yellow, in branched clusters. Fr pappus with simple hairs, not forming a clock. Hybrids not infrequent.

1* RAGWORT Senecio jacobaea. Medium/tall biennial, often hairless; little branched. Lvs pinnately lobed, end lobe *small,* blunt. Flhds 15-25 mm, in dense flat-topped clusters; outer sepal-like bracts few, much shorter than *dark-tipped* inner bracts; June-Nov. Dry grassy places. T. **1a* Hoary Ragwort** S. eruci-folius □ is grey with down and has lvs more narrowly lobed, the end lobe more pointed and longer outer bracts; July-Sept. Southern. **1b* Marsh Ragwort** S. aquaticus is more widely branched and has lower lvs varying from oval to pinnately lobed, the end lobe larger and blunt, and looser clusters of larger (25-30 mm) flhds. Damp grassland. **1c** S. erraticus is very like 1b, but has lvs always pinnately lobed, the lobes at right angles to axis instead of directed forwards. F, G, eastern.

2* OXFORD RAGWORT Senecio squalidus. Medium perennial, almost hairless; well branched. Lvs lanceolate to pinnately lobed, the end lobe *sharply* pointed, the upper unstalked. Flhds 15-20 mm, sepal-like bracts black-tipped, the outer row much shorter; Apr-Nov. Walls, bare and waste ground. (B, F), from S Italy. **2a** S. vernalis is annual with matted hairs and larger (20-30 mm) flhds. F, G, S.

3 MARSH FLEAWORT Senecio palustris. Medium/tall annual/perennial, covered with *white* woolly hairs. Lvs *lanceolate,* toothed or not. Flhds pale yellow, 20-30 mm; June-July. Marshes. T, but extinct in B.

4* FEN RAGWORT Senecio paludosus. Medium/tall hairy perennial. Lvs lanceolate, *coarsely* saw-toothed, unstalked, shiny above, cottony *white* beneath. Flhds 30-40 mm, in a flat-topped cluster; July-Aug. Wet places. T, but rare in B. **4a* Broad-leaved Ragwort** S. fluviatilis is taller, to 1½ m, and stouter, with less deeply toothed lvs hairless beneath and smaller (30 mm) flhds. Damp places. (B), F, G, southern.

5 ALPINE RAGWORT Senecio fuchsii. Medium/tall perennial, almost hairless. Lvs lanceolate, *finely* saw-toothed, short-stalked. Flhds with *few* rays, 15-25 mm, in a flat-topped cluster; July-Sept. Woods and rocks, mainly in hills and mountains. T, but (B).

6* GROUNDSEL Senecio vulgaris. Low/short annual, often downy, not foetid. Lvs pinnately lobed, hairless above. Flhds usually *rayless,* brush-like, but occasionally rayed, yellow, in loose clusters; sepal-like bracts *black-tipped,* the outer very short; all year. A weed of cultivation, disturbed ground. T. **6a* Sticky Groundsel** S. viscosus is taller, foetid and greyish with sticky hairs and has larger paler yellow flhds with very short rolled-back rays and bracts not black-tipped, the outer half as long as the inner; July-Oct. Bare and waste places, fens. **6b* Heath Groundsel** S. sylvaticus is greyish with sticky hairs (but less so than 6a) and somewhat foetid, with flhds like 6a and bracts purple-tipped, the outer still smaller; July-Oct. Dry sandy and heathy places.

7* SILVER RAGWORT Senecio cineraria. Medium undershrub, covered with a dense *silvery* white felt; well branched. Lvs oval to oblong, toothed to pinnately lobed, green above. Flhds 8-12 mm; June-Aug. Cliffs and rocks by sea. (B, F), from Mediterranean.

8* FIELD FLEAWORT Senecio integrifolius. Low/short perennial, often whitely downy; *unbranched.* Lvs oval, wrinkled, toothed or not, mainly in a basal rosette, narrower up stems. Flhds *few,* 15-25 mm, in a loose cluster; May-June. Short dry turf. T. **8a*** S. spathulifolius is stouter, with lvs more coarsely toothed and larger and more numerous lvs, almost clasping the stem. Grassy places. T, rare in B. **8b** S. crispatus is like 8a, but lower lvs ace-of-spades shaped, and flhds 25-35 mm; Apr-Aug. Damp places in hills, on lime. G.

242

1a* Hoary Ragwort

Daisy Family *(contd.)*

1* GLOBE THISTLE *Echinops sphaerocephalus.* Medium/tall thistle-like perennial, to 2 m, stickily hairy and white-woolly. Lvs lanceolate, pinnately lobed, spiny. Flhds rayless, *globular,* pale blue, spiny; June-Sept. Dry bare stony places. T, but (B). **1a** *E. commutatus* is less spiny and not stickily hairy, with whiter fls. (F, G), from Mediterranean.

2* CARLINE THISTLE *Carlina vulgaris.* Low/short erect biennial, spiny. Lvs thistle-like, oblong, pinnately lobed, prickly, the lower cottony, not in a rosette. Flhds rayless, yellow-brown, but the conspicuous spreading *yellow* sepal-like bracts appear like rays, folding up in wet weather; July-Sept. Dead plants survive through winter. Grassland, dunes; on lime. T.

3 STEMLESS CARLINE THISTLE *Carlina acaulis.* Low biennial/perennial, spiny. Lvs all in a *rosette,* pinnately lobed, more or less prostrate. Flhds rayless, whitish, but appearing rayed from the conspicuous spreading *silvery* sepal-like bracts, solitary, unstalked; May-Sept. Grassy, rocky places in mountains. F, G, S.

4* ALPINE SAWWORT *Saussurea alpina.* Short stout hairy perennial. Lvs lanceolate, more or less *toothed,* shaggily hairy beneath, spineless. Flhds rayless, purple, fragrant, in compact clusters; Aug-Sept. Mountain grassland. B, G, S.

5* LESSER BURDOCK *Arctium minus.* Medium/tall stout downy spineless biennial; stems arching. Lvs broadly heart-shaped, to 30 cm, longer than broad, with hollow stalks. Flhds 15-30 mm, rayless, egg-shaped, purple, in short-stalked spikes; hooked bracts forming *burs* when dried and lasting through winter; July-Sept. Shady and waste places. T. **5a* Greater Burdock** *A. lappa* has blunter lvs as long as broad with solid stalks and larger (35-40 mm), more globular flhds. **5b** *A. tomentosum* has lf undersides and smaller (20-25 mm) flhds covered with white cottony down. Extensive hybridisation occurs.

6* COTTON THISTLE *Onopordum acanthium.* Tall spiny biennial, to 1½ m, *white* with cottony down; stems *broadly* winged. Lvs oblong, spiny. Flhds rayless, pale purple, solitary; sepal-like bracts cottony at base, ending in yellow spines; July-Sept. Bare and waste ground, often sandy. T.

7* MILK THISTLE *Silybum marianum.* Medium/tall spiny annual/biennial; stems downy, unwinged. Lvs oblong, wavy-edged, dark green with conspicuous *white veins,* spiny. Flhds rayless, purple, solitary; sepal-like bracts ending in sharp yellow spines; June-Aug. Bare and waste places. (B), F, G.

8* CABBAGE THISTLE *Cirsium oleraceum.* Medium/tall perennial, more or less hairless. Lvs lanceolate, pinnately lobed to toothed. Flhds rayless, egg-shaped, *straw-yellow,* with narrow spiny sepal-like bracts, in dense clusters almost hidden by topmost lvs; July-Aug. Damp places. T, but (B, rare). **8a** *C. spinosissimum* is shorter and much spinier, with all lvs pinnately lobed and spiny and flhds whitish. Mountains. F, G. **8b Yellow Melancholy Thistle** *C. eristithales* □ has lemon-yellow flhds not surrounded by upper lvs, usually solitary and nodding. Hill districts. F, G.

8b Yellow Melancholy Thistle

Daisy Family (contd.)

THISTLES: *Carduus* and *Cirsium.* Stems usually winged and spiny. Lvs oblong or lanceolate, and usually pinnately lobed, spiny and wavy-edged. Flhds rayless, brush-like, purple (except 244/8); sepal-like bracts usually spine-tipped. Fr with a pappus, feathery in *Cirsium* but undivided in *Carduus,* not forming a clock.

1* CREEPING THISTLE *Cirsium arvense.* Short/tall *creeping* perennial, usually *hairless;* little branched, stems spineless. Lvs sometimes cottony beneath. Flhds *lilac,* 15-25 mm, fragrant, in clusters; bracts purplish, scarcely spiny; June-Sept. Grassy and waste places, a weed of cultivation. T. **1a** *C. rivulare* is hairier, with lvs scarcely spiny and much larger (25-35 mm) and rounder purple unstalked flhds. G.

2* SPEAR THISTLE *Cirsium vulgare.* Medium/tall biennial, downy, *sharply* spiny. Lvs deeply pinnately lobed. Flhds 20-40 mm across, often *solitary;* bracts often hairless, with yellow-tipped spines; July-Sept. Bare and waste places. T.

3* WOOLLY THISTLE *Cirsium eriophorum.* Tall stout biennial, very spiny, covered with *white wool;* stem unwinged. Lvs almost pinnate. Flhds red-purple, *globular,* 40-70 mm across, usually solitary; bracts covered with cobwebby wool; July-Sept. Bare and grassy places; on lime. B,F,G.

4* MEADOW THISTLE *Cirsium dissectum.* Medium perennial, cottony all over but *scarcely* spiny; little branched, stems unwinged; has creeping runners. Lvs whiter beneath, the upper undivided but clasping the stem. Flhds red-purple, 20-25 mm across, usually solitary; June-Aug. Damp grassy places, fens. B,F, G. **4a* Melancholy Thistle** *C. heterophyllum* □ is taller and more thickly white-felted, with larger (30-50 mm) flhds. T. **4b* Tuberous Thistle** *C. tuberosum* □ is taller and has tuberous roots, no runners, all lvs pinnately lobed and green on both sides, and larger (25-30 mm) flhds. Drier grassland, on lime. B(rare), F,G.

5* DWARF THISTLE *Cirsium acaule.* Low/short, usually prostrate perennial, almost hairless. Lvs all in a basal *rosette,* rather spiny. Flhds red-purple, 20-50 mm across, usually *unstalked* but occasionally stalked to 15 cm, usually solitary; bracts purplish, scarcely spiny; June-Sept. Dry grassland; on lime. T.

6* MUSK THISTLE *Carduus nutans.* Medium/tall biennial, with cottony white hairs. Lvs deeply pinnately lobed. Flhds bright red-purple, 35-50 mm, often solitary, slightly *nodding,* on spineless upper stalks; bracts conspicuous, purple; June-Sept. Bare and grassy places; on lime. T. **6a** *C. defloratus* has lfless unwinged fl-stems and erect bracts. Dry grassy places in hills. G.

7* SLENDER THISTLE *Carduus tenuiflorus.* Short/tall slender annual/biennial, with cottony white hairs. Lvs cottony white beneath. Flhds pinkish, *8 mm* across, in clusters; May-Aug. Grassy and waste places, often by sea. B,F,G.

8 GREAT MARSH THISTLE *Carduus personata.* Tall perennial, to 2 m, with cottony white hairs. Lvs *softly* spiny, the upper undivided, white beneath. Flhds 15-20 mm, unstalked, in tight clusters; bracts *not* spine-tipped, the outer recurved, often blackish; July-Aug. Damp places in mountains. F,G.

9* WELTED THISTLE *Carduus acanthoides.* Tall biennial, to 2 m. Lvs *weakly* spined, cottony white beneath. Flhds red-purple, 10-15 mm across, in a cluster, with stalks spineless *at top;* bracts weakly spined; June-Sept. Grassy places, hedge-banks. T. **9a** *C. crispus* is very similar, but has stronger spines and fl stalks spiny to top. F,G,S. **9b* Marsh Thistle** *Cirsium palustre* □ is less branched, and usually purple-tinged, especially on bracts, and has more cottony down and flhds less reddish, occasionally white, on stalks spiny to top. More often in marshes and woods.

9b* Marsh Thistle   **4b*** Tuberous Thistle   **4a*** Melancholy Thistle

Daisy Family *(contd.)*

KNAPWEEDS and STAR THISTLES *Centaurea*. Spineless (except 5, 6). Stems stiff, downy. Lvs spirally arranged. Flhds rayless but with enlarged sterile outer florets often giving appearance of rays, deeply lobed to appear star-like. Sepal-like bracts with toothed or cut appendages. Pappus not forming a clock. Numerous casuals and hybrids also occur.

1* BLACK KNAPWEED *Centaurea nigra*. Short/medium perennial. Lvs *lanceolate,* the lower sparsely toothed. Flhds more often brush-like than apparently rayed (1a), solitary, *20-40 mm* across; sepal-like bracts blackish-brown with long fine teeth; June-Sept. Hybridises with 1b. Grassy places. T. **1b* Brown Knapweed** *C. jacea* is shorter and slenderer, with smaller (10-20 mm) paler flhds always apparently rayed and browner bracts with a paler irregularly cut margin. T, but (B). **1c** *Jurinea cyanoides* has lvs white-felted beneath and inrolled at edges. G, southern.

2* GREATER KNAPWEED *Centaurea scabiosa*. Medium perennial. Lvs *pinnately lobed,* sparsely bristly. Flhds apparently rayed, solitary, *30-50 mm* across; sepal-like bracts *green,* with long slender teeth; June-Sept. Grassy places; on lime. T. **2a* Jersey Knapweed** *C. paniculata* has cottony stems and much smaller fls in a long branched spike. (B, rare), F. **2b** *C. stoebe* has lf-lobes linear, flhds smaller, more numerous, and bracts green. F, G.

3* PERENNIAL CORNFLOWER *Centaurea montana*. Medium creeping downy perennial. Lvs *lanceolate,* unstalked. Flhds apparently rayed, blue, solitary, *60-80 mm* across; May-Aug. Grassy places. (B), F, G.

4* CORNFLOWER *Centaurea cyanus*. Medium annual, grey with down. Lvs *pinnately lobed,* stalked, the upper lanceolate, unstalked. Flhds apparently rayed, bright blue, solitary, *15-30 mm* across; June-Aug. A weed of cultivation, especially cornfields. T.

5* YELLOW STAR-THISTLE *Centaurea solstitialis*. Medium/tall annual, grey-white with down; well branched. Lvs pinnately lobed, the upper linear. Flhds rayless, pale *yellow,* solitary; sepal-like bracts ending in a long stiff yellow *spine,* with smaller spines below; July-Sept. Bare and waste ground. (T, casual), from S Europe. **5a* Cockspur Star-Thistle** *C. melitensis* has pinnate bracts and sticky hairs on fls. **5b*** *Carthamus lanatus* has spiny lvs and lf-like spiny bracts. **5c* Safflower** *Carthamus tinctorius* is like 5b but fls bright reddish-orange. (B, F, G, casual), from Asia.

6* RED STAR-THISTLE *Centaurea calcitrapa*. Short/medium perennial, almost hairless; widely branched. Lvs pinnately lobed, the lobes *bristle-pointed.* Flhds rayless, thistle-like, red-purple, solitary, 8-10 mm across; sepal-like bracts ending in a long stout yellow *spine,* with shorter spines below; July-Sept. Dry bare places. (B), F, G. **6a* Rough Star-Thistle** *C. aspera* has larger (25 mm) flhds and very short, palmately arranged, spreading or down-turned reddish spines. (B), F.

7* SAWWORT *Serratula tinctoria*. Slender medium perennial, hairless, thistle-like but *spineless;* stems stiff. Lvs pinnately lobed, lobes *finely* saw-toothed. Flhds rayless, purple, in a branched cluster; sepal-like bracts purplish; July-Oct. Pappus yellowish. Damp grassy places. T.

5b* *Carthamus lanatus*

5a* Cockspur Star-Thistle   **5c*** Safflower

Daisy Family (contd.)

All plants on this page have dandelion-like flhds, with no disc florets, and all except 5 and 8 have milky juice.

1* GOATSBEARD *Tragopogon pratensis.* Medium hairless annual/perennial; little branched. Lvs *linear,* grass-like. Flhds yellow, solitary, florets either shorter or longer than sepal-like bracts, opening fully only on sunny mornings, whence folk-name of Jack-go-to-bed-at-noon; May-Aug. Pappus makes a large clock. Grassy places. T. **1a** *T. dubium* is taller with stem very swollen beneath flhd. Dry woods. F, G.

2* SALSIFY *Tragopogon porrifolius.* Resembles 1, but has much larger dull purple flhds, whose florets are about *as long as* sepal-like *bracts;* Apr-June. Grassy places, formerly cultivated. (T), from Mediterranean.

3 PURPLE VIPER'S GRASS *Scorzonera purpurea.* Short/medium hairless perennial; little branched. Lvs *linear,* keeled. Flhds *lilac-purple,* solitary, florets much longer than sepal-like bracts; May-June. Grassy and rocky places. F, G. **3a** *Podospermum laciniatum* □ has lvs with linear lobes, the end lobe broader, and fls brownish. G, southern.

4* VIPER'S GRASS *Scorzonera humilis.* Low/short perennial, almost hairless; little branched. Lvs narrow *lanceolate,* untoothed, becoming small and scale-like up stems. Flhds pale *yellow,* solitary, florets much longer than sepal-like bracts; May-July. Damp grassy places. T, southern, rare in B. **4a** *S. austriaca* has a ring of scales at base of stem and brighter yellow fls. F, G. **4b** *S. hispanica* has much branched stems. G, southern.

5* CHICORY *Cichorium intybus.* Medium/tall perennial, hairy or not; stems stiff, well branched, *no* milky juice. Lvs pinnately lobed, the upper undivided. Flhds *clear* blue, 25-40 mm, in lfy spikes; June-Sept. Grassy and waste places. T.

6* ALPINE SOW-THISTLE *Cicerbita alpina.* Medium/tall perennial; stems unbranched, with *reddish* hairs. Lvs pinnately lobed, hairless, clasping the stem. Flhds mauvish blue, 20 mm, in lfy spikes; July-Sept. Grassy places in *mountains.* T, rare in B. **6a* Blue Sow-thistle** *C. macrophylla* is taller, stouter and patch-forming, with lvs less lobed but more toothed, and larger (30 mm) flhds. Waysides, waste places. (T), from the Caucasus. **6b*** *C. plumieri* is hairless, with more lf lobes and fls in a cluster. (B), F, G. **6c*** *C. tatarica* is shorter and almost hairless, with lvs more lobed but less toothed, and compacter clusters of darker flhds. (B), G, S. **6d** *C. sibirica* is like 6c with untoothed lanceolate lvs. S.

7 BLUE LETTUCE *Lactuca perennis.* Medium hairless perennial. Lvs pinnately lobed, the upper clasping the stem, greyish. Flhds 30-40 mm, blue-purple or white, long-stalked from *arrow-shaped* bracts, in a loose cluster; May-Aug. Bare and grassy places. F, G.

8 PURPLE LETTUCE *Prenanthes purpurea.* Medium/tall greyish perennial, not unlike Wall Lettuce (p.252), but with purple fls. Lvs oblong, often *waisted* or lobed, clasping the stem. Flhds occasionally white, each with very few florets, 20 mm, in a loose cluster; July-Sept. Pappus white. *Shady places.* F, G.

3a *Podospermum laciniatum*

Daisy Family *(contd.)*

All plants on this page have yellow dandelion-like flhds with no disc florets. All except 7, 8 and 9 have milky juice.

1* SMOOTH SOW-THISTLE *Sonchus oleraceus.* Short/tall greyish annual, hairless. Lvs pinnately lobed, the end lobe the largest, with *softly* spiny margins, clasping the stem with arrow-shaped points. Flhds pale yellow, *20-25 mm,* in a loose cluster; May-Nov. Pappuses making a clock. Bare and waste ground, a weed of cultivation. T. **1a* Prickly Sow-thistle** *S. asper* □ has lvs less lobed, often undivided, with sharper spines and clasping the stem firmly with rounded lobes.

2* PERENNIAL SOW-THISTLE *Sonchus arvensis.* Tall patch-forming perennial, almost hairless. Lvs pinnately lobed, greyish beneath, softly spiny on margins, clasping the stem with rounded lobes. Flhds *40-50 mm,* rich yellow, in clusters; stalks and sepal-like bracts with numerous sticky yellow hairs; July-Sept. Bare and waste places. T.

3* MARSH SOW-THISTLE *Sonchus palustris.* Very tall, to 3 m, tufted perennial, hairless below. Lvs *deeply* pinnately lobed, greyish, edged with softly spiny teeth, clasping the stem with arrow-shaped lobes. Flhds pale yellow, *30-40 mm,* in tight clusters; stalks and sepal-like bracts with numerous sticky blackish-green hairs; July-Sept. Fresh and brackish marshes, ditches. T, southern, rare in B.

4* WALL LETTUCE *Mycelis muralis.* Slender short/tall hairless perennial, often purple-tinged. Lvs pinnately lobed, toothed, the end lobe large, *triangular* and sharply cut, clasping the stem with rounded toothed lobes. Flhds 7-10 mm, with only five florets, in a loose cluster; June-Sept. Shady places, walls, rocks. T.

5* PRICKLY LETTUCE *Lactuca serriola.* Tall foetid biennial, hairless, stems prickly. Lvs greyish, oblong, sometimes irregularly lobed, alternate, weakly spiny on margins, *prickly* on midrib beneath, clasping the stem with arrow-shaped points; in the sun the upper lvs are held in the vertical plane and oriented north/south, whence name Compass Plant. Flhds 11-13 mm, numerous, in branched spikes; sepal-like bracts often purplish; July-Sept. Fr pale brown to *grey-green.* Bare and waste ground. T, southern. **5a* Great Lettuce** *L. virosa* holds upper lvs horizontally and has fr maroon to blackish. B,F,G.

6* LEAST LETTUCE *Lactuca saligna.* Medium/tall hairless, *spineless* annual/biennial; little branched. Lvs greyish, pinnately lobed, the lobes curved *backwards,* with a white midrib, the upper narrow, undivided and held almost vertically against the stem, which they clasp with arrow-shaped points. Flhds 10 mm, in a slender stalked spike, sepal-like bracts green; July-Aug. Fr beak white. Bare grassy or shingly places, often near the sea. B,F,G. **6a, 6b,** p.305.

7* NIPPLEWORT *Lapsana communis.* Short/tall hairy annual; *no* milky juice. Lvs pointed oval, toothed, often pinnately lobed *at base.* Flhds 10-20 mm, in branched clusters, not opening in dull weather; June-Oct. Shady, bare and waste places. T.

8* LAMB'S SUCCORY *Arnoseris minima.* Low/short hairless annual; *no* milky juice. Lvs all from roots, lanceolate, toothed. Flhds very shortly rayed, 7-10 mm, solitary, on lfless stems much *swollen* at top; June-Aug. No pappus. Bare places, especially sandy. T, southern, rare in B.

9 CHONDRILLA *Chondrilla juncea.* Medium/tall biennial, hairy only on lower stem. Lower lvs lanceolate, lobed, withering before flg time; upper lvs narrower, not lobed, sometimes finely toothed. Flhds 10 mm, unstalked, in small clusters up stems, sepal-like bracts *downy;* July-Sept. Dry sandy or stony places. F,G. **9a** *Willemetia stipitata* has stem black hairy, lvs less toothed, mostly in a rosette, and larger flhds. Wet places. G, southern.

1a* Prickly Sow-thistle

Daisy Family (contd.)

Yellow dandelion-like fls, with no disc florets. Lvs all in a basal rosette (except 5a, 8), oblong, broadly toothed or shallowly pinnately lobed (except 9).

DANDELIONS *Taraxacum*. A variable, difficult group divided into eight Sections, each with many microspecies, typical examples of three being shown here. Perennials, mostly hairless, with *milky juice*. Lvs usually well lobed. Flhds solitary, on hollow lfless stalks, often downy at top. Pappus making a clock.

1* DANDELION *Taraxacum* Sect. *Vulgaria*. Low/short perennial. Flhds 35-50 mm, outer florets usually *grey-violet* beneath, sometimes purple, never red; sepal-like bracts broader than 2 but narrower than 3, erect or *recurved;* all year, but especially Apr-June. Fr colour variable, but not purplish-red. Grassy and waste places. T. **1a,** p.305.

2* LESSER DANDELION *Taraxacum* Sect. *Erythrosperma*. Low delicate perennial. Flhds 15-25 mm, sepal-like bracts narrow, appressed, tips of inner ones appearing *forked;* Apr-June. Fr usually *purplish-red.* Dry grassy and bare places, especially on lime. T.

3 RED-VEINED DANDELION *Taraxacum* Sect. *Spectabilia*. Low/short perennial. Lvs often dark-spotted and with reddish stalk and midrib. Flhds 20-35 mm, outer florets sometimes reddish or purplish beneath, and sepal-like bracts broader and often closely pressed but not recurved; Apr-Aug. Moist and wet places, often in mountains. T. **3a Marsh Dandelion** *T.* Sect. *Palustria*, has lvs narrower, very finely toothed or untoothed, not spotted or reddish; sepal-like bracts broad, closely pressed, with a broad pale margin; Apr-June. Fr usually olive-green. Marshes, fens, streamsides.

4* COMMON CATSEAR *Hypochaeris radicata*. Short/medium perennial, little branched. Lvs usually roughly hairy, the end lobe blunt. Flhds solitary, lfless, with scattered *scale-like* bracts; June-Sept. Drier grassy places. T. **4a* Autumn Hawkbit** *Leontodon autumnalis* is usually hairless and branched, with shiny, more deeply lobed lvs, the end lobe pointed, and smaller (10-35 mm) flhds, outer florets usually reddish beneath and scale-like bracts mainly near top of stalk; July-Oct.

5* ROUGH HAWKBIT *Leontodon hispidus*. Short/medium perennial, shaggily hairy; unbranched. Lvs shallowly lobed, the end lobe rounded. Flhds solitary, 25-40 mm, outer florets often *reddish* or orange beneath; stalks lfless but with *0-2* small scale-like bracts; June-Oct. Pappus making a clock. Grassland; on lime. T. **5a, 5b, 5c,** p.305.

6* LESSER HAWKBIT *Leontodon taraxacoides*. Low/short perennial, hairy or not; unbranched. Flhds solitary, 12-20 mm, outer florets *grey-violet* beneath; stalks lfless, *bractless* and often hairless; June-Oct. Pappus forming a clock. Dry grassy places, dunes. T, southern. **6a* Smooth Catsear** *Hypochaeris glabra* □ is a usually hairless annual, with shiny lvs and smaller (12-15 mm) flhds, opening fully only in bright sunshine, the florets no longer than the sepal-like bracts; stalks with a few scale-like bracts; especially on sandy soils.

7* SPOTTED CATSEAR *Hypochaeris maculata*. Short/medium hairy perennial. Lvs lanceolate, usually heavily *spotted* purplish-black. Flhds pale yellow, 40-50 mm, on stalks with a few tiny lvs; June-Aug. Grassland, especially on lime. T, but rare in B.

8* SMOOTH HAWKSBEARD *Crepis capillaris*. Short/medium perennial, mostly hairless; *branched.* Lvs shiny, not all in a basal rosette, the upper narrower and clasping the stem with arrow-shaped points. Flhds 10-15 mm, in loose clusters, outer florets often *reddish* beneath; outer sepal-like bracts *half-spreading;* June-Nov. Grassy and waste places. T.

9* MOUSE-EAR HAWKWEED *Pilosella officinarum* agg. Low/short creeping perennial, shaggy with *white* hairs, unbranched, with lfy runners. Lvs elliptical, untoothed, whitish beneath. Flhds lemon yellow, solitary, 20-30 mm, outer florets often *reddish* beneath, on lfless, bractless stalks; May-Oct. Bare and grassy places. T. **9a*,** p.305.

Daisy Family (contd.)

HAWKWEEDS *Hieracium*. An exceptionally variable and difficult group, with some hundreds of microspecies. Here only four major groupings can be illustrated. Milky juice, stems unbranched, more or less lfy. Lvs lanceolate, toothed, often sparsely, alternate. Pappus hairs usually pale brownish.

1* FEW-LEAVED HAWKWEED *Hieracium murorum* agg. Short/medium perennials, hairy. Lvs *few,* both on stems and at base. Flhds 20-30 mm, rather *few* in the cluster; June-Aug. Grassy places, walls, rocks. T. **1a* Spotted Hawkweed** *H. maculatum* is the most frequent of a few forms with distinctively purple-blotched lvs. B, F, G.

2* LEAFY HAWKWEED *Hieracium umbellatum* agg. Medium/tall perennials, softly hairy. Lvs *numerous,* all up stems. Flhds 20-30 mm, *numerous,* in clusters; June-Nov. Grassy and heathy places. T.

3* ALPINE HAWKWEED *Hieracium alpinum* agg. Low/short perennials, shaggily hairy. Lvs mostly in a basal *rosette,* few and very small up stems. Flhds usually *solitary,* 25-35 mm; July-Aug. Grassy and rocky places in mountains. T.

4* ORANGE HAWKWEED *Pilosella aurantiacum* agg. Short/medium perennials, covered with *blackish* hairs. Lvs mainly in a basal rosette. Flhds *orange-red* (if yellow, see *P. lactucella,* p.305), 15 mm, in a tight cluster; June-Aug. Grassy and waste places. T.

5* HAWKWEED OX-TONGUE *Picris hieracioides.* Medium perennial, roughly hairy; *branched.* Lvs lanceolate, *wavy-edged.* Flhds 20-35 mm, in a cluster; sepal-like bracts *narrow,* the outer spreading, covered with blackish hairs; July-Oct. Pappus creamy. Grassy and bare places. T.

6* BRISTLY OX-TONGUE *Picris echioides.* Medium perennial, covered with rough *bristles;* well branched. Lvs oblong, wavy-edged, covered with *whitish pimples.* Flhds pale yellow, 20-25 mm, in a cluster; sepal-like bracts *broadly* triangular; June-Nov. Pappus white. Rough grassy places. B, F, G.

HAWKSBEARDS *Crepis.* Distinguished from other dandelion-like fls by the outer row of sepal-like bracts being *much shorter* and usually spreading. Pappus usually white.

7* BEAKED HAWKSBEARD *Crepis vesicaria.* Medium downy biennial. Lvs *pinnately lobed,* dandelion-like, clasping the stem with pointed lobes. Flhds 15-25 mm, erect in bud, in a cluster, outer florets usually *orange* beneath; May-July. Grassy and waste places. B, F, G. **7a* Rough Hawksbeard** *C. biennis* □ is taller and hairier, with fewer, larger (25-35 mm) richer yellow flhds with outer florets yellow beneath. T. **7b* Bristly Hawksbeard** *C. setosa* has paler yellow flhds with numerous spiny yellow bristles. (B, rare), F, G. **7c* Stinking Hawksbeard** *C. foetida* smells of bitter almonds when bruised and has fewer flhds, drooping in bud. Rare in B. **7d* French Hawksbeard** *C. nicaeensis* has lvs clasping stem with arrow-shaped points and florets often red-tipped. T, but (B, rare, S). **7e** *Lagoseris sancta* has stems lfless, lvs more coarsely lobed and no pappus. F, G, range expanding.

8* MARSH HAWKSBEARD *Crepis paludosa.* Medium perennial, almost hairless; stems lfy. Lvs lanceolate, *sharply* toothed, clasping the stem with *pointed* bases. Flhds dull orange-yellow, 15-25 mm, rather few to the cluster; sepal-like bracts downy, with numerous sticky *blackish* hairs; July-Sept. Pappus *brownish white.* *Damp* grassy places, stream-sides. T. **8a** *C. tectorum* is a shorter, much branched annual with pappus pure white. Dry sandy grassy places. F, G, S. **8b** *C. pulchra* is like 8a but taller, less branched and stickily hairy with hairless bracts. Dry sunny slopes and woods. F, G, southern. **8c** *C. praemorsa* has lvs scarcely toothed, few or none on the stem, and dense yellow flhds. F, G, S, eastern.

9* NORTHERN HAWKSBEARD *Crepis mollis.* Slender medium perennial, hairy or not. Lvs lanceolate, toothed or not, clasping the stem with *rounded* bases. Flhds 20-30 mm, rather few in the cluster; July-Aug. Pappus *pure white.* Woods, streamsides, in hill districts. B (rare), F, G.

Water-Plantain Family Alismataceae

The first of the Monocotyledon families (see p.10). Hairless aquatic plants, with lvs all from roots, 3-petalled fls and fr (nutlets) in a rounded head.

1* COMMON WATER-PLANTAIN *Alisma plantago-aquatica.* Medium/tall perennial, with *broad* lanceolate lvs, rounded or heart-shaped at base. Fls in whorls, pale *lilac,* 8-10 mm; June-Sept. Fr head a close ring of nutlets with long beak arising at or below middle. By fresh water. T. **1a* Narrow-leaved Water-Plantain** *A. lanceolatum* □ has narrower lvs tapering into stalk, pinker fls and nutlets with short beak arising near top. **1b* Ribbon-leaved Water-Plantain** *A. gramineum* has often submerged, ribbon-like lvs, slightly broader at tip, smaller fls and nutlets with coiled beak. T, eastern, rare in B.

2* LESSER WATER-PLANTAIN *Baldellia ranunculoides.* Low/short perennial, with *narrow* lanceolate lvs. Fls pale *pink,* 10-15 mm; in whorls of 10-20; June-Sept. In and by shallow, often peaty, fresh water. T, western. **2a** *B. repens* is creeping, rooting at the lf-junctions, with fls 15-22 mm in whorls of 2-6. F,G, western.

3* FLOATING WATER-PLANTAIN *Luronium natans.* Floating perennial, with elliptical lvs, also narrow tapering submerged ones. Fls *white* with a yellow spot, 12-15 mm; May-Aug. Still and slow fresh water. T, western.

4* STAR-FRUIT *Damasonium alisma.* Low/short annual, with blunt oval lvs, heart-shaped at base, floating or submerged. Fls white, 6 mm; June-Sept. Fr heads with long-beaked nutlets spreading like a 6-pointed *star.* In or by shallow fresh water. B(rare),F.

5 PARNASSUS-LEAVED WATER-PLANTAIN *Caldesia parnassifolia.* Medium/tall perennial, with conspicuously *heart-shaped* long-stalked lvs. Fls white, in whorls; July-Sept. By fresh water. F,G.

6* ARROWHEAD *Sagittaria sagittifolia.* Medium/tall perennial, with lvs *arrow-shaped* and in fast streams also oval or lanceolate (floating) or ribbon-like (submerged). Fls white with a large purple spot, 20 mm; July-Aug. Still and flowing shallow fresh water. T. **6a* Canadian Arrowhead** *S. rigida* is shorter, with pointed oval lvs, all aerial, and fls all white, sometimes with a yellow spot. (B), from N America. **6b** *S. natans* has basal lobes of lvs directed downwards and smaller fls. S(rare).

Flowering Rush Family Butomaceae

7* FLOWERING RUSH *Butomus umbellatus.* Tall hairless perennial. Lvs long, rush-like, *three-cornered,* all from roots. Fls 6-petalled, bright pink, in an umbel; July-Aug. Fr egg-shaped, purple when ripe. In and by shallow fresh water. T, southern.

Frog-Bit Family Hydrocharitaceae

8* FROGBIT *Hydrocharis morsus-ranae.* Hairless floating perennial, with bronzy-green, *kidney-shaped* lvs. Fls 3-petalled, white with a yellow spot, 20 mm; July-Aug. Still fresh water. T. **8a Water Chestnut** *Trapa natans* □ (Trapaceae; dicotyledonous) is rooted, with toothed quadrangular lvs and all-white 4-petalled fls; June-July. F,G.

9* WATER SOLDIER *Stratiotes aloides.* Floating perennial, submerged except at flg time, when rosettes of *spine-toothed* lanceolate lvs rise to surface to reveal 3-petalled white 30-40 mm fls, stamens and styles on separate plants; June-Aug. Still fresh water. T.

1a* Narrow-leaved Water-Plantain

8a Water Chestnut

Lily Family   Liliaceae

pp.260-68. Lvs undivided. Fls with three petals and three sepals, often the same colour and so appearing 6-petalled.

1* LILY OF THE VALLEY *Convallaria majalis.* Low/short patch-forming hairless perennial. Lvs *broad* elliptical, two on each stem, nearly opposite. Fls creamy white, *fragrant,* bell-shaped, drooping in a one-sided spike; May-June. Fr a red berry. Drier woodland. T.

2* MAY LILY *Maianthemum bifolium.* Low/short patch-forming hairless perennial; unbranched. Lvs *heart-shaped,* shiny, two on each stem, not opposite. Fls white, very small, *not* fragrant, with prominent stamens, in a spike; May-July. Fr a red berry. Woods. T, rare in B.

3* SNOWDON LILY *Lloydia serotina.* Low slender hairless perennial; unbranched. Lvs grasslike, almost thread-like. Fls white, *purple-veined,* 20 mm, bell-shaped, *solitary;* May. Grassy places, rocks in hills or mountains. B, rare.

4* KERRY LILY *Simethis planifolia.* Low/short hairless perennial. Lvs narrow, grass-like, often *curled,* greyish, all at the base. Fls white, purplish outside, 20 mm, 6-petalled, in a loose head; May-June. Heathy and rocky places. Ireland (rare), F.

5 ST BERNARD'S LILY *Anthericum liliago.* Slender medium hairless perennial, *little branched.* Lvs all at base, long and grasslike. Fls white, starlike, with conspicuous yellow stamens, 30-50 mm, 6-petalled, in a loose *elongated* cluster; bracts long, lanceolate; May-June. Dry grassland, especially in hills. F, G, S.
5a *A. ramosum* □ has smaller (25 mm) fls with shorter bracts in widely branched clusters; June-July. F, G.

6 FALSE HELLEBORINE *Veratrum album.* Medium/tall stout hairy perennial. Lvs broad elliptical, *pleated* lengthwise, unstalked, arranged in *threes.* Fls yellowish-green, often white inside, starlike, 6-petalled, unstalked, in branched spikes; July-Aug. Grassy places in hills and mountains. F, G, S.

7* SCOTTISH ASPHODEL *Tofieldia pusilla.* Low/short, rather slender hairless perennial; unbranched. Lvs flat, sword-shaped, usually *all at base,* occasionally a few small up stem. Fls greenish-yellow, with a *green* 3-lobed bract, in a *short* spike; June-Aug. Wet places on mountains and in tundra. B, F, S.

8 GERMAN ASPHODEL *Tofieldia calyculata.* Short hairless perennial. Lvs flat, sword-shaped, both at base and smaller *up stem.* Fls greenish-yellow, occasionally reddish, with both a green lanceolate and a *papery* 3-lobed bract, in an *elongated* or tight cluster; July-Aug. Damp grassy places, bogs. G, S (Gotland).

Lily Family continued on p.262.

Pipewort Family   Eriocaulaceae

9* PIPEWORT *Eriocaulon aquaticum.* Short/medium hairless aquatic perennial; unbranched. Lvs *all submerged* at base of stem, narrow, pointed, translucent. Fls white, with *grey* bracts, in flat button-like heads, on lfless stems projecting a few cm out of the water; July-Sept. Shallow still fresh water, bare wet ground. B (W Ireland, Inner Hebrides).

2* May Lily

5a *Anthericum ramosum*

Lily Family (contd.)

1* FRITILLARY *Fritillaria meleagris.* Short hairless perennial. Lvs grass-like, greyish, up the stem. Fls *bell-like,* solitary, nodding, *varying* from dull purple to creamy white, chequered with darker blotches; Apr-May. Damp meadows, forming extensive colonies when these are not ploughed. T, southern, scattered.

2* BOG ASPHODEL *Narthecium ossifragum.* Low/short hairless perennial. Lvs sword-shaped, iris-like, all from roots, in rather flattened tufts, often tinged orange. Fls *orange*-yellow, starlike, 6-petalled, in spikes on lfless stems; July-Aug. Fr deep orange. Bogs, wet heaths. T, western.

3* MARTAGON LILY *Lilium martagon.* Medium/tall hairless perennial; stem red-spotted. Lvs elliptical, dark green, in whorls up stem. Fls in a close cluster, dull pink spotted dark purple, with petals *curled back,* making the pinkish stamens prominent; June-July. Woods, scrub, mountain grassland. T, southern, but (B). **3a* Pyrenean Lily** *L. pyrenaicum* □ is shorter and foetid, with narrower alternate lvs and larger dark-spotted yellow fls. (B), from the Pyrenees. **3b** *L. bulbiferum* has erect, sometimes yellowish fls with petals not curled back. With (ssp. *bulbiferum*) or without (ssp. *croceum*) bulbils at base of lvs. G.

4* YELLOW STAR OF BETHLEHEM *Gagea lutea.* Low slender perennial, almost hairless. Lvs like Bluebell (p.266), but solitary, yellower green, more hooded at tip, and with 3-5 prominent ridges on the back, 5-12 mm broad. Fls yellow, 6-petalled, with a *green band* on the outer side of each petal, starlike; in an umbel-like cluster with a pair of large lflike bracts at its base on an otherwise lfless stem; Apr-May. Open woods, scrub, grassland. T.

5 LEAST GAGEA *Gagea minima.* Like 4, but has a very delicate root-lf *1-2 mm* wide, and fls long-stalked, 1-7 in a head; Mar-May. Woods, grassy places, on lime. F,G,S, eastern.

6 MEADOW GAGEA *Gagea pratensis.* Low slender hairy perennial. Lvs linear, *2-5 mm* broad, *greyish,* sharply keeled, solitary. Fls pale greenish-yellow, starlike, in a long-stalked umbel-like cluster with a pair of broad lflike bracts on an otherwise lfless stem; Mar-Apr. Grassy places. F,G. **6a** *G. bohemica* has fls usually solitary and bracts not opposite; Feb-Mar.

7 BELGIAN GAGEA *Gagea spathacea.* Short perennial with two very narrow lvs. Fls yellow-green, 2-5 on each spike, with hairless stalks and very broad bracts; Apr-May. Damp grassy places in woods. F,G. **7a** *G. arvensis* has downy stems and two more shaggily hairy bracts; Feb-Apr. Not on lime; also in cultivated places. F,G,S.

8* WILD TULIP *Tulipa sylvestris.* Slender short/medium hairless perennial. Lvs linear, grass-like, hooded at tip, with *no* prominent veins, grooves or ridges, all from roots. Fls yellow, tinged green or red outside, like a slender garden tulip, fragrant, solitary on lfless stalk; Apr-May. Dry grassy and bare places. (B),F,G.

262   1* Fritillary, colour varieties

Lily Family *(contd.)*

LEEKS and GARLICS *Allium.* Hairless unbranched bulbous perennials, smelling *strongly* of garlic or onion. Fls usually bell-shaped, often mixed with bulbils, in a head or loose umbel, enclosed by one or more papery bracts, which may fall before flowers open.

1* RAMSONS *Allium ursinum.* Short/medium *carpeting* perennial; stem triangular. Lvs *broad* elliptical, all from roots, recalling Lily of the Valley (p.260), but brighter green and garlic-smelling. Fls white, *starlike,* in an umbel; Apr-June. Woods, shady banks. T.

2* THREE-CORNERED LEEK *Allium triquetrum.* Short/medium perennial; stem 3-sided. Lvs linear, sharply keeled, all at base. Fls white, with a narrow *green line* on the back.of each petal, no bulbils, in a drooping one-sided umbel-like head; Apr-June. Hedge-banks, stream-sides, shady places. (B), from Mediterranean. **2a* Few-flowered Leek** *A. paradoxum* is often colonial and has a single narrower lf and bulbils mixed with the smaller fls, which have unequal stalks; Apr-May. (B,F,G), from the Caucasus.

3* CROW GARLIC *Allium vineale.* Medium stiff perennial. Lvs semi-cylindrical, grooved, hollow. Fls greenish or pinkish, stamens *protruding,* bract shorter than fls; either long-stalked in a loose umbel, usually mixed with bulbils, or in a tight head of bulbils only; June-Aug. Grassy and cultivated places. T. **3a* Field Garlic** *A. oleraceum* □ has flatter lvs, looser umbels of fls with unequal stalks and stamens not protruding; form with bulbils only is rare. T.

4* CHIVES *Allium schoenoprasum.* Short/medium *tufted* perennial; a herb used for flavouring. Lvs cylindrical, hollow, greyish, *all from roots.* Fls purple-pink, short-stalked, in a head, stamens not protruding, papery bract short; June-Aug. Grassy places, rocks. T, rare in B.

5* KEELED GARLIC *Allium carinatum.* Medium perennial. Lvs linear, grooved, keeled beneath, up to middle of stem. Fls pink-purple, *blunt*-petalled, with *protruding* purple stamens and long bracts; long-stalked in a loose umbel, usually mixed with bulbils; July-Aug. Open woods, rocks, grassy places. T, but (B). **5a* Rosy Garlic** *A. roseum* has lvs only at base of stem, pinker fls, stamens not protruding, and shorter bracts; Apr-June. (B), from Mediterranean.

6* WELSH ONION *Allium fistulosum.* Medium/tall perennial. Lvs cylindrical, hollow. Fls yellowish-white, numerous, in umbel on stem swollen and hollow *above middle;* June-Sept. Escape from cultivation, occasionally naturalised. (T). **6a***, p.305.

7* SAND LEEK *Allium scorodoprasum.* Tall stiff perennial. Lvs linear, rough-edged. Fls red-purple, stamens *not* protruding, long-stalked in an umbel, mixed with *bulbils;* June-Aug. Sandy, grassy and cultivated places, hedge-banks. T. **7a** *A. rotundum* has lvs not finely toothed and a denser head of darker purple fls without bulbils. F,G.

8* WILD LEEK *Allium ampeloprasum.* Tall stout perennial, *to 2m.* Lvs linear, greyish, keeled, finely toothed. Fls pale purple to whitish, with yellow anthers, slightly protruding stamens, and papery bract falling before flg; in large compact umbels *50-70 mm* across, sometimes with a few small bulbils; July-Aug. Hedge-and other banks, rocks, often near the sea. B(rare),F. **8a* Babington's Leek** *A. babingtonii* has loose umbels with fewer fls, numerous bulbils and sometimes also secondary umbels. B(rare). **8b***, p.305.

9 GERMAN GARLIC *Allium senescens.* Short/medium perennial; stem angled. Lvs variable, linear, sometimes *twisted,* all from roots. Fls rose-pink, numerous in a dense umbel on a lfless stem; June-July. Dry, often grassy or sandy places. F,G,S, south-eastern. **9a, 9b, 9c, 9d,** p.305.

10* ROUND-HEADED LEEK *Allium sphaerocephalon.* Medium perennial. Lvs semi-cylindrical, grooved, hollow. Fls pink-purple, with *protruding stamens,* no bulbils and short papery bracts; in *globular heads;* June-Aug. Grassy places, rocks, dunes. B(rare),F,G.

1

2

3

3a

4

5

6

7

8

9

10

Lily Family *(contd.)*

1* BLUEBELL *Endymion non-scriptus.* Short hairless carpeting perennial. Lvs linear, keeled, with a hooded tip, all from roots. Fls azure blue, occasionally white, elongated bell-shaped, fragrant, in a *long* one-sided spike, drooping at the tip, on a lfless stem; anthers creamy; Apr-June. Fr egg-shaped, seeds black. Woods, scrub, hedge-banks, also on sea cliffs and mountains. B, F, G.
1a* Spanish Bluebell *E. hispanicus* □ is stouter, with broader lvs, larger paler fls in an erect not one-sided spike, and blue anthers; rather later flg. (B, F), from S W Europe.

2* SPRING SQUILL *Scilla verna.* Low/short hairless perennial. Lvs narrow, grass-like, often curly, *all from roots,* appearing before fls. Fls sky-blue, rarely white. 6-petalled, starlike, with bluish *bracts,* in a cluster on a lfless stem; *Apr-June.* Grassy places, especially near the sea. B, F, S.

3 ALPINE SQUILL *Scilla bifolia.* Low/short hairless perennial. Lvs narrow lanceolate, usually two, less often 3-5, *all on stems.* Fls bright blue, rarely pink or white, 6-petalled, starlike, with *no* bracts, in a loose cluster; Mar-Aug. Woods, scrub, grassland, also on mountains. F, G.

4* AUTUMN SQUILL *Scilla autumnalis.* Low/short hairless perennial. Lvs narrow, grass-like, *all from roots,* appearing *after* fls. Fls purplish-blue, starlike, with *no* bracts, in a cluster on a lfless stem; *Aug-Oct.* Dry grassy places, often near the sea. B, F.

5* GRAPE HYACINTH *Muscari atlanticum.* Short hairless perennial. Lvs *1-3 mm* wide, grass-like, *grooved* and semi-cylindrical, rather limp, all from roots. Fls dark blue, *egg-shaped,* with small whitish petal-lobes, in a tight elongated head on a lfless stem; Apr-May. Dry grassy and cultivated places. B (rare), F, G. **5a** *M. neglectum* is larger overall with more fls. G, southern, rare.

6 SMALL GRAPE HYACINTH *Muscari botryodes.* Short hairless perennial. Lvs 3-7 mm wide, grass-like, broader *at tip,* all from roots. Fls pale blue-violet, *globular,* with small whitish petal-lobes, in a loose, more conical head than 5, on a lfless stem; Mar-May. Grassy places, woods. F, G, S. **6a* Tassel Hyacinth** *M. comosum* □ is larger and has broader lvs and larger spikes of longer stalked fls, with a "tassel" of purple sterile fls at the top. (B), F, G, southern.

7* MEADOW SAFFRON *Colchicum autumnale.* Short hairless perennial. Lvs *oblong-lanceolate,* bright green, in clumps, appearing in spring and dying *before* flg time. Fls pale rosy mauve, rarely white, *crocus-like,* on long weak white stalks (which are actually tubular and part of the fl), solitary; *Aug-Sept.* Fr egg-shaped, appearing in spring. Differs from Autumn Crocus (p.272) both in its lvs and in its *orange* anthers and six stamens. Woods, damp meadows. T, southern.

Lily Family continued on p. 268.

Yam Family   Dioscoreaceae

8* BLACK BRYONY *Tamus communis.* Hairless perennial climber to 4 m, twining clockwise; no tendrils (unlike the completely unrelated White Bryony (p.148)). Lvs *heart-shaped,* dark green, shiny, alternate. Fls yellow-green, tiny, *6-petalled,* in loose, sometimes branched spikes, stamens and styles on separate plants; May-Aug. Fr a red berry. Woods, scrub, hedges. B, F, G.

6a* Tassel Hyacinth, fruits × ½

7* Meadow Saffron, fruits × ½

Lily Family (contd.)

1* COMMON SOLOMON'S SEAL *Polygonatum multiflorum.* Medium patch-forming hairless perennial; stems rounded, arching. Lvs elliptical, alternate, along the stems. Fls greenish-white, bell-shaped, waisted, unscented, in hanging *clusters* of 1-3 at base of lvs; May-June. Fr a *blue-black* berry. Woods, scrub. T. **1a* Angular Solomon's Seal** *P. odoratum* □ has stems angled and fls fragrant, 1-2 together and not waisted; on lime. **1b*** *P.* × *hybridum,* the hybrid between 1 and 1a, is commonly grown in gardens and sometimes escapes. (T). **1c** *Streptopus amplexifolius* is like 1a but fls small, star-like, on long stalks. Mountain woods. G, southern, rare.

2* WHORLED SOLOMON'S SEAL *Polygonatum verticillatum.* Medium/tall hairless perennial. Lvs narrow lanceolate, in *whorls* of 4-5 up the stem. Fls white, tipped green, bell-shaped, unscented, 1-3 together at each whorl; June-July. Fr a *red* berry. Woods in hill and mountain districts. T, rare in B.

3* HERB PARIS *Paris quadrifolia.* Low/short hairless colonial perennial. Lvs pointed oval, unstalked, in a *whorl* of usually four (less often 3-8) together at top of otherwise lfless stem. Fls yellow-green, 4-6-petalled, starlike, with prominent stamens, solitary; May-June. Fr a black berry. Woods, on lime. T.

4* SPIKED STAR OF BETHLEHEM *Ornithogalum pyrenaicum.* Medium/tall hairless perennial; unbranched. Lvs linear, grass-like, greyish, all from roots, usually withered by flg time. Fls *greenish-white* or yellowish, 6-petalled, starlike, in a *stalked spike* on a lfless stem; May-July. Woods, hedge-banks, grassland. B, F, G.

5* DROOPING STAR OF BETHLEHEM *Ornithogalum nutans.* Short/medium hairless perennial. Lvs linear, grass-like, grooved and with a central *white stripe,* all from roots. Fls white, with a green stripe on the back of each petal, bell-shaped, nodding, in a one-sided spike on a lfless stem; Apr-May. Grassy and cultivated places. (T), southern, from S Europe.

6* COMMON STAR OF BETHLEHEM *Ornithogalum umbellatum.* Low/short hairless perennial. Lvs linear, grass-like, limp, grooved and with a central *white stripe,* all from roots. Fls white, with a green stripe on the back of each petal, 6-petalled, starlike, in an *umbel-like* cluster, on a lfless stem; May-June. Grassy and cultivated places. T, southern.

7* WILD ASPARAGUS *Asparagus officinalis.* Medium/tall perennial, hairless; well branched. Lvs (actually reduced stems) short, *needle-like,* in tufts. Fls yellowish or greenish-white, bell-shaped, 1-2 together at base of branches; stamens and styles on separate plants; June-Aug. Fr a red berry. Grassy and waste places; often by the sea, when may be prostrate, with fleshy greyish 'lvs' and male fls tinged red at base. Widely cultivated. T.

8* BUTCHER'S BROOM *Ruscus aculeatus.* Medium evergreen hairless bush. Lvs (actually flattened branches) oval, ending in a sharp *spine.* Fls whitish or greenish, tiny, 6-petalled, solitary or paired on upper surface of 'lvs; stamens and styles on separate plants; Jan-Apr. Fr a red berry. Woods, scrub, hedge-banks. B, F.

fruits of: **1a*** Angular Solomon's Seal

2* Whorled Solomon's Seal

1* Common Solomon's Seal

Daffodil Family   Amaryllidaceae.

Bulbous perennials, hairless. Lvs linear, all from roots. Fl buds enclosed in a sheath (spathe), which splits at flg; petals in two rings, each usually of three petals or petal-like sepals each. Fr a capsule.

1* WILD DAFFODIL *Narcissus pseudonarcissus.* Medium perennial. Lvs greyish, grooved. Fls yellow, the erect *trumpet-like* inner ring as long as but darker yellow than the spreading outer ring, nodding, solitary on lfless stem; Mar-Apr. Woods, grassland; numerous garden forms, including double ones, escape and appear on waysides and waste ground. T, southern.   **1a* Tenby Daffodil** *N. obvallaris* has erecter uniformly deep yellow fls, the trumpet longer than the outer petals. B (Pembrokeshire), rare.

2* PRIMROSE PEERLESS *Narcissus* × *biflorus.* Probably originally a natural hybrid between 1 and 3. Fls *paired,* fragrant, the short trumpet deep yellow with a whitish crisped margin, the outer petals creamy or whitish; Apr-May. Widely naturalised in grassy places. (B, F).

3* POET'S NARCISSUS *Narcissus poeticus.* Medium perennial. Lvs greyish, grooved. Fls with a short yellow trumpet, the margin *red and crisped,* and white outer petals, fragrant, nodding, solitary; Apr-May. Grassy places. (B, F, G), from Mediterranean.   **3a** *N. stellaris* has lvs less than 5 mm wide. G, southern, rare.

4* SPRING SNOWFLAKE *Leucojum vernum.* Short perennial. Lvs *bright* green. Fls white tipped green, bell-shaped, nodding, petals 20-25 mm, *solitary* or occasionally paired on lfless stems; anthers orange; Feb-Mar. Damp woods, copses, meadows. B (rare), F, G.

5* SUMMER SNOWFLAKE *Leucojum aestivum.* Short/medium *tufted* perennial. Lvs bright green. Fls like 4 but smaller (petals 15-20 mm) and 3-6 in a *cluster* on unequal stalks; Apr-May. Damp grassy places, by fresh water. B (rare), F, G.

6* SNOWDROP *Galanthus nivalis.* Low/short perennial. Lvs *greyish,* narrower than 4 and 5, grooved, keeled. Fls white, inner petals tipped green, outer row green on the back, bell-shaped, nodding, petals 20-25 mm long, *solitary* on lfless stems, anthers green; Jan-Mar. Damp woods, shady streamsides, meadows, and extensively naturalised. B, F, G.

Arum Family   Araceae

7* LORDS AND LADIES *Arum maculatum.* Low/short hairless perennial. Lvs bluntly *arrow-shaped,* dark green, often *dark-spotted,* all from roots, appearing in January. Fls tiny, male above female, in dense whorls topped by the conspicuous purple finger-like spadix, fls and spadix both being enveloped by the even more conspicuous pale green hooded spathe, whose base conceals the fls; Apr-May. Fr a spike of bright orange berries. Woods, copses, shady banks. B, F, G.   **7a* Large Cuckoo Pint** *A. italicum* ☐ has larger, more triangular lvs with creamy veins, appearing in autumn, and a yellow spadix. B, F.   **7b* Bog Arum** *Calla palustris* ☐ is aquatic, with unspotted heart-shaped lvs, a shorter greenish spadix in a smaller white spathe, and red berries; June-Aug. Swamps, lake-sides. T, but (B).

N.B. **Adderstongue** *Ophioglossum vulgatum* ☐ and **Moonwort** *Botrychium lunaria* ☐ are two rather atypical ferns, whose spathe-like fronds respectively oval and pinnate, together with their spadix-like spike of spores, make them look not unlike diminutive arums.

Adderstongue

270   **7*** Lords and Ladies   **7a*** Large Cuckoo Pint   **7b*** Bog Arum   Moonwort

Iris Family   Iridaceae

IRISES *Iris.* Hairless perennials, often showing conspicuous rhizomes (underground stems) above ground. Lvs sword-shaped (except 5), flat, mostly from roots. Fls showy, with three spreading, sometimes bearded, outer petals (falls), three more or less erect and twisted inner ones (standards), all narrow at the base, and three large, almost petal-like stigmas with branched tips (crests). Fl buds enclosed in a sheath (spathe) which splits at flg time. Fr a 3-sided capsule.

N.B. Illustrations for this page are painted at half natural size.

1 SIBERIAN IRIS *Iris sibirica.* Slender tufted perennial. Lvs narrow, grass-like. Fls blue-violet, 60 mm, with an *all-brown* spathe. Grassy places. F,G,(S), southeastern.

2* PURPLE FLAG *Iris versicolor.* Medium/tall; little branched. Lvs broader than 3, slightly greyish and without raised midrib. Fls pale *pinkish*-purple, blade of falls as long as haft, crests nearly *white,* 1-3 together; June-July. Seeds dark brown. Wet places. (B,rare), from N America.

3* YELLOW IRIS *Iris pseudacorus.* Tall; stems branched. Lvs with a raised midrib. Fls *yellow,* 80-100 mm across, 1-3 together; June-Aug. Seeds brown. Marshes, by fresh water. T.

4* GARDEN IRIS *Iris germanica.* Stout medium/tall; branched. Lvs *greyish.* Fls fragrant, falls blue-violet with *yellow* beard, standards deep lilac, up to 100 mm across, 2-3 together; spathe green and brown; May-June. Grassy places, waysides. (T), from Mediterranean. **4a** *I. aphylla* is shorter, with lvs longer than stems. G, eastern. **4b** *I. sambucina* has lvs smelling strongly of Elder (p.222) and fls darker. (G, southern).

5* BUTTERFLY IRIS *Iris spuria.* Medium; little branched. Lvs rather narrow, *unpleasant*-smelling when crushed. Fls pale blue-violet, falls with rounded blade half as long as haft and keel of haft *yellow,* crests violet, 40-50 mm across, unstalked, 1-3 together; May-June. Damp grassland and chalk downs. (B,rare), F,G.

6* STINKING IRIS *Iris foetidissima.* Medium, tufted. Lvs dark green, smelling *sickly sweet* when crushed. Fls grey-purple, occasionally yellowish, 80 mm across, 2-3 together; May-July. Seeds bright orange. Open woods, scrub, hedge-banks, sea cliffs and dunes. B,F.

7* GLADIOLUS *Gladiolus illyricus.* Medium/tall hairless perennial. Lvs sword-shaped, all from roots. Fls red-purple, 6-petalled, 3-8 in a spike on a lfless stem; June-July. Grassy and heathy places, scrub, marshes. B(rare),F. Several similar species occur as garden escapes. **7a** *G. palustris* has fls marked with white. G. **7b* Montbretia** *Crocosmia × crocosmiiflora* is shorter and patch-forming with a one-sided spike of smaller orange fls; July-Aug. A man-made hybrid widely established on cliffs and waste ground. (B,F).

8* SAND CROCUS *Romulea columnae.* Low hairless perennial, to 5 cm. Lvs threadlike, *curly,* all from roots. Fls pale purple, greenish outside, much smaller than 9 *(10 mm* across) and opening only in full sunshine; Apr. Sandy grassland, dunes. B(rare),F.

9* SPRING CROCUS *Crocus albiflorus.* Low hairless perennial. Lvs linear, grooved, keeled, with a *white midrib,* tufted. Fls like garden crocuses, purple, white or both, solitary on lfless stalk; styles orange; Mar-Apr. Grassy places. (B,F),G. **9a* Autumn Crocus** *C. nudiflorus* has lvs appearing in spring and dying before fls appear in Sept-Oct. Cf. Meadow Saffron (p.266), which has 6 (not 3) stamens and much broader lvs. From S W Europe.

10* BLUE-EYED GRASS *Sisyrinchium bermudiana.* Low/short hairless perennial. Lvs linear, tufted, all from roots. Fls blue with a yellow centre, 6-petalled, *starlike,* 2-4 together on a *winged* lfless stem, opening only in sunshine; July. Grassy places. B,(F,G).

Orchid Family   Orchidaceae

pp. 274-86. Unbranched perennials, hairless or almost so, except *Epipactis* (p. 284). Lvs undivided, untoothed, often linear/lanceolate, keeled and rather fleshy. Fls of rather diverse shape, but two-lipped, the lower lip often developed to appear the predominant part of the fl; often spurred; in a stalked spike with a lflike bract at base of each fl. Fr egg-shaped to cylindrical.

1* LADY'S SLIPPER *Cypripedium calceolus*. Short/medium perennial. Lvs broad lanceolate, strongly ribbed, pale green. Fls maroon with a large hollow *yellow lip*, spotted red inside, solitary, the largest orchid fl of the region; May-June. Woods, scrub, in hill and mountain districts; mainly on lime. T, rare in B.

2* BEE ORCHID *Ophrys apifera*. Short/medium perennial. Lvs elliptical, pointed. Fls with sepals *pink*, petals *green* and lip red-brown, *rounded at tip*, patterned to appear like the rear of a small bumblebee apparently visiting the fl; June-July. Grassland, open scrub, dunes; only on lime. B, F, G. **2a* Wasp Orchid** *O. apifera* var. *trollii* □ has a point to the lip (in 2 this point is bent back under the lip) and also lacks the pale U-shaped mark usually found on the bee orchid's lip.

3* LATE SPIDER ORCHID *Ophrys fuciflora*. Shorter than 2 and differs in having both petals and sepals pink and a more elaborate pattern on the lip, supposedly resembling a spider, also a *heart-shaped* appendage at the tip; June-July. Grassy places; on lime. B (rare), F, G.

4* EARLY SPIDER ORCHID *Ophrys sphegodes*. Shorter than 2 and differs in having both sepals and petals *yellow-green*, and an *H- or X-shaped* mark on the lip, sometimes also likened to the Greek letter pi (π); Apr-June. Grassy places; on lime. B, F, G.

5* FLY ORCHID *Ophrys insectifera*. Slender short/medium perennial. Lvs narrow elliptical, shiny. Fls brown with green sepals, the narrow lip 3-lobed, the middle lobe forked, and with a *bluish patch* at its base; May-June. Woods, scrub, fens, grassy places; on lime. T.

6 BLACK VANILLA ORCHID *Nigritella nigra*. Low/short perennial. Lvs linear, pointed. Fls *blackish-purple*, rarely pink, spurred, *vanilla-scented,* in a tight head; June-Aug. Mountain meadows. F, G, S.

7 CALYPSO *Calypso bulbosa*. Low/short perennial. Lf elliptical, markedly veined, *solitary.* Fl purplish-pink with a large *hollow* yellow-blotched pale pink lip, solitary; May. Marshes, marshy woodland. S, a high northern plant.

5* Fly Or

2* Bee Orchid   2a* Wasp Orchid   3* Late Spider   4* Early Spider
                                                   Orchid   Orchid

274

Orchid Family (contd.)

1* EARLY PURPLE ORCHID *Orchis mascula.* Short/medium perennial. Lvs narrow oblong, pointed, usually *blotched* purplish-black. Fls purple, less often pink-purple or white, smelling of tom-cats; petals with one sepal forming a hood, the two other sepals *erect,* lip shallowly 3-lobed, spur long; Apr-June. Woods, scrub, grassland. T. **1a Pale-flowered Orchid** *O. pallens* □ has broader lvs and pale yellow fls. On lime. F, G, eastern.

2* GREEN-WINGED ORCHID *Orchis morio.* Short perennial. Lvs narrow oblong, pointed, unspotted. Fls purple, pink or white, fragrant, hood formed by all sepals (which are *green-veined)* and petals, lip shallowly 3-lobed, spur long; fewer and in more open spike than 1; May-June. Grassland, open scrub. T.

3* LADY ORCHID *Orchis purpurea.* Medium/tall perennial. Lvs broad oblong, shiny, unspotted, mostly in a basal rosette. Fls shaped like a *manikin,* all sepals and petals forming the hood or 'head', and the long lip being lobed narrowly to make the 'arms' and *broadly* to make the 'legs'; hood *dark* purple, and lip *pale* pink, fragrant, spur short; May-June. Woods, scrub; on lime. B (rare), F, G.

4* MILITARY ORCHID *Orchis militaris.* Short/medium perennial. Lvs broad oblong, unspotted. Fls manikin-shaped, like 3, but with a small central lobe between the 'legs', pinkish-grey with darker pink markings, especially inside the hood; spur short; May-June. Wood edges, scrub, grassland; on lime. B (rare), F, G.

5* BURNT ORCHID *Orchis ustulata.* Low/short perennial. Lvs oblong, pointed, unspotted. Fls like miniatures of 3, the hood *dark maroon* at first, becoming paler, the lip less markedly 4-lobed, white with pink dots, short-spurred, fragrant; May-June. Grassland; on lime. T, southern. **5a Toothed Orchid** *O. tridentata* □ has the 'legs' of the manikin poorly developed, the central lobe of the lip being itself shallowly 3-lobed, and having purplish-red spots. G.

6* MONKEY ORCHID *Orchis simia.* Short/medium perennial. Lvs broad oblong, shiny, unspotted. Fls like 4 but more delicate, and 'arms' and 'legs' of manikin *much thinner* with the central tooth accentuated to form the monkey's 'tail'; May-June. Grassland, open scrub; mainly on lime. B (rare), F, G.

7 BUG ORCHID *Orchis coriophora.* Short perennial. Lvs narrow lanceolate. Fls red-brown, all sepals (which are *green-veined)* and petals joined in a hood, the lip 3-lobed and dark wine-purple tipped red and green, spur short, *smelling of bed-bugs* (but a more southern darker red form with white bracts and a large central lobe to the lip is vanilla-scented); Apr-June. Damp grassland. F, G.

8 LOOSE-FLOWERED ORCHID *Orchis laxiflora.* Medium perennial. Lvs lanceolate, *unspotted.* Fls purple, in a rather loose spike, hood formed from petals and one sepal, the two other sepals *spreading,* lip often only *2-lobed,* spur short; bracts and stem often purplish; May-June. Damp meadows. F, G. **8a** *O. palustris* □ has paler fls with tip clearly 3-lobed. G, S.

5a Toothed Orchid   **8a** *Orchis palustris*

Orchid Family *(contd.)*

1* PYRAMIDAL ORCHID *Anacamptis pyramidalis*. Medium perennial. Lvs narrow lanceolate, unspotted. Fls *bright pink* or rose-purple, sepals spreading, petals hooded, lip 3-lobed, spur very *long*, slender and curved, often foxy-smelling, in a densely flattened *pyramidal* spike; June-Aug. Dry grassland, dunes; mainly on lime. T, southern.

MARSH AND SPOTTED ORCHIDS *Dactylorhiza*. Fls with sepals spreading outwards and forwards like a bird's wing, and a broad lip. Frequently hybridise with each other.

2* EARLY MARSH ORCHID *Dactylorhiza incarnata*. Short/medium perennial. Lvs oblong, keeled, *hooded at tip, yellowish-green*, unspotted. Fls very variable in colour, pink, purple, brick-red or yellow; lip with a small central tooth and looking very narrow because the sides soon fold *backwards*, spur straight; May-July. Damp grassland, marshes, fens, dune slacks. T. **2a* Flecked Marsh Orchid** *D. cruenta* is shorter and has stem, lvs and bracts streaked dark purple on both sides, and fls dark red-purple with untoothed lip. T, rare. **2b* Pugsley's Marsh Orchid** *D. traunsteineri* has narrower, usually spotted lvs and fewer dark rose-purple fls with a 3-lobed lip less markedly folded back but strongly marked with dark purple.

3* SOUTHERN MARSH ORCHID *Dactylorhiza praetermissa*. Short/medium perennial. Lvs dark green, not hooded at tip, unspotted. Fls rose-purple, lip broad, streaked and spotted darker, with a small central tooth, the sides *not* folding backwards; spur short, stout; bracts and upper stem often purplish; June-July. Damp grassland, marshes, fens, dune slacks. T, but rare in S. **3a* Northern Marsh Orchid** *D. purpurella* is shorter, with occasionally spotted lvs and wine-purple fls with lip either untoothed or shallowly 3-lobed, and a tapering spur. B, G, S.

4* BROAD-LEAVED MARSH ORCHID *Dactylorhiza majalis*. Medium/tall perennial. Lvs broad oblong, *bluish-green*, sometimes dark-spotted. Fls lilac-purple (bright purple or reddish in mountains), the shallowly 3-lobed lip with darker lines and spots, spur short; May-July. Damp meadows, marshes, on lime. T.

5* COMMON SPOTTED ORCHID *Dactylorhiza fuchsii*. Short/medium perennial. Lvs narrow lanceolate, keeled, usually *dark-spotted*. Fls pale pink, pale purple or white, dotted and lined with crimson or purple, lip markedly 3-lobed, spur long; June-Aug. Grassy places, open scrub; on lime. T. **5a* Heath Spotted Orchid** *D. maculata* □ has the lip with a wavy edge and a single small tooth. Heathy places on acid soils, including bogs.

6 ELDER-FLOWERED ORCHID *Dactylorhiza sambucina*. Short perennial. Lvs pale green, shiny, unspotted. Fls either *yellow* with a purple-spotted lip and pale *green* bracts or *red-purple* with a purple lip and *reddish* bracts, often mixed in same colony; lip very shallowly 3-lobed; not fragrant; Apr-June. Mountain grassland. F, G, S.

7* FRAGRANT ORCHID *Gymnadenia conopsea*. Medium perennial. Lvs narrow lanceolate, unspotted. Fls *pale* purplish-pink, *fragrant* (sometimes clove-scented), with spreading sepals, hooded petals, 3-lobed lip and *long* slender spur; June-July. Grassland, especially on lime, fens. T. **7a** *G. odoratissima* has narrower greyish lvs, pale pink or white vanilla-scented fls in a shorter tighter spike, and a shorter stouter spur. F, G, S.

Orchid Family *(contd.)*

1* MAN ORCHID *Aceras anthropophorum.* Short perennial. Lvs oblong lanceolate, keeled, shiny. Fls *greenish-yellow,* in long dense spikes, petals and sepals forming a hood and lip shaped as a manikin (cf. Lady Orchid, 276/3), often tinged red-brown, unspurred; May-June. Grassland, scrub; on lime. B, F, G.

2* LIZARD ORCHID *Himantoglossum hircinum.* Short/medium greyish perennial. Lvs oblong lanceolate, soon withering. Fls purplish grey-green, petals and sepals forming a hood, the *long straplike lip* 3-lobed, spur short; smelling strongly of billy goat; in a tangled, often elongated spike; June-July. Grassy banks, scrub, wood edges, dunes. B (rare), F, G.

3* DENSE-FLOWERED ORCHID *Neotinea intacta.* Short perennial. Lvs oblong elliptical, blunt, either unspotted or with *lines* of brownish spots. Fls either greenish-white or pink, sepals and petals hooded, lip shortly 3-lobed, spur short, vanilla-scented, in a short *tight* spike; Apr-June. May be more conspicuous in fr. Sandy grassland, dunes, limestone rocks. B (W Ireland, Isle of Man).

4 FALSE MUSK ORCHID *Chamorchis alpina.* Low perennial, hard to detect among other vegetation. Lvs linear, *grasslike.* Fls green tinged with violet or purple, sepals and petals hooded, lip yellowish, obscurely 3-lobed, unspurred, few in the spike; July-Aug. Damp grassland in *mountains;* on lime. S.

5* MUSK ORCHID *Herminium monorchis.* Low perennial. Lvs oval, yellow-green, 2-3 near *base* of stem. Fls greenish-yellow, petals and sepals spreading, lip 3-lobed, the middle lobe much the longest, spur very short or none; *honey-scented* (not musky); June-July. Grassland; on lime; in Britain prefers dry turf, on Continent damp places. T.

6* FROG ORCHID *Coeloglossum viride.* Low/short perennial. Lvs lanceolate, narrower up stem. Fls yellow-green tinged red-brown, sepals and petals hooded, lip strap-shaped, *forked* near tip with a *tooth* in the middle, spur very short; June-July. Grassland, often on mountains. T.

7* LESSER TWAYBLADE *Listera cordata. Low* slender perennial, sometimes only 5 cm. Lvs *heart-shaped,* shiny, *a single pair* at base of stem. Fls reddish-green, petals and sepals half-hooded, lip obscurely 3-lobed, the middle lobe conspicuously forked, unspurred; June-July. Moors, bogs, coniferous woods, often half-hidden among heather and moss or under bracken. T.

8* COMMON TWAYBLADE *Listera ovata.* Medium perennial. Lvs *broad oval,* not shiny, *a single pair* at base of stem. Fls yellow-green, petals and sepals half-hooded, lip forked, unspurred, in a long spike; May-July. Woods, scrub, grassland. T.

Military Orchid (p. 276) Lady Orchid (p. 276)

Monkey Orchid (p. 276)

6* Frog Orchid

1* Man Orchid

2* Lizard Orchid

Orchid Family (contd.)

1* GREATER BUTTERFLY ORCHID *Platanthera chlorantha.* Medium perennial. Lvs broad elliptical, shiny, *a single pair* at base, much smaller and lanceolate up stem. Fls greenish-white, *vanilla-scented,* sepals and petals spreading, lip long, narrow and undivided, spur very long, in a rather loose spike; pollen masses diverging; June-July. Woods, open scrub, grassland. T. **1a* Lesser Butterfly Orchid** *P. bifolia* □ is shorter and smaller, with less broad basal lvs and pollen masses parallel. Also on moors and in marshes.

2* RED HELLEBORINE *Cephalanthera rubra.* Short/medium perennial, slightly tinged purple. Lvs narrow lanceolate, sometimes almost grasslike in deep shade. Fls bright *purple-pink,* not opening widely, with an unspurred whitish lip; June-July. Woods, especially of beech, and scrub; on lime. T, but rare in B.

3* WHITE HELLEBORINE *Cephalanthera damasonium.* Short/medium perennial. Lvs *broad lanceolate,* slightly bluish-green, narrower up stem. Fls creamy white, the yellow base of the unspurred lip usually hidden because fl scarcely opens, bracts conspicuous and *lflike;* May-July. Woods, especially of beech, shady banks; on lime. T.

4* NARROW-LEAVED HELLEBORINE *Cephalanthera longifolia.* Slenderer than 3. Lvs longer, *narrower,* parallel-sided and darker green. Fls pure white, opening more widely to reveal a much smaller orange spot at base of lip, with bracts too *small* to appear lflike; May-June. Woods. T.

5* WHITE FROG ORCHID *Pseudorchis albida.* Short perennial. Lvs narrow oblong, keeled, shiny, unspotted. Fls *creamy white, fragrant,* sepals and petals hooded, lip 3-lobed, short-spurred; May-June. *Upland grassland. T.*

6* AUTUMN LADY'S TRESSES *Spiranthes spiralis.* Low perennial. Lvs pointed oval, bluish-green, in a basal rosette which withers before flg, leaving only a few scale-like lvs up stem. Fls white, fragrant, the lip greenish, unspurred, with a frilled recurved margin, *spirally* in a *single* row up the spike; Aug-Sept. *Dry* grassland. B, F, G. **6a Summer Lady's Tresses** *S. aestivalis* □ has no basal rosette but linear-lanceolate lvs up the stem, fls all-white, slightly larger and less fragrant, and longer bracts; July-Aug. Marshy or boggy places, often by streams. Extinct in B; rare in F, G.

7* IRISH LADY'S TRESSES *Spiranthes romanzoffiana.* Low/short perennial. Lvs linear-lanceolate up the stem, no basal rosette. Fls white or creamy white, greenish at the base, fragrant, unspurred, in *three spiral* rows on a spike consequently much broader than 6; bracts conspicuous; Aug-Sept. *Wet* grassy places, peat marshes, bogs. B, western.

8* CREEPING LADY'S TRESSES *Goodyera repens.* Low/short creeping perennial, with *runners.* Lvs pointed oval conspicuously net-veined, in basal *rosettes,* smaller and scale-like up the stems. Fls white, fragrant (to some people unpleasantly so), in a *spiral* spike, the unspurred lip neither frilled nor recurved; July-Aug. *Woods,* especially coniferous, mainly in hills and mountains. T.

6a Summer Lady's Tresses

Orchid Family (contd.)

HELLEBORINES *Epipactis.* Lvs usually broad elliptical, pointed. Fls in one-sided stalked spikes; lip unspurred, in two parts joined by a narrow waist, the base cupped, the tip more or less triangular.

1* MARSH HELLEBORINE *Epipactis palustris.* Medium perennial. Lvs narrow elliptical, keeled, folded. Fls with sepals purple or purple-brown, petals *crimson and white,* and lip white with cup crimson-streaked and tip with a yellow spot and a *notched frilly margin;* also sometimes fls yellowish-white with lip white; July-Aug. Fr downy. *Fens,* marshes, dune slacks. T.

2* DARK RED HELLEBORINE *Epipactis atrorubens.* Medium perennial, *tinged red-purple;* stem downy. Lvs in two rows up stem. *Fls dark wine-red,* fragrant, opening fully; June-July. Fr downy. Woods, grassy places, rocks, dunes; on lime. T.

3* BROAD-LEAVED HELLEBORINE *Epipactis helleborine.* Medium/tall perennial, often tinged purplish. Lvs sometimes short and almost rounded, alternating *spirally* up stem. Fls very variable, greenish-yellow to purple-red, unscented, opening fully, lip appearing rounded because point of tip recurved *underneath;* July-Sept. Woods, shady banks, less often open hillsides, dune slacks. T. **3a** *E. microphylla* □ has much smaller lvs and fls greenish with purple margin, the lip showing a small point. Woods. F,G,S, rare. **3b** *E. muelleri* □ has pale green fls with the inside of the lip cup reddish. F,G.

4* NARROW-LIPPED HELLEBORINE *Epipactis leptochila.* Short/medium perennial, slightly downy, tinged *yellow.* Lvs in *two rows* up stem. Fls green, usually tinged white or yellow, *drooping,* sometimes not opening fully, the lip with cup mottled red and tip extended to a *point;* July-Aug. Woods, especially of beech. B,F,G.

5* VIOLET HELLEBORINE *Epipactis purpurata.* Medium perennial, purple-tinged; stems usually in a *clump.* Lvs narrow elliptical, spirally up stem. Fls pale greenish-white (purplish-green outside), lip with cup usually mottled violet and tip recurved; bracts often longer than fls; Aug-Sept. Woods, often of beech. B,F,G.

6* DUNE HELLEBORINE *Epipactis dunensis.* Medium perennial; stem downy above, purplish below. Lvs in *two rows* up stem. Fls pale yellow-green, *not fully opening,* lip with cup mottled red inside and tip recurved as in 3; lowest bracts *longer than fls;* June-July. Dune slacks, pine plantations. B (Lancashire, Anglesey).

7* GREEN-FLOWERED HELLEBORINE *Epipactis phyllanthes.* Medium perennial; stems hairless. Lvs variable, in *two rows* up stem. Fls green, often tinged yellow or purple, drooping, *scarcely opening,* lip cup usually whitish; lowest bracts longer than fls; July-Sept. Woods, dunes. B,F,G. **7a** *E. confusa* □ is slenderer and has fls opening fully with a pink lip. G,S.

2* Dark Red Helleborine

3* Broad-leaved Helleborine

3a *Epipactis microphylla*

1* Marsh Helleborine

3b *E. muelleri*

5* Violet Helleborine

6* Dune Helleborine

7* Green-flowered Helleborine

7a *Epipact confusa*

Orchid Family (contd.)

1* BIRDSNEST ORCHID *Neottia nidus-avis.* Short/medium perennial; stem *honey-coloured. No lvs* but scale-like bracts on stem. Fls honey-brown like stem, with a somewhat sickly fragrance, petals and sepals small, lip large, unspurred, *2-lobed* (Broomrapes (p.218) have a 3-lobed lip); May-July. Woods, especially of beech, in deep shade. T.

2* CORALROOT ORCHID *Corallorhiza trifida.* Low/short perennial; stem yellow-green. *No lvs* but scale-like bracts on stem. Fls yellow-green or yellowish-white, sepals and petals spreading, lip obscurely 3-lobed, unspurred and marked with *red;* June-July. Woods, especially coniferous, in mountains, dune slacks. T.

3* GHOST ORCHID *Epipogium aphyllum.* Low perennial; stem *mauvish-yellow. No lvs* but a few tiny pale brown bracts. Fls pale mauve and pale yellow, petals and sepals spreading, lip recurved, undivided, spur whitish, solitary or 2-3 in a spike; June-Sept, flg most freely after a wet spring. Broad-leaved woods, especially of beech, in deep shade, often in damp places. T, but rare in B.

4* BOG ORCHID *Malaxis paludosa.* Low perennial, sometimes only 5 cm high. Lvs oval, *2-4* up the stems, usually with tiny bulbils on their margins. Fls yellow-green, sepals and petals spreading, the narrow lanceolate lip twisted round to appear at the top; July-Sept. Bogs and other wet *acid* places, almost always among *sphagnum* moss. T.

5 ONE-LEAVED BOG ORCHID *Malaxis monophyllos.* Low/short perennial. Lf *solitary,* more or less oval. Fls greenish-yellow, sepals and petals spreading, the narrowly triangular unspurred lip twisted round to appear at the top; July. Damp places, usually among *sphagnum* moss. G, S, rare.

6* FEN ORCHID *Liparis loeselii.* Low/short perennial. Lvs broad lanceolate, shiny, yellow-green, a *single pair* at base of stem. Fls yellow-green, petals and sepals spreading, the broad unspurred lip twisted round to appear at the top; June-July. *Fens,* marshes, dune slacks; *on lime.* T.

7 VIOLET BIRDSNEST ORCHID *Limodorum abortivum.* Medium perennial; stem violet. No lvs but scales on stem. Fls violet with a yellow tinge, *40 mm* across, petals and sepals spreading, lip undivided, spurred; May-July, flg most freely after a wet spring. Woods, especially coniferous, shady banks. F, G.

Ghost Orchid, Coralroot Orchid and the two Birdsnest Orchids, together with Yellow Birdsnest (p.168) are all *saprophytes,* plants without chlorophyll (green colouring matter) which feed on rotting vegetation with the aid of a fungus partner. They are not to be confused with the other group of plants with no chlorophyll, the Broomrapes and Toothworts (p.218), which are parasites on the roots of other plants.

2* Coralroot Orchid

Common Broomrape (p.218)

1* Birdsnest Orchid   **3*** Ghost Orchid   Yellow Birdsnest (p.168)   Toothwort (p.218)

Key to Waterweeds

(based on leaf-shape)

Includes all flowering plants in the book that grow actually in or under the water, but not waterside plants that just "get their feet wet". Those with conspicuous flowers are also included in the Main Key (p.13). The present key is based on leaf-shape, subdivided into floating or emergent and submerged leaves. Aquatic plants can be very confusing because their leaves may be of different shapes, depending on whether they are above the surface or submerged, and if submerged also on the speed of the water in which they grow.

Leaves oval or pointed oval

floating: Floating Water-plantain, Water Chestnut 258; Purslanes 290; Pondweeds 294

submerged: Purslanes 290; Ivy-leaved Duckweed 292; Pondweeds 294

Leaves pointed oblong

floating: Amphibious Bistort 40

submerged: Pondweeds 294

Leaves roundish

floating: Water-lilies 66; Fringed Water-lily 178; Frogbit 258; Duckweeds 292

submerged: Yellow Water-lily 66

Leaves short and narrow

emergent or floating: Blinks 42; Water Tillaea 102; Waterworts, Mudworts 290; Marestail, Water Starwort 292

submerged: Blinks 42; Water Tillaea 102; Waterworts, Mudworts 290, Marestail, Water Starwort, Canadian Waterweed 292; Naiads 294

Leaves quill-like, tufted (usually submerged)

Water Soldier 258: Pipewort 260; Awlwort, Shoreweed, Water Lobelia, Quillworts 290

Leaves long and grass like

floating: Bur-reeds 290

submerged: Ribbon-leaved Water-plantain 258; Bur-reeds 290; Tape-grass 292; Pondweeds, Eel-grasses, Tasselweeds 294; Common Clubrush 296

Compound leaves

emergent or floating: Water Crowfoots 70; Watercress 90; Fool's Watercress 162; Bogbean 72

submerged: Water Crowfoots 70; Watercress 90; Fool's Watercress, River Water Dropwort 162; Water Violet 176; Bladderworts 290; Hornworts, Water Milfoils 292

The waterweeds in the key opposite are those with inconspicuous flowers that are largely or completely submerged. Their text and illustrations are on pages 290-295. Water plants with conspicuous flowers will be found in the main text and in the main key on p.13. The diagram below shows the zonation of water plants from those growing on damp or marshy ground, through those that like to "keep their feet wet" to those that are completely submerged. The order in which plants are shown does not necessarily indicate the precise order in which they will occur at any given pond or lake, nor, of course, is the selection of plants in any way comprehensive.

Plants rooted on bottom with leaves floating

Water-lilies
Broad-leaved Pondweed
Amphibious Bistort
Water Crowfoots
Water Starworts
Water Chestnut
Unbranched Bur-reed

Plants of river or pond banks

Meadowsweet
Common Meadow-rue
Purple Loosestrife
Great Willowherb
Yellow Loosestrife
Orange Balsam
Gipsywort
Water Mint
Marsh Woundwort
Hemp Agrimony

Plants of bare mud at water's edge

Celery-leaved Buttercup
Ivy-leaved Crowfoot
Brooklime, Shoreweed
Water Purslane
Common Water Starwort
Creeping and Marsh Yellowcress
Bur Marigolds, Blinks
Mudwort

Plants growing with their "feet wet"

Watercress, Monkey Flower
Fool's Watercress, Flowering Rush
Yellow Iris, Water Plantains
Water Dock, Arrowhead
Greater Spearwort, Great Yellowcress
Water Speedwell, Great Water Parsnip
Lesser Water Parsnip

Plants rooted on bottom with leaves submerged

Pondweeds
Water Starworts
Water Crowfoots (especially fast streams)
Water Milfoils
Hornwort
Awlwort
Water Lobelia
Pipewort

Plants not rooted on bottom with leaves floating

Frogbit
Duckweeds
Water Soldier (summer)
Bladderworts

Plants not rooted on bottom with leaves submerged

Ivy-leaved Duckweed
Water Soldier (winter)

BLINKS *Montia fontana* (Portulacaceae): see p.42.

1 AWLWORT *Subularia aquatica* (Cruciferae: see p.84). Low annual, usually submerged. Lvs narrow, grass-like, pointed, all in a basal tuft. Fls small, white, with 4 narrow petals, in a *stalked spike;* June-Sept. Fr egg-shaped. Shallow pools in hills and mountains, not on lime. T, northern.

Crassula aquatica, C. vaillantii (Crassulaceae): see p.102.

2* WATERWORT *Elatine alsinastrum* agg. (Elatinaceae). Low/medium annual/perennial, sometimes creeping, often submerged. Lvs narrow oval, linear when submerged, opposite or *in whorls.* Fls tiny, pinkish or whitish, 3- or 4-petalled, *solitary* at base of lvs. In or by shallow, often peaty pools, sometimes drying out in summer. T.

3* WATER PURSLANE *Lythrum portula* (Lythraceae, see p.150). More or less prostrate creeping hairless annual; stems often reddish. Lvs oval, tapering to stalk, opposite, sometimes reddish. Fls small, with 6 pinkish or no petals, *solitary,* unstalked at base of lvs, July-Aug. Fr roundish. In and by shallow water, often drying out in summer, and on damp mud in woods, not on lime. T. **3a,** p.306.

4* HAMPSHIRE PURSLANE *Ludwigia palustris* (Onagraceae, see pp.148-50). Prostrate hairless perennial; stems reddish. Lvs pointed oval, glossy, red-veined, opposite. Fls petalless, green tinged red, unstalked *in pairs* at base of upper lvs; June-July. Fr surmounted by sepals. In and by fresh water. B(rare),F,G.

5* MUDWORT *Limosella aquatica* (Scrophulariaceae, see p.208). Prostrate hairless annual with creeping *runners.* Lvs narrower elliptical, long-stalked, all in a basal rosette. Fls small, pink or whitish, 4-5-petalled, stalked, *solitary,* at base of lvs; June-Sept. Fr oval, on down-turned stalks. Wet mud. T. **5a*,** p.306.

6* GREATER BLADDERWORT *Utricularia vulgaris* agg. (Lentibulariaceae, see p.218). Floating *rootless* insectivorous perennial. Lvs *pinnately* divided into numerous thread-like green segments, with small *bladders* and tiny bristles. Fls *yellow, 2-lipped* with a blunt spur, on lfless stalks above the water; June-Aug. Fr roundish. Still fresh water, usually fairly deep. T. **6a* Intermediate Bladderwort** *U. intermedia* agg. is smaller, with toothed lflets and bladders not on the feathery green lvs but on colourless ones, often sunk in mud; spur of fl pointed. Prefers peaty water. **6b* Lesser Bladderwort** *U. minor* agg.☐ is like 6a but has untoothed lflets, bladders on green lvs as well, and smaller pale yellow fls with a short blunt spur.

7* SHOREWEED *Littorella uniflora* (Plantaginaceae, see p.220). Low hairless perennial, with rooting *runners,* often forming submerged swards. Lvs *semi-cylindrical,* flat on one side, all in a basal tuft. Fls whitish, 4-petalled, solitary, not usually appearing under water; male stalked and with *prominent stamens* separate from unstalked female; June-Aug. In and by shallow fresh water, not on lime. T.

8* WATER LOBELIA *Lobelia dortmanna* (Campanulaceae, see p.226). Hairless perennial with submerged rosettes of untoothed linear lvs. Fls *pale lilac, 2-lipped,* in lfless spikes above surface; July-Aug. Lakes with acid water. T, western.

PIPEWORT *Eriocaulon aquaticum:* see p.260.

9* UNBRANCHED BUR-REED *Sparganium simplex* (Sparganiaceae). Medium hairless perennial. Lvs long, linear, keeled, the numerous floating ones longer and *ribbon-like.* Fls small, green, in *rounded heads,* male and female separate, in an unbranched spike (Cf. Branched Bur-reed, p.296); June-Aug. Fr beaked. In and by shallow fresh water. T. **9a*, 9b*,** p.306.

10* QUILLWORT *Isoetes lacustris* (Isoetaceae). Not a flowering plant, but its submerged tuft of cylindrical pointed dark green lvs could be confused with 1, 5a, 7, 8 or Pipewort (p.260). Spores in capsules at base of lvs. Still water in hill districts. T, northern. **10a* Spring Quillwort** *I. setacea* is smaller with paler green, more often spreading or recurved lvs. Also in lowlands.

1

2

3

4

5

6

6b

7

8

9

10

1* **RIGID HORNWORT** *Ceratophyllum demersum* (Ceratophyllaceae). Submerged rootless perennial. Lvs dark green, rather stiff, *forked*, with linear toothed segments, in whorls. Fls minute, *solitary*, unstalked, with 8 or more linear petals, whitish male and green female at base of separate lf-whorls; July-Sept. Fr a nut with 3 spines. Still or slow fresh water. T. **1a***, p. 306.

2* **SPIKED WATER MILFOIL** *Myriophyllum spicatum* (Haloragidaceae). Submerged rooted perennial. Lvs *pinnate*, feathery, with linear segments, *in whorls* usually of 4. Fls small, pinkish or red, with 4 or no petals, male (anthers yellow) above female, in a *whorled spike* projecting above water; bracts small, undivided; June-Sept. Still or slow, fresh or brackish water. T. **2a* Whorled Water Milfoil** *M. verticillatum* □ is easily told by its conspicuous pinnate bracts; fls greenish-yellow. 2a and 2b are only in fresh water. N.B. Some long trailing Water Crowfoots (p. 70) have rather similar feathery lvs. **2b***, p. 306.

3* **MARESTAIL** *Hippuris vulgaris* (Hippuridaceae). Unbranched perennial, erect but largely submerged in fast streams. Lvs *strap-like*, dark green, *in whorls*, longer and paler green when submerged. Fls petalless, pink, male and female separate, not or very shortly stalked at base of lvs; June-July. Fr a greenish nut. Fresh water. T. Not to be confused with Horsetails (p. 296), which are not flowering plants.

4* **WATER STARWORT** *Callitriche stagnalis* agg. (Callitrichaceae). A difficult group, lf shape varying with presence or depth of water and the ripe fr needed for sure identification often hard to find. Annuals/perennials, submerged, partly submerged or on wet mud. Lvs varying from oval to linear (**4a** □), opposite, often forming a *rosette* when not submerged. Fls tiny, green, petalless, male (anthers yellow) and female separate, solitary or paired at base of lvs; Apr-Sept. Fr 4-lobed. Fresh water, wet mud. T.

5* **CANADIAN WATERWEED** *Elodea canadensis* (Hydrocharitaceae, see p. 258). Submerged perennial. Lvs oblong, blunt, dark green, in overlapping *whorls of 3*. Fls white tinged purple, long-stalked and floating, but infrequent; male (rare) and female on separate plants; May-Sept. Still or slow fresh water. (T) from N America. **5a* Esthwaite Waterweed** *E. nuttallii* □ has much narrower pointed toothed lvs, in whorls of 3-5; only female fls in Europe. Deep fresh water. B (Ireland, rare), F & G, increasing. **5b* Curly Water-thyme** *Lagarosiphon major* □ has curled, bluntly toothed lvs in interrupted whorls; female fls only (rare). (B) from Africa. **5c* Large-flowered Water-thyme** *Egeria densa* □ has lvs up to 5 cm, usually in whorls of 4, and much larger white fls (male only). (B) from S America. **5d***, p. 306.

6* **TAPE-GRASS** *Vallisneria spiralis*. Submerged perennial. Lvs all from roots, long and *ribbon-like*, with 3-5 veins, slightly toothed at tip. Fls small, pinkish-white, male (several in a head) separate from female (solitary, with stalks twisted after flg); June-Oct. Fresh water, usually *artificially heated* canals. (B, F), increasing, from tropics.

Duckweed Family   Lemnaceae

Small floating annuals, with no stems or apparent lvs, but a thallus, which may be a modification of one or both. Fls minute, petalless, rare in B; male and female separate but within same sheath. Still water, often making a complete green carpet. N.B. Water Fern *Azolla filiculoides* (**10***), which appears like a feathery duckweed and turns red in autumn, is a fern, not a flowering plant.

7* **COMMON DUCKWEED** *Lemna minor*. Thallus *rounded*, 1½-4 mm, nearly flat on both sides, 1-5-veined, floating *on surface;* one root. Fls May-July. T. **7a*, 7b***, p. 306.

8* **IVY-LEAVED DUCKWEED** *Lemna trisulca*. Thalli floating just *below surface*, lanceolate, translucent, 7-12 mm, *at right angles* to each other, each with one root; fls May-July. T.

9* **ROOTLESS DUCKWEED** *Wolffia arrhiza*. The smallest European flowering plant, with a rootless egg-shaped thallus ½-1 mm across. Fl not found in N W Europe. B, F, G.

Pondweed Family   Potamogetonaceae

Submerged perennials, a few with floating lvs and emergent fl spikes; lvs alternate. Variable and often hard to identify because lf-shape varies with depth and speed of water. Fls small, green, petal-less, 4-sepalled, in stalked spikes at base of lvs. Fr with 4 nutlets. Usually in still or slow fresh water. Hybrids are frequent. A very simplified treatment is given, mainly based on lf characters.

1* BROAD-LEAVED PONDWEED *Potamogeton natans.* Floating lvs *broad oval,* apparently (unlike 1a-1f) jointed at junction with long stalk; submerged lvs linear. Fls in stout long-stalked emergent spikes; May-Sept. T.   **1a* Bog Pondweed** *P. polygonifolius* has floating lvs often reddish and submerged lvs lanceolate. Often in very shallow water, not on lime.   **1b* Fen Pondweed** *P. coloratus* □ has all lvs broad, translucent, net-veined and often reddish. Usually on lime. **1c*** □, **1d*** □, **1e*, 1f*,** p. 306.

2* SHINING PONDWEED *Potamogeton lucens.* All lvs submerged, *oblong lanceolate,* scarcely stalked, translucent, *wavy-edged,* shiny. Fl stalks thickened at top; June-Sept. Fairly deep water, especially on lime. T.   **2a* Long-stalked Pondweed** *P. praelongus* □ has greener, less wavy, blunter-tipped lvs and longer fl-stalks not thickened at top. Cf. 1d and 1e when without floating lvs.

3* PERFOLIATE PONDWEED *Potamogeton perfoliatus.* Lvs all submerged, dark green, oval, *clasping* the stem. Fl. June-Sept. T.

4* CURLED PONDWEED *Potamogeton crispus.* Stems 4-angled. All lvs submerged, narrow oblong, markedly *wavy,* translucent, finely toothed. Fl spikes small; May-Sept. T.

5* SMALL PONDWEED *Potamogeton berchtoldii.* Stems slightly flattened. All lvs submerged, narrow linear, *grass-like,* 3-veined; June-Sept. Not on lime, unlike 5a-5e. T.   **5a*, 5b*, 5c*** □, **5d*, 5e*, 5f*** □, **5g*,** p. 306.

6* FENNEL PONDWEED *Potamogeton pectinatus.* Lvs all submerged, dark green, narrow, often *rounded* like a bootlace. Fl spikes short; June-Sept. Also in fast fresh, brackish and slightly polluted water; likes lime. T. **6a*,** p. 306.

7* OPPOSITE-LEAVED PONDWEED *Groenlandia densa.* All lvs submerged, *opposite,* narrow oval, unstalked, sometimes very short. Fls in heads; May-Sept. Also in fast and lime-rich water. T.

Eel-Grass Family   Zosteraceae

The only flowering plants that grow completely submerged in the sea.

8* EEL-GRASS *Zostera marina.* Perennial. Lvs more than 20 cm long, *grass-like,* 5-10 mm wide, blunt or pointed at tip, with 3 or more parallel veins. Fls minute, petalless, green, male and female separate, in *branched spikes;* June-Sept. Muddy, sandy and gravelly sea shores, near and below spring low water mark. T.   **8a*, 8b*,** p. 306.

9* SPIRAL TASSELWEED *Ruppia cirrhosa* (Ruppiaceae). Slender submerged perennial. Lvs narrow, *thread-like,* dark green, alternate or opposite. Fls petalless, green, with 2 anthers, rising to surface on long stalks, *in pairs,* the individual stalks much shorter than the common one, which coils in fr; July-Sept. Shallow flowing salt or brackish water. T.   **9a*,** p. 306.

10* HORNED PONDWEED *Zannichellia palustris* (Zannichelliaceae). Slender submerged perennial. Lvs *thread-like,* usually opposite. Fls tiny, petalless, green, male (1 stamen) and female separate, *solitary,* short- or unstalked at base of lvs; May-Aug. Fr beaked. Still or slow fresh or brackish water. T.

11* SLENDER NAIAD *Najas flexilis* (Najadaceae). Slender submerged perennial. Lvs narrow, grass-like, *in whorls* of 2-3, translucent. Fls small, green, 2-lipped, unstalked, *1-3 together* at base of lvs, male and female separate; Aug-Sept. Lakes. B, G, S.   **11a* Holly-leaved Naiad** *N. marina* □ has stems slightly and lvs strongly toothed, and male and female fls on separate plants. Brackish water. B (rare), F, G.   **11b,** p. 306.

A selection of tall waterside species from families mostly not otherwise covered. All T.

Rush Family: Juncaceae.

Typical common waterside species are **1* Soft Rush** *Juncus effusus,* with smooth glossy stems, no proper leaves and compact brownish fl-heads, and **2* Jointed Rush** *J. articulatus* with narrow curved segmented lvs and spreading greenish or brownish fl-heads. N.B. Wood-rushes *Luzula,* growing mainly in drier habitats, are easily told by their long whitish hairs.

Grass Family: Gramineae.

Typical common waterside species are **3* Reed** *Phragmites australis,* forming extensive beds, with stiff stems surviving as hard canes through the winter, broad greyish lvs and spreading purplish fl-heads; and **4* Reed Sweetgrass** *Glyceria maxima,* forming less extensive beds, with stems not overwintering, softer green lvs, sometimes crinkled at tip, and pale green fl-heads.

Arum Family: Araceae.

5* Sweet Flag *Acorus calamus* has lvs like Yellow Iris (p.272), but with one or both edges crinkled, sweet-smelling when crushed, and tiny green fls tightly packed into a phallic spike.

Bur-reed Family: Sparganiaceae.

6* Branched Bur-reed *Sparganium erectum* also has iris-like lvs, but 3-sided near the base, and small greenish fls in unstalked globular heads, arranged in lfy branched spikes. Other bur-reeds, p.290.

Reed-mace Family: Typhaceae.

7* Bulrush or **Cattail** *Typha latifolia* grows in large patches and has long narrow greyish lvs 10-20 mm wide, and chocolate-brown fls tightly packed into a sausage-shaped spike 10-15 cm long. **8* Lesser Bulrush** *T. angustifolia* is less common, and has narrower lvs, *c.* 5 mm wide, and two sausage-shaped spikes because of the gap between the male and female flheads.

Sedge Family: Cyperaceae.

A very large family; four typical waterside species are: **9* Common Club-rush** or **Bulrush** *Scirpus lacustris* has soft rounded rush-like stems, lvs (if present) only floating or submerged and strap-shaped, and fls in red-brown rush-like tufts at top of stem. **10* Lesser Pond Sedge** *Carex acutiformis* has narrow grass-like lvs, greyish when young, blackish catkin-like fl-spikes and conspicuous yellow stamens. **11* Greater Tussock Sedge** *C. paniculata* makes substantial tussocks of narrow rough-edged grass-like lvs; fls greenish in a single tight spike. **12* Bottle Sedge** *C. rostrata* has narrow rough-edged 3-sided lvs and fl-spikes short, sausage-like, stalked, often growing out into shallow lakes.

Horsetail Family: Equisetaceae.

These are not flowering plants; cf Marestail (p.292). **13* Water Horsetail** *Equisetum fluviatile* is the often completely unbranched stem, sometimes also with whorls of branches, that grows in quantity in shallow fresh water, sometimes also in marshes. **14* Marsh Horsetail** *E. palustre,* always branched, and growing only in marshes, differs in having 4-8 instead of 10-30 furrows on the stem and black teeth at the stem joints.

Appendix 1 Additional species

To allow adequate space for description of the illustrated species on certain plates, some of the subsidiary species have been removed to this Appendix. These are mostly non-British species. As with the subsidiary species in the main text, however, descriptions refer back to the main numbered species. This, and where appropriate also the family and genus description, must therefore also be taken into account.

Page 24
3a* Japanese Larch *Larix kaempferi* has twigs often red-brown, lvs bluer green than *L. decidua* and female fls green. (T) from Japan. **3 × 3a*** is also widely planted.
5a* Western Red Cedar *Thuja plicata* has lvs resin-scented, leading shoots erect and larger cones than *Chamaecyparis lawsoniana*. Less often naturalised.
8 Joint Pine *Ephedra distachya* (Gnetales: Ephedraceae)□. Well-branched undershrub with green twigs, appearing lfless because lvs so small. Fls small, yellow-green, often unstalked; male and female fls on different plants; May-June. Fr fleshy, red. Sandy shores, rocks, walls. F, western.

Page 30
4a* Bayberry *Myrica carolinensis* has downy twigs and lvs, catkins with the lvs, and waxy white berry-like fr. (B,G) from N America.
7a *Betula humilis* is taller than *B. nana*, with twigs and narrower lvs often downy. Bogs, fens. G.
8b Green Alder *Alnus viridis* is always a shrub, with narrower lvs than *A. glutinosa*, sticky when young, and fls with lvs. Mountains. F,G.

Page 44
1a *Rumex thyrsiflorus* has lvs 4-14 times as long as wide; fl-head much branched, dense; fls up to six weeks later than *R. acetosa*. Dry places. F,G,S, eastern.
1b* French Sorrel *R. scutatus* is shorter, with lvs often greyish, about as long as broad. Mountain screes in the south, also on walls and in waste places. T, but (B). **1c** *R. arifolius* has lvs about twice as long as wide, the lobes very short; ochreae not fringed. Mountain meadows. F,G,S.
3c* Monk's Rhubarb *Rumex alpinus* is less tall than *R. obtusifolius* with much broader roundish lvs; valves oval, not toothed or swollen. Usually near houses, upland. (B),F,G. **3d* Willow-leaved Dock** *R. triangulivalvis* has narrower pointed lvs, widely branching fl-stems and triangular untoothed valves, all swollen; July/Aug. (T), from N America. **3e* Argentine Dock** *R. frutescens* is short and creeping, with lvs wavy-edged and leathery, and valves triangular, untoothed and swollen; July-Aug. (B,G), from S America. **3f** *R. pseudonatronatus* is like *R. longifolius*, but has much narrower lvs. S(rare), northern.
6b* Patience Dock *Rumex patientia* has more rounded valves than *R. hydrolapathum* with often only one swollen; May-June. (B) from S E Europe.

24/8 Joint Pine

Page 46
2c* Grey Goosefoot *Chenopodium opulifolium* has lvs broader than long though sometimes with small lateral lobes, as in *C. glaucum*. Scattered. **2d* Green Goosefoot** *C. suecicum* is brighter green than *C. album*, the broader lvs with sharp, ascending teeth. B, S. 2a-2d have stems always green and are less common.
6c *Atriplex calotheca* is like *A. hastata*, but with bracts deeply toothed. G, S.
6d *A. oblongifolia* is like *A. patula*, but has narrower lvs and bracts always untoothed. F (rare), G, southern.
9a *Polycnemum arvense* has warty, spirally twisted stems, softer spines and longer bracts than *P. majus*. G.

Page 54
1c *Stellaria longifolia* resembles *S. graminea* with rough stems and fls 2-4 mm; June-July. Damp woods. G, S. **1d** *S. crassifolia* has slightly fleshy lvs, fls 5-8 mm, petals longer than sepals; July. Wet places. G, S, mainly coastal.
1e *S. humifusa* is like 1d, but has smaller lvs and larger fls. S, northern.
2e *Stellaria crassipes* is perennial, with thick unstalked lvs, fls usually solitary and sepal-teeth recurved in fr; June-July. S, northern. **2f** *S. calycantha* is perennial, with square stems, lvs yellowish and sometimes narrower, and smaller fls; June-July. S, northern.
5e *Cerastium dubium* is an annual resembling *C. cerastoides* but larger and with slightly hairy stems. Damp places. F, G, lowland.

Page 56
1a* Pearlwort Spurrey *Spergula morisonii* is more erect than *S. arvensis*, with unchannelled lvs and blunt petals; Apr-June. Bare sandy ground. T, but (B, rare).
1b *S. pentandra* resembles 1a but has pointed, narrower petals and longer, less densely clustered lvs. F, G.
2d *Spergula segetalis* is low, with minute white fls, the petals shorter than the pointed sepals; May-July. Arable fields. F, G. **2e Greek Sea Spurrey** *S. bocconii* has stipules triangular and not silvery and 2 mm white or pink and white fls. Coastal sand and rocks. F, (G). **2f** *S. echinosperma* is like 2c with black seeds. F, G.
7b* Hairy Rupturewort *Herniaria hirsuta* is a slightly hairy perennial; fr blunt, shorter than sepals; June-Aug. Maritime sand and rocks. B (rare), F, south-western.

Page 70
1a *Ranunculus nivalis* is smaller than *R. auricomus*, the 5 petals never distorted. Mountains. S. **1b** *R. sulphureus* is stouter than 1a, with dense brown hairs on sepals. Arctic.
2a *Ranunculus pygmaeus* is very much smaller than *R. sceleratus*, its fr-head not conical. Mountains. S.
5a *Ranunculus aconitifolius* is altogether smaller than *R. platanifolius*, with 3-5-lobed lvs divided to the midrib; fl-stalks hairy. Mountains. F, G.

Page 86
1a *Erysimum crepidifolium* has hairs not flattened, lvs often lobed or toothed, smaller (18 mm) yellow fls than Wallflower and pods 4-angled; June-Sept. Dry grassland, often on lime. G. **1b** *E. odoratum* is like 1a, but taller, with more deeply toothed lvs and fls brighter yellow. F, G, eastern. **1c* Hare's-ear Cabbage** *Conringia orientalis* is like 1b but lvs clasp stem and fls smaller, greenish. Pods to 140 mm. (T), from E Europe.
6a *Erysimum repandum* has fls 10 mm with a longer style than *E. cheiranthoides*, and pods 45-100 mm, rounded. G. **6b** *E. hieracifolium* is biennial/perennial, with fls 10 mm on stalks equalling sepals, and pods 30-35 mm, parallel to stem. Grassland. G.
8a* Austrian Yellowcress *Rorippa austriaca* has lvs clasping stem and fls half the size of *R. amphibia;* pod spherical, shorter-stalked. (B, F, S), G.

Page 98

2d* *Lepidium perfoliatum* has root lvs 2-pinnate, stem lvs rounded and enclosing the stem, pale yellow petals longer than sepals, and round pods. (T) from E Europe. **2e* Tall Pepperwort** *L. graminifolium* is perennial with lvs not clasping, sepals white-edged and pods not notched. (B),F,G.

Page 102

10a *Sempervivum tectorum* has fls flat, the 12 or more petals with smooth edges. (G), scattered.

Page 104

2a* Pyrenean Saxifrage *Saxifraga umbrosa* □ is very tufted, lvs rounded with blunt teeth, on short, hairy stalks. (B,rare). **2b* Kidney Saxifrage** *S. hirsuta* □ has hairy, kidney-shaped lvs on long cylindrical stalks. **2 × 2a* London Pride** *S. × urbium* (B,F), **2 × 2b*** *S. × polita* (Ireland), and **2a × 2b*** *S. × geum* (B,F,G) are all intermediate between the parents. **2c*** *S. rotundifolia* has larger rounder lvs than *S. spathularis* and yellow spots on fls. (B,F),G. **2d* Pick-a-back Plant** *Tolmiea menziesii* □ is taller and has yellowish heart-shaped lvs, with buds at their base in autumn, and red-purple fls with narrow petals. (B), from N America. **2e* Fringe Cups** *Tellima grandiflora* □ is taller still, with darker green, more rounded lvs, and greenish-pink fls with jaggedly toothed petals. (B) from N America.

2b* Kidney Saxifrage

2a* Pyrenean Saxifrage

2d* Pick-a-back Plant

2e* Fringe Cups

3a *Saxifraga tenuis* is slenderer and less hairy than *S. nivalis,* with fls stalked, in a looser head. S, northern.

4a *Saxifraga adscendens* is larger than *S. tridactylites* and not stickily hairy, with all lvs toothed, the lower in a rosette; June-July. Mountains. S. **4b** *S. osloensis* is intermediate between 4 and 4a. S, rare.

11 Hawkweed Saxifrage *Saxifraga hieracifolia* □. Low/short, with a rosette of hairy oval lvs. Fl stem lfless with a spike of small reddish-green fls. Mountains, S.

Page 112

7a *Potentilla supina* does not root at lf-junctions; lvs always green; petals shorter than sepals. Dry grassy places. F,G,eastern. **7b** *P. multifida* □ does not creep and has lvs 2-pinnate with narrow segments, silvery beneath only, and smaller fls than *P. anserina* in loose clusters. S, northern.

104/11 Hawkweed Saxifrage

112/7b *Potentilla multifida*

2b *Pyrus salvifolia* is smaller than *P. pyraster,* with narrower untoothed lvs, grey downy beneath. F,G. **2c* Plymouth Pear** *P. cordata* may be a shrub to 8 m, and has purplish twigs, oval lvs, young lvs and sepals thickly downy, smaller fls, often pink on the back, and rounder fr with no dead sepals. B(rare),F.
3b* Wild Service Tree *Sorbus torminalis* □ has lvs to 90 mm, green on both sides, deeply lobed; fr 12-18 mm long, brown. **3c* Swedish Whitebeam** *S. intermedia* □ is intermediate between 3 and 3a, with pinnately lobed lvs. G,S, and planted T. **3d* Broad-leaved Whitebeam** *S. latifolia* agg.□, intermediate between 3 and 3b, has shallower, more triangular lobes than 3c, fls 20 mm, and an orange or brown fr. Woods, limestone rocks. B,F,G. Several hybrids are known in this group and a tree similar to 3d is often planted and naturalised.
4b *Crataegus calycina* has minutely toothed lf-lobes, narrower sepals than *C. monogyna,* and brighter haws. F,G, eastern, S.
6a* Dwarf Cherry *Prunus cerasus* is a shrub with lvs dark green and glossy above. Hedges. (T), from S W Asia. **6b** *P. fruticosa* is like 6a but shorter and has smaller fls. Scrub. G. **6c St Lucie's Cherry** *P. mahaleb* has smaller fls and black fr. F,G,(S).
7a* Rum Cherry *Prunus serotina* is larger than *P. padus* to 25 m, with shiny lvs to 120 mm; fls 5-10 mm, and fr rather purplish. (T), from N America.
8c *Cotoneaster nebrodensis* has even sepals hairy. F,G,rare.

2b *Chamaecytisus ratisbonensis* has one-sided spikes of red-spotted fls. Dry places. G, eastern.

1b Mountain Zigzag Clover *Trifolium alpestre* differs from *T. medium* in its narrower lflets and stipules, and unstalked flheads. Uplands. F,G, eastern.
2b *Dorycnium pentaphyllum* may have woody stems; keel dark red. G.
4b* Reversed Clover *Trifolium resupinatum* is a hairless annual with fls twisted upside down so that heads appear flatter than *T. fragiferum,* and similarly swollen fr heads. Bare ground; sometimes cultivated. (B,casual),F,G, southern.
5a* Clustered Clover *Trifolium glomeratum* is hairless and often semi-prostrate, and has lflets often with a pale patch, flheads globular, and calyx-teeth turned back in fr. B,F. **5b* Twin-flowered Clover** *T. bocconei* has pale pink fls in often paired but unequal heads. B(rare),F. **5c* Rough Clover** *T. scabrum* is more often prostrate than *T. striatum,* with white fls and calyx-teeth curving outwards in fr; May-June. B,F,G, southern.
7a *Trifolium rubens* is hairless with longer heads of purpler fls than *T. incarnatum.* G.

114/3b* Wild Service Tree

114/3c* Swedish Whitebeam

114/3d* Broad-leaved Whitebeam

Page 132

2b* *Oxalis stricta* differs from *O. europaea* in having tiny stipules and fr stalks bent back. (B,F,G).

3a* *O. corymbosa* is smaller than *O. articulata,* with a cluster of bulbils at the stem base, spots mainly near lf edges and fewer, more purplish-pink fls.

3b* *O. latifolia* is almost hairless, with lvs unspotted but sometimes purplish beneath and pale to deep pink fls.

5b *Tribulus terrestris* (Zygophyllaceae), creeping on dry ground, has pale yellow fls, smaller than *Erodium cicutarium* and spiny star-like fr. F.

7c *Linum viscosum* is downy and has broader lvs than *L. perenne* and pink fls, blue when dry. G, southern. **7d** *L. tenuifolium* has smaller pink or white fls with all sepals pointed. F,G. **7e White Flax** *L. suffruticosum* may be woody at base and has white fls with all sepals pointed. F.

Page 134

5b *Geranium divaricatum* is annual with petals pink, less notched than *G. pyrenaicum* and equalling sepals. G.

Page 136

5a Hairy Spurge *Euphorbia villosa* is taller, with lvs downy on both sides and fr scarcely warty; June-July. F,G, southern. **5b Marsh Spurge** *E. palustris* is taller still than *E. villosa,* with greyish lvs hairless on both sides and fr with short warts only. F,G,S. **5c* Sweet Spurge** *E. dulcis* has lvs almost hairless and glands green, turning purple. (B),F,G. 5a, 5b and 5c all have lvs sometimes finely toothed at tip, bracts green and glands oval. **5d** *E. brittingeri* is woody at base and downier than *E. hyberna,* with lvs all toothed and fr slightly curved. F,G, southern.

Page 138

4a* Chalk Milkwort *Polygala calcarea* has blunt lvs broadest above the middle, the lower stalked and in a rosette; fls bright gentian blue or white; and fr broader than inner sepals; Apr-June. Only on lime. B,F,G, western. **4b Tufted Milkwort** *P. comosa* has lower lvs blunt and fls usually pink. F,G,S, eastern. **4c* Heath Milkwort** *P. serpyllifolia* has at least the lower lvs opposite and more crowded than *P. vulgaris,* and shorter spikes of usually dark blue or dark pink fls. Not on lime. Southern. **4d* Dwarf Milkwort** *P. amarella* has bitter-tasting lvs, the lower in a rosette, the stem ones blunt, broadest near the tip, and fls blue, violet, pink or white, 2-4 mm; May-July. Northern; rare in B.

Page 140

1c *Ribes petraeum* may be taller than *R. rubrum,* with larger wrinkled lvs, pinkish fls and purpler fr. F,G, southern.

5a *Rhamnus saxatilis* has smaller, rounder lvs than *R. catharticus* on short stalks. G, southern, rare.

8a *Myricaria germanica* has longer fatter spikes of larger fls than *Tamarix gallica.* River gravels. F,G,S, eastern.

Page 146

1b White Violet *Viola alba* □ is hairless and has non-rooting runners, more oval lvs persisting through the winter, narrower stipules and larger (15-20 mm) fls than *V. odorata.* F,G, southern. **1c** *V. collina* has no runners, narrower lvs and stipules, and pale blue fls with a whitish spur. On lime. F,G,S, southern. **1d** *V. suavis* has shorter runners, very large (to 200 mm) long-stalked summer lvs and fls violet with white throat. F,(G).

2c *Viola mirabilis* is larger than *V. riviniana* with a line of hairs on stem, untoothed stipules turning brown, and fls 20 mm, pale violet with whitish spur. Woods, on lime. F,G,S.

4b *Viola elatior* is taller than *V. pumila,* with much longer stipules, at least equalling leaf-stalks, heart-shaped lvs and 20-25 mm fls. Damp woods and grassland. F,G,S.

6a *Viola epipsila* is larger than *V. palustris,* with lvs in pairs, some hairs beneath lvs. 15-20 mm fls and bracts above middle of stalk. G,S. **6b** *V. uliginosa* has heart-shaped lvs and larger violet fls. G,S, northern, rare.

1b White Violet

Page 178
1a Jersey Thrift *Armeria arenaria* is taller and stouter than *A. maritima*, with broader 3-5-veined lvs, hairless fl-stalks, fls usually deep pink, sepal-teeth with long bristles, and outer bract enfolding the head. F, G.
4a *Limonium auriculae-ursifolium* is taller and stouter than *L. binervosum*, with broader greyish, often sticky lvs, lf-stalks 5-9-veined and inner bracts red-edged. F.

Page 184
4b Flax Dodder *Cuscuta epilinum* differs from *C. europaea* in its yellowish fls with curved petal-lobes; parasitic on Common Flax (p. 132). G, S, rare.
5a *Polemonium acutiflorum* has bell-shaped fls. On lime. S, northern.
7a Dyer's Woodruff *Asperula tinctoria* has lower lvs usually in whorls of 6, whiter fls than *A. cynanchica* with three petal-lobes, and oval bracts. F, G, S.
8a *Galium glaucum* is taller, stouter, more branched and greyish than *G. odoratum*, with whitish stems, swollen at the joints, narrower lvs with inrolled margins and minute points, and fr rough but not bristly; hybridises with Lady's Bedstraw (p. 186); May-July. Sandy, grassy places. F, G.

Page 188
1e* Dwarf Comfrey *Symphytum grandiflorum* is much shorter than *S. officinale* and creeping, with shorter broader lvs, creamy fls, buds tipped brick-red, and sepal-teeth cut nearly to base; an early (sometimes winter) flowering garden escape. (B), from S W Asia.

Page 190
1b *Pulmonaria angustifolia* is like *P. longifolia*, but has linear-lanceolate lvs 8-10 times as long as broad. G, S, south-eastern. **1c** *P. mollis* is stickily hairy with winged lf stalks. F, G, southern.
3a* Alpine Forgetmenot *Myosotis alpestris* is smaller than *M. sylvatica*, with deeper blue fragrant fls, sepals with few hooked hairs, and fr black on erect stalks with spike scarcely elongating. Mountains. B (rare), F, G.
4c *Myosotis sparsiflora* is shorter than *M. arvensis*, and almost hairless, with broader lvs and fewer fls; June. G, S.
8a *Hackelia deflexa* has larger (5-7 mm) fls and sepal-teeth than *Lappula myosotis* and fr-stalks turned down. G, S.

Page 204
3a* Round-leaved Mint *Mentha rotundifolia* □ is apple-scented and always thickly downy, with rounder lvs and pinker fls than *M. spicata*. B, F, G. The hybrid **3 × 3a*** *M.* × *villosa* var. *alopecuroides* is a large coarse escape from cultivation with sharply toothed rounded lvs and stamens not protruding. (B, F, G).

Page 210
9a *Lindernia pyxidaria* □ has unstalked lvs and tubular white fls, petals shorter than sepals. G, southern.

Page 214
1a* *Veronica longifolia* is taller than *V. spicata* and hairless, with longer, more strongly toothed lvs, and looser, sometimes branched spikes of larger fls with pointed sepals. (B), F, G.
2a *Veronica prostrata* is half-prostrate with lvs less deeply toothed than *V. teucrium* and fls 6-8 mm with hairless sepals. On lime. **2b** *V. austriaca* has short-stalked lvs. G (rare).

204/3a* Round-leaved Mint

210/9a *Lindernia pyxidaria*

3c *Veronica urticaefolia* is like *V. montana* with pale pink fls in longer spikes. Mountain woods, on lime. G.

8a* American Speedwell *Veronica peregrina* has narrowly elliptical, often untoothed lvs, and lilac fls in even lfier spikes. A garden weed. (B,F,G), from N America. **8b* Spring Speedwell** *V. verna* is shorter than *V. arvensis* and has lvs pinnately lobed. Rare in B. **8c* Fingered Speedwell** *V. triphyllos* is half-prostrate, with palmately lobed lvs and long-stalked dark blue fls longer than sepals; Mar-May. Rare in B. **8d* Breckland Speedwell** *V. praecox* has more deeply toothed lvs than *V. arvensis,* often purple below, and petals longer than sepals. T but (B,rare). **8e* French Speedwell** *V. acinifolia* is like 8d but has lvs less toothed and fr clearly 2-lobed. (B,rare),F,G. **8f** *V. dillenii* is like 8b but larger, with fleshy lvs and larger bright blue fls. G.

10c *Veronica anagalloides* has narrower lvs than *V. anagallis-aquatica* and fr egg-shaped. F,G, southern.

Page 218

2b *Orobanche flava* is like 2a, but fls browner and lower lip unequally lobed. On Butterbur (p.238) and Coltsfoot (p.240). F,G. **2c** *O. alsatica* is like *O. elatior* but hairless, with fls lilac-veined. F,G, eastern.

3e *Orobanche gracilis* is like *O. alba,* but fls yellow, red inside, anthers yellow; on peaflowers. F,G. **3f** *O. teucrii* is like 3e, but fls bent almost in a right angle; on gemanders (p.196). **3g* Carrot Broomrape** *O. maritima* has stem and bracts always purple; on Wild Carrot (p.156) by the sea.

4a* Branched Broomrape *Orobanche ramosa* is usually branched and sometimes yellowish, with smaller (10-17 mm) fls than *O. purpurea;* on Hemp (p. 38), Tomato (p.206), Potato, Tobacco and other plants. B(rare),F,G. **4b** *O. arenaria* has fls 25-35 mm. On Field Wormwood (p.238). F,G,rare.

5a *Orobanche amethystea* is reddish-purple, with smaller, much less hairy fls than *O. caryophyllacea;* on Field Eryngo (p.154). F,G, western.

8a Alpine Butterwort *Pinguicula alpina* has larger (8-10 mm) white fls than *P. lusitanica* with two yellow spots in throat and a conical spur. Damp places on mountains. S. **8b** *P. villosa* is smaller than *P. lusitanica* and has brownish lvs and pale violet fls with two yellow spots on lower lip and straight spur. S.

Page 222

2b *Valeriana tripteris* differs from *V. montana* in its greyish lvs, heart-shaped at roots, 3-lobed on stems, and narrower white-edged bracts.

4a *Valerianella coronata* has narrower lvs than *V. locusta,* bluer fls in rounded heads, and fr conspicuously crowned by enlarged sepals. F,G.

10a Blue Honeysuckle *Lonicera caerulea* is less hairy than *L. xylosteum,* with blunter lvs, bell-shaped fls and black berries fused together. Mountains. G,S. **10b Black-berried Honeysuckle** *L. nigra* is hairless with long-stalked pinkish-white fls and a distinct pair of black berries. F,G. **10c** *L. alpigena* has fls brownish and red berries fused. Mountains. G, southern.

Page 226

1a *Campanula uniflora* is shorter than *C. rotundifolia* with lower lvs oblong, and smaller (5-10 mm) solitary nodding fls. Mountains, arctic heaths. S. **1b** *C. scheuchzeri* has unstalked lvs. G,S. **1c** *C. cochlearifolia* is smaller than *C. rotundifolia* and white with hairs below. F,G. **1d** *C. rhomboidalis* with oval stem lvs and **1e** *C. baumgartenii* with lanceolate stem lvs, both have fl stalks longer than fls. F,G. 1b-1e are all rare in mountains.

Page 230

9a* Rough Cocklebur *Xanthium strumarium* is greyish with roughly hairy lvs broader than *X. spinosum.* **9b* Stinking Cocklebur** *X. echinatum* is yellowish-green and strong-smelling, the lvs covered with short stiff hair and yellow dots, and male flhds conical. 9a and 9b both have lvs spineless, less lobed, toothed and not white beneath.

Page 232
1b *Bellidiastrum michelii* is taller and stouter than *Bellis perennis,* with larger fls (to 40 mm), the rays either all white or pink; Mar-Sept. Mountains. G, southern.
4a *Aster amellus* is roughly hairy and not fleshy, and has fls always rayed, solitary on each branch of the cluster, and bracts curved and coloured at the edges. Dry grassy places. F,G, southern.
6a *Stenactis annua* has fls 15-25 mm, with longer spreading white rays than *Erigeron acer.* Lvs toothed. **6b** *S. strigosa* has rays white or purple and much narrower lvs than *Erigeron acer.* Both 6a and 6b are found in damp and waste places. (F,G), from N America.

Page 238
1e* *Petasites japonicus* has smaller lvs and creamy flhds whose large pale green bracts make the whole spike resemble a small cauliflower. (B), from E Asia.
2a *Achillea nobilis* has lvs flat, not feathery, and pale green with winged stalks, and rays very pale yellow; July-Aug. Bare ground. F,G. **2b** *A. collina* has narrower, leathery lvs. **2c** *A. pannonica* is woolly. **2d** *A. setacea* has weak stems and crowded, pointed lflets. 2b-2d are all G, south-eastern.
4d *Artemisia pontica* is taller than *A. vulgaris* and shrubby, with lvs downy white on both sides and pale yellow flhds. G. **4e Tarragon** *A. dracunculus,* the culinary herb, is very aromatic and has green linear-lanceolate lvs, the lower 3-lobed at tip, and greenish flhds. (F,G), from Asia. **4f** *A. rupestris* is shorter than *A. vulgaris* and woody at base, with hairless green lvs. S. **4g** *A. austriaca* is like 4d with lvs silky below and fls sometimes red. G, eastern.
4h *A. laciniata* is like *A. campestris,* but stem hairier and often purplish, and segments of lower lvs blunter. S (Oland).

Page 252
6a Pliant Lettuce *Lactuca viminea*□ has lower lvs deeply pinnately cut with linear lobes, and unstalked flhds often in threes in a branched spike. F,G. **6b** *L. quercina* □ has a weak green stem, lvs varying from toothed to pinnate, and fr beak black. G,S(rare).

252/6a

252/6b

254/1b

254/5a

Page 254
1b *Aposeris foetida* □ is stouter than *Taraxacum* Sect. *Vulgaria* with foetid juice, sparser florets and no pappus. G, south-eastern.
5a *Leontodon hyoseroides* □ has lvs deeply pinnately lobed, all lobes long and narrow. F,G. **5b** *L. incanus,* grey-hairy, and **5c** *L. helveticus,* with clearly stalked lvs, are both rare in mountains. G, southern.
9a* *Pilosella lactucella* agg. is similar to *P. officinarum* but with flhds 2-5 together in a cluster, like Orange Hawkweed (p.256). T, but (B).

Page 264
6a* Onion *Allium cepa* is taller and stouter than *A. fistulosum,* with greenish-white fls on a stem swollen and hollow below middle.
8b* Cultivated Leek *Allium porrum* has broader lvs than *A. ampeloprasum,* white fls, reddish anthers and a persistent green bract. Escape from cultivation. (T).
9a *Allium angulosum* has keeled lvs. Marshy places. G. **9b** *A. suaveolens* and **9c** *A. strictum* have bright purple fls and round stems; 9b in damp meadows, 9c on rocks. G(rare). **9d** *A. victorialis* has broader lvs than 9c and yellowish-white fls. G, southern, rare.

Page 290

3a *Lythrum borysthenicum* is usually hairy and erect, with lvs unstalked and sometimes alternate. F.

5a* Welsh Mudwort *Limosella australis* has smaller, more linear lvs than *L. aquatica,* larger fls, white with an orange tube, and rounder fr. Near the sea. B (Wales, rare).

9a* Floating Bur-reed *Sparganium angustifolium* is smaller than *S. simplex,* with all lvs floating and not keeled, their sheaths inflated, and fewer flhds, the lflike bract of the lowest well overtopping the top head; Aug-Sept. Peaty pools. **9b* Least Bur-reed** *S. minimum* is smaller than 9a, with still narrower lvs, some not floating, and still fewer flhds, the lowest bract scarcely overtopping the top head; June-July. Still water.

Page 292

1a* Soft Hornwort *Ceratophyllum submersum* has laxer, paler green lvs than *C. demersum* and fr with 1 or no spines. Less frequent, often in brackish water. N.B. Some species of Stonewort (Characeae) could be confused with Hornworts, but they are flowerless plants, Algae, recognisable by their grey or translucent green colour, and by their foetid smell.

2b* Alternate Water Milfoil *Myriophyllum alterniflorum* is slenderer than *M. spicatum* and has red-streaked yellow fls in a shorter spike, the upper solitary or paired. Not on lime.

5d* Greater Water-thyme *Elodea callitrichoides* is larger than *E. canadensis,* with longer, paler green lvs, sometimes in pairs, and female fls only. (B) from S America.

7a* Greater Duckweed *Lemna polyrrhiza* has thalli 5-8 mm, 5-11-veined, often purplish beneath, with a tuft of roots. **7b* Fat Duckweed** *L. gibba* has thalli 3-5 mm and markedly swollen. Also in brackish and polluted water.

Page 294

1c* Loddon Pondweed *Potamogeton nodosus* (□ on p.295) has all lvs broad, the submerged ones narrower and net-veined. B,F,G. **1d* Various-leaved Pondweed** *P. gramineus* (□ on p.295) may have floating lvs like *P. polygonifolius,* but smaller, thinner and blunter; submerged lvs narrow, wavy-edged, unstalked; fl stalks thickened at top. Not on lime. **1e* Reddish Pondweed** *P. alpinus* may have narrow elliptical net-veined floating lvs; submerged lvs similar but unstalked, often reddish brown. Not on lime. **1f* American Pondweed** *P. epihydrus* has flattened stems, narrow elliptical floating lvs and linear unstalked submerged lvs. B(rare).

5a* Lesser Pondweed *Potamogeton pusillus* is very similar to *P. berchtoldii* but prefers lime, also in brackish water. **5b* Hair-like Pondweed** *P. trichoides* has 1-veined lvs. **5c* Shetland Pondweed** *P. rutilus* (□ on p.295) has more finely pointed lvs. Rare in B. **5d* Flat-stalked Pondweed** *P. friesii* has slightly broader 5-veined lvs. **5e* Sharp-leaved Pondweed** *P. acutifolius* has sharply pointed lvs twice as broad as *P. berchtoldii,* with 3 main and many smaller veins; fl stalks very short. **5f* Blunt-leaved Pondweed** *P. obtusifolius* (□ on p.295) has bluntly pointed lvs twice as broad as *P. berchtoldii* and rather short fl-stalks. **5g* Grasswrack Pondweed** *P. compressus* has blunt 5-veined lvs twice as broad as *P. berchtoldii* and much longer fl stalks. 5b-5e are in still water only.

6a* Slender-leaved Pondweed *Potamogeton filiformis* is smaller than *P. pectinatus* with flattened stems and long-stalked fl spikes, the individual whorls widely spaced.

8a* Narrow-leaved Eel-grass *Zostera angustifolia* has lvs 15-30 cm long and 2-3 mm wide, 1-3-veined and notched at tip when mature. Muddy shores, mainly from half to low tide mark. B,G,S. **8b* Dwarf Eel-grass** *Z. noltii* is like *Z. angustifolia* but with 1-veined lvs less than 20 cm long and 1 mm wide, and fl-spikes unbranched.

9a* Beaked Tasselweed *Ruppia maritima* has paler green lvs, longer fl stalks and the common stalk not coiling. Still water.

11b *Najas minor* is like *N. marina,* but has lvs narrower and less sharply toothed and male and female fls on same plant. F,G.

Appendix 2 **Introduced Shrubs**

So many ornamental or cultivated shrubs have been introduced to Europe during the past three or four centuries and have been self-sown, bird-sown or otherwise naturalised that it is not possible to deal with them in as great detail as the native trees or shrubs. Many will anyway be familiar from gardens. The following list gives brief details of the most important.

As elsewhere in the book, an asterisk indicates a plant to be found wild in Britain.

*Fig *Ficus carica* (Moraceae). Lvs palmately lobed. Fls small, green. Fr pear-shaped.

*Shrubby Orache *Atriplex halimus* (Chenopodiaceae). Lvs broad, silvery. Fls small, green.

*Flowering Currant *Ribes sanguineum* (Grossulariaceae). Lvs palmately lobed. Fls reddish-pink. Fr blue-black.

*Escallonia *Escallonia macrantha* (Escalloniaceae). Lvs oval. Fls pink.

*Pittosporum *Pittosporum crassifolium* (Pittosporaceae). Lvs oval, evergreen. Fls purple-black.

Sorbaria *Sorbaria sorbifolia* (Rosaceae). Lvs pinnate. Fls white.

*Physocarpus *Physocarpus opulifolius* (Rosaceae). Lvs roundish. Fls white.

*Spiraea *Spiraea salicifolia* agg. (Rosaceae). Lvs elliptical. Fls pink.

*Purple-flowering Raspberry *Rubus odoratus* (Rosaceae). Lvs palmately lobed. Fls pink.

*Wineberry *R. phoenicolasius.* Lvs trefoil. Fls pink.

*Rubus *R. spectabilis.* Lvs trefoil. Fls bright pink.

Roses **Rosa sempervirens, R. foetida, *R. virginiana, *R. rugosa* (Rosaceae). Lvs pinnate. Fls pink.

Quince *Cydonia oblonga* (Rosaceae). Lvs pointed oval. Fls pink.

*Juneberry *Amelanchier grandiflora* (Rosaceae). Like *A. ovalis* (p.114), but young lvs purplish and fl spikes drooping. *A. spicata* has lvs thickly white-downy when young.

*Himalayan Cotoneaster *Cotoneaster simonsii* (Rosaceae). Like *C. integerrimus* (p.114), but reaches 4 m with most twigs downy and lvs grey-green with fewer shorter hairs beneath.

*Wall Cotoneaster *C. horizontalis* is like *C. microphyllus* (p.114) but lvs smaller and branches spreading horizontally.

*Firethorn *Pyracantha coccinea* (Rosaceae). Spiny. Lvs elliptical. Fls white.

*Medlar *Mespilus germanica* (Rosaceae). Sometimes a small tree and spiny. Lvs lanceolate. Fls white. Fr brown.

Choke Cherry *Prunus virginiana* (Rosaceae). Like *P. serotina* (p.114), but usually a shrub, with matt lvs and dark red fr.

*Cherry Laurel *P. laurocerasus.* Lvs oblong, dark green, glossy. Fls in spikes, white.

Aronia *Aronia melanocarpa* (Rosaceae). Lvs pointed oval, dark glossy green. Fls white. Fr black.

Box Elder *Acer negundo* (Aceraceae). Lvs trefoil or pinnate. Fls green, in hanging clusters, before lvs.

*Vine *Vitis vinifera* (Vitaceae). Woody climber. Lvs palmately lobed. Fls green.

Virginia Creepers **Parthenocissus quinquefolia, P. inserta* (Vitaceae). Woody climbers. Lvs with 5 elliptical lflets. Fls green.

*Fuchsia *Fuchsia magellanica* (Onagraceae). Lvs oval, toothed. Fls bell-shaped, purple and red.

*Rhododendron *Rhododendron ponticum* (□ on p.170) (Ericaceae). Lvs lanceolate, dark green. Fls large, purple.

*Yellow Rhododendron *R. luteum.* Similar but fls yellow.

*Shallon *Gaultheria shallon* (Ericaceae). Carpeting evergreen undershrub. Lvs broad, leathery. Fls pink, bell-shaped. Fr black.

*Prickly Heath *Pernettya mucronata* (Ericaceae). Suckers. Lvs toothed, sharp-tipped. Fls pink, bell-shaped. Fr pink.

*Lilac *Syringa vulgaris* (Oleaceae). Lvs pointed oval. Fls in spikes, lilac or white.
*Butterfly Bush *Buddleja davidii* (Buddlejaceae). Lvs lanceolate. Fls mauve, in long dense spikes.
*Snowberry *Symphoricarpos rivularis* (Caprifoliaceae). Suckering. Lvs rounded or lobed. Fls small, pink. Fr large, white.
*Flowering Nutmeg *Leycesteria formosa* (Caprifoliaceae). Lvs ace-of-spades. Bracts lf-like. Fls purplish.

Appendix 3   **Aggregates**

Where the abbreviation "agg." follows a scientific name, several closely similar species or microspecies are included under the one heading, as follows:

Salix caprea (p.26) inc. *S. coaetanea.*
S. cinerea inc. *S. atrocinerea.*
S. nigricans inc. *S. borealis.*
S. phylicifolia inc. *S. bicolor, S. hibernica.*
S. starkeana inc. *S. xerophila.*
S. glauca inc. *S. stipulifera.*
S. lanata inc. *S. glandulifera.*
S. repens inc. *S. arenaria, S. rosmarinifolia.*
Polygonum minus (p.40) inc. *P. foliosum.*
Atriplex hastata (p.46) inc. *A. glabriuscula, A. longipes.*
Salicornia europaea (p.48): too numerous to list.
Sagina intermedia (p.56) inc. *S. caespitosa.*
Ranunculus montanus (p.68) inc. *R. oreophilus.*
Fumaria officinalis (p.78) inc. *F. rostellata.*
F. vaillantii inc. *F. schrammii.*
Papaver radicatum (p.80) inc. *P. dahlianum, P. laestadianum, P. lapponicum.*
Draba alpina (p.84) inc. *D. glacialis.*
Biscutella laevigata inc. *B. apricorum, B. divoniensis, B. guilloni, B. neustriaca,* all very local in F.
Cardamine pratensis (p.90) inc. *C. nymanii, C. palustris, C. matthiolii.*
Cakile maritima (p.92) inc. *C. edentula.*
Cochlearia officinalis (p.92) inc. *C. pyrenaica, C. alpina, C. micacea, C. scotica.*
Arabis hirsuta (p.94) inc. *A. brownii.*
Draba norvegica (p.96) inc. *D. cacuminum.*
D. daurica inc. *D. cinerea.*
Saxifraga rosacea (p.104) inc. *S. hartii.*
Alchemilla vulgaris (p.106): too numerous to list.
Rosa canina (p.108): too numerous to list.
R. tomentosa inc. *R. mollis, R. scabriuscula, R. sherardii, R. villosa.*
Rubus fruticosus (p.108): too numerous to list.
Potentilla argentea (p.112) inc. *P. neglecta.*
P. nivea inc. *P. chamissonis.*
Sorbus aria (p.114): too numerous to list.
S. latifolia: too numerous to list.
Linum perenne (p.132) inc. *L. austriacum, L. leonii.*
Geranium bohemicum (p.134) inc. *G. lanuginosum.*
Ribes rubrum (p.140) inc. *R. spicatum.*
Hypericum perforatum (p.144) inc. *H. elegans.*
Oenothera biennis (p.148) inc. *O. chicagoensis, O. rubricaulis, O. strigosa, O. suaveolens.*
O. parviflora inc. *O. ammophila.*
Epilobium montanum (p.150) inc. *E. duriaei.*
Anthriscus sylvestris (p.156) inc. *A. nitida.*
Monotropa hypopitys (p.168) inc. *M. hypophegea.*
Empetrum nigrum (p.172) inc. *E. hermaphroditum.*
Limonium binervosum (p.178) inc. *L. paradoxum, L. recurvum, L. transwallianum.*
Gentianella campestris (p.182) inc. *G. baltica.*

Myosotis caespitosa (p.188) inc. *M. laxa.*
Lamium amplexicaule (p.200) inc. *L. moluccellifolium.*
Galeopsis tetrahit (p.200) inc. *G. bifida.*
G. angustifolia inc. *G. ladanum.*
Thymus serpyllum (p.204): too numerous to list.
Euphrasia officinalis (p.216): too numerous to list.
Rhinanthus minor (p.216): too numerous to list.
Valeriana officinalis (p.222) inc. *V. sambucifolia.*
Valerianella locusta (p.222) inc. *V. carinata, V. dentata, V. eriocarpa, V. rimosa.*
Artemisia campestris (p.238) inc. *A. borealis.*
Taraxacum spp. (p.254): too numerous to list.
Pilosella spp. (p.254): too numerous to list.
Hieracium spp. (p.256): too numerous to list.
Elatine alsinastrum (p.290) inc. *E. hexandra, E. hydropiper, E. triandra.*
Utricularia vulgaris (p.290) inc. *U. neglecta.*
U. intermedia inc. *U. ochroleuca.*
U. minor inc. *U. bremii.*
Callitriche stagnalis (p.292) inc. *C. hermaphroditica, C. truncata, C. intermedia,*
  C. obtusangula, C. platycarpa, C. palustris.
Spiraea salicifolia (App.2) inc. *S. alba, S. douglasii, S. tomentosa.*

Plant Ecology

Anyone who becomes seriously interested in identifying plants will soon notice that most species are restricted to a particular type of habitat, and will want to know more about the reasons for their distribution. We have here tried to indicate briefly the factors that control plant distribution, and to give some idea of the sorts of plants found in different habitats.

Climate determines the overall distribution of plants, that is the total geographical range of a species. Within that range a species may favour, for example, a particular soil type or may grow only in wet places.

The major climatic factors are **temperature** and **rainfall. Temperature** splits areas up into northern and southern parts, though in the south of an area the mountains, with a northern climate, often harbour many northern species. Thus Mountain Avens *Dryas octopetala*, a mountain plant in the Alps and Central Scotland, is found at sea-level in North Scotland and Scandinavia.

Rainfall divides our area into eastern and western parts, owing to the prevailing wet westerly winds, so that Spring Speedwell *Veronica verna*, for example, is not found in most of northern and western Europe, the wettest part, and in Britain occurs only in East Anglia, the part with the most continental climate and flora and only 60 cm (25 inches) of rain a year. Conversely the Pale Butterwort *Pinguicula lusitanica*, is found only in wet places on the Atlantic seaboard of Europe.

Soil. Within the range defined by climate, many plants are confined to a particular soil. The plants of salty soils are a striking example, confined to the sea-shore, except for occasional patches of salty ground inland, as for example Lesser Sea Spurrey *Spergularia marina*. Plants confined to limestone and chalk soils are perhaps the biggest such group, including many orchids, such as the striking Man Orchid *Aceras anthropomorphum*. The distribution between lime- and acid-soil loving plants is so important that on the map on page 6 we have shown the major areas of limestone soils in Europe.

Soils that are usually lime-free and hence acid are formed on sandstone, producing well-drained soils, and on clays, where they are sticky and wet. In many wet districts the dead plants decay very slowly and their remains form peat, which is usually acid, with a characteristic flora of plants such as Common Sundew *Drosera rotundifolia*. In fens alkaline peat is found, with a different flora.

Dependent largely on these factors, therefore, characteristic habitats are found, each dominated by different plants. In the lowlands, forest normally forms unless controlled by man, when various types of grassland commonly result. Natural grassland in our area is found on salt-marshes and mountains, and on mountain tops a form of tundra is found.

Woodland formed of deciduous trees has a flora of plants that come into leaf and flower early and catch the light before the leaves on the trees cut it out. Plants such as Early Purple Orchid *Orchis mascula* and Bluebell *Endymion non-scriptus* are characteristic. Most of the low-lying part of the area covered in this book would naturally be covered by oakwood. Beech trees often intercept so much light that in some beechwoods no plants can survive except those such as the Birdsnest Orchid *Neottia nidus-avis*, living on dead leaves. On the Continent beechwood is found mainly in the hills and has a richer flora, but in Britain it is on the edge of its range and is found mainly in the south.

Pinewoods have leaves on the trees all year, but natural pinewoods have well-scattered trees and ample light, and seem very different from the sombre plantations. The flora normally comprises dwarf shrubs, mainly in the heather family (Ericaceae), of which the commonest is Heather *Calluna vulgaris*. When the trees are cleared, the same areas form the vast tracts of **heather moorland**, so characteristic of the British hills. These moors have very few plants in their flora.

Grasslands have very different flowers depending on whether they are on acid, neutral, or lime-rich soils, and whether they are cut for hay or grazed. Chalk and limestone grassland have especially rich floras, with many rare or decreasing plants. In grazed pastures the plants are often low-growing, with their buds low down and protected. Hay-meadows, usually neutral grassland, contain taller plants, flowering in spring and early summer, but they too have their buds low down.

Salt-marshes formed in estuaries are often grass-covered, but they have a very distinctive flora, including many members of the Goosefoot family. Many of the plants, such as the Glassworts *Salicornia*, are succulent and flower late in the year. Succulence enables them to counteract the salt in the soil.

Similarly the plants of other coastal habitats are often succulent. The floras of **shingle, sandy, and rocky shores** are all quite different, and *Glaucium flavum*, the Yellow Horned Poppy, is a typical plant of shingle banks. Behind the shore, dunes of lime-rich shell sand are often found and these have a flora somewhat resembling that of chalk grassland. Sea-shore plants are often covered with a thick greyish layer of protective wax, to prevent them drying out in their very exposed habitat. For the same reason they are also often cushion-shaped, and these characteristics are shared by many plants of mountain tops, another exposed habitat. Indeed several plants, such as Thrift *Armeria maritima* are found both on sea-cliffs and on mountain tops.

In addition there are many other habitats, some of which are widespread, such as those of **freshwater** or those, such as **arable fields,** that are directly the result of human activity. These habitats have a flora of annual plants, but many of the typical cornfield weeds such as Cornflower *Centaurea cyanus* are now scarce owing to the widespread use of modern herbicides.

Finally one can mention the more specialised habitats which have a particular fascination for botanists as they are the home of many rare and beautiful plants. These are the **limestone rocks** and inland cliffs, especially in the mountains; these habitats are protected from grazing, and often from collectors too. In addition to the mountain plants such as Alpine Gentian *Gentiana nivalis*, many lowland woodland plants also occur here.

The following lists are intended to give an impression of the sorts of plants to be found in certain typical habitats. They are based on a grid system so that reading across the page increasingly wet soils are found, while down the page the lists proceed from the pioneer species of open ground to the plants associated with the climax woodland community. They are divided firstly by the different soils – chalk and limestone, neutral soils (clays, for example), and acid sandy and peaty soils. In addition a list for salty soils by the sea, divided on the distance from the sea and whether the ground is muddy, sandy or rocky is included.

By identifying first the soil type and then finding the correct position on the grid in terms of wetness and the stage of development of the community (open, scrub or wooded) the typical species of that habitat can be located. Few localities will contain all of those mentioned, nor is the list exhaustive, but they may be regarded as indicator species, suggesting the others that are likely to be found as well.

Reading down the columns, as the habitats become more wooded, groups of species are separated by gaps. Species from one stage may often survive through to later stages, but the later plants are unlikely to appear earlier. In addition species in one column are likely to appear in the immediately adjacent column.

Calcareous Soils

	Dry *Rocks*	**Moister** *Grassland*	**Wet** *Fen*
Open	Sticky catchfly Whitlow grasses Rock stonecrop Buckler mustard Wallflower Dark red helleborine Pellitory of the wall Spring sandwort Alpine penny-cress Spiked speedwell Hairy rock-cress Stone bramble	Wild candytuft Autumn gentian Rockroses Clustered bellflower Common spotted orchid Pyramidal orchid Fragrant orchid Bee orchid Salad burnet Horseshoe vetch Fairy flax Hairy violet Greater hawkbit Dwarf thistle	Marsh pea Marsh valerian Marsh helleborine Early marsh orchid Milk parsley Common comfrey Marsh cinquefoil Greater spearwort Fen orchid Marsh bedstraw Yellow loosestrife Cabbage thistle Water chickweed
Scrub	Traveller's joy Hazel Blackthorn	Hawthorn Dogwood Wayfaring tree Wild privet Spindle tree Sweet briar	Alder buckthorn Buckthorn Grey willow Goat willow Guelder rose Blackcurrant
Wooded	Whitebeams Yew Wych elm	Beech Ash Lime Crab apple Wild cherry Field maple	Alder Birches Hawthorn

Understory	Herb robert
	Mountain
	  St. John's wort
	Wall lettuce
	Jacob's ladder
	Enchanter's
	  nightshade
	Herb bennet
	Wild strawberry
	Broad-leaved
	  willowherb
		Spurge laurel
		Dog's mercury
		Sanicle
		Woodruff
		Wood dog violet
		Fly orchid
		Birdsnest orchid
		White helleborine
		Hellebores
		Alpine ragwort
		Wood sorrel
			Hedge bindweed
			Tufted vetch
			Yellow iris
			Dewberry
			Comfrey
			Hop
			Hemp agrimony

Neutral Soils

	Dry *Waste Ground*	**Moister** *Grassland*	**Wet** *Marsh*

Open	Red dead-nettle
	Broad-leaved dock
	Common field
	  speedwell
	Groundsel
	Shepherd's purse
	Dandelions
	Common stitchwort
	Poppies
	Fumitories
	Charlock
	Hoary cinquefoil
		Birdsfoot trefoil
		Great burnet
		Pepper saxifrage
		Dyer's greenweed
		Saw-wort
		Greater butterfly
		  orchid
		Common mouse-ear
		Knapweeds
		Meadow buttercup
		Ribwort plantain
		Yarrow
			Broad-leaved
			  pondweed
			Water-lilies
			Amphibious bistort
			
			Marsh marigold
			Branched bur-reed
			Yellow iris
			Purple loosestrife
			Meadowsweet
			Water forgetmenots
			Lesser spearwort
			Great willow-herb

Scrub	Goat willow Elder	Hazel Midland hawthorn Holly Blackthorn	Goat willow Grey willow
Wood	Common oak	Durmast oak Wych elm Wild service tree Field maple	Black poplar Crack willow White willow Alder

Understory	Nettle (and see
	  next column)
	Barren strawberry
	Bramble
	Rosebay willowherb
		Bugle, Wood anemone
		Hepatica, Ramsons
		Primrose, Common
		  dog violet
		Yellow pimpernel
		Red campion
		Early purple orchid
		Wolfsbane
		Purple lettuce
			Cuckoo flower
			Ragged robin
			Water avens
			Creeping buttercup
			Bugle
			Hedge woundwort
			Creeping jenny
			Marsh thistle

Acid soils

	Dry *Heath*	**Moister** *Moor*	**Wet** *Bog*

Open	Spring beauty
	Sand spurrey
	Changing
	  forgetmenot
	Annual knawel
	Cudweeds
	
	Thyme
	Harebell
	Petty whin
	Maiden pink
	Tormentil
	Heath milkwort
	Heath dog violet
		Shepherd's cress
		Sheep's sorrel
		
		Lousewort
		Heath spotted orchid
		Lesser butterfly
		  orchid
		Coralroot orchid
		Foxgloves
		Heath bedstraw
		Dwarf cornel
			Bog pondweed
			Bogbean
			
			Marsh asphodel
			Sundews
			Grass of Parnassus
			Butterworts
			Bog orchids
			Marsh lousewort
			Marsh pennywort
			Marsh St John's wort
			Marsh violet

Dwarf Shrub	Bell heather Gorse Broom Bilberry	Heather Bearberry Crowberry	Cross-leaved heath Bog rosemary Cranberry Bog myrtle Creeping willow

	Dry *Heath*	**Moister** *Moor*	**Wet** *Bog*
Scrub	Rowan Downy rose	Juniper Aspen	Gorse
Wood	Birches Durmast oak	Scots pine	Norway spruce Birch

Understory	Common cow-wheat
	Lesser stitchwort
	Trailing St John's
	  wort
	Lesser periwinkle
	Climbing corydalis
		Wintergreens
		Twinflower
		Bitter vetch
		Cowberry
		Creeping lady's
		  tresses
			Labrador tea
			Chickweed
			  wintergreen

Marine

	Muddy *Dry saltmarsh*	**Sandy** *Fixed dunes*	**Rocky** *Cliffs*
Upper Shore	Grass-leaved orache Sea plantain Sea aster Sea arrow-grass Sea wormwood Scurvy-grasses Marsh mallow Wild celery Sea clover Dittander	Sea buckthorn Burnet rose Bloody cranesbill Purple milk-vetch Yellow-wort Wild pansy Centauries Crow garlic Common rest-harrow	Thrift Rock samphire Squills Goldilocks aster Rock spurrey English stonecrop Rock sea-lavender Wild cabbage Stocks

	Lower saltmarsh
	Sea milkwort
	Common sea
	  lavender
	Sea purslane
	Common seablite
	Sea spurreys
		Slack
		Creeping willow
		Marsh helleborine
		Yellow iris
		Bog pimpernel
		Marsh pennywort
		Dewberry
		
		Mobile dune
		Houndstongue
		Sea bindweed
		Viper's bugloss
		Ragwort
		Sea spurge
		Common whitlow-
		  grass
		Common storksbill
		Rue leaved saxifrage
			Fixed shingle
			Tree mallow
			Nottingham catchfly
			Sea campion
			Curled dock
			Goosegrass
			Wall-pepper
			Lady's fingers
			Yellow vetch
			Bittersweet
			Herb robert

Lower Shore	*Mud* Glassworts Eelgrasses	*Foreshore* Frosted orache Sea rocket Saltwort Sea beet Sea sandwort	*Shingle banks* Sea kale Yellow horned poppy Sea holly Sea mayweed Sea pea Shrubby seablite Oyster plant

Glossary

Alternate

Anther

Some Basic Terms to Learn

The following botanical terms are used very frequently in the text, since it is not possible to "translate" them. If they are unfamiliar to you, it will simplify your use of this book to memorise them:

Anther, Bract, Petal, Pinnate, Sepal, Stamen, Stigma, Stipule, Style.

Appressed

Acid soils have very few basic minerals and are formed on rocks such as sandstone. Peaty soils are usually acid since plant humus is often so. See also Ecology (p. 309)

Aggregate: group of closely related species, often microspecies (q.v.): see Appendix 3 (p. 308).

Alternate: neither opposite nor whorled.

Anther: the part of the flower producing the male pollen.

Annual plants live for a year or less. They are usually shallow rooted and never woody.

Appressed: flattened against the stem.

Bracts

Bulbil

Berry: fleshy fruit.

Biennials: live for two years. Usually the first year's growth produces a leaf-rosette, the second the flowers.

Bog: a habitat on wet, acid peat.

Bracts are small, usually leaf-like organs just below the flowers, and sometimes, as in Daisies, numerous and overlapping.

Bulbs are underground storage organs, composed of fleshy leaves.

Bulbils are small bulb-like organs at the base of the leaves or in place of the flowers, breaking off to form new plants.

Calyx

Catkins

Calyx refers to the sepals as a whole, usually used when they are joined.

Casual: plant appearing irregularly, without fixed localities.

Catkins are hanging flower-spikes, the individual flowers of which are usually rather inconspicuous.

Cluster: loose group of flowers.

Composite: member of the Daisy Family, Compositae.

Corms are bulb-like underground storage organs, comprising a swollen stem.

Coppice: trees or shrubs cut to the ground and growing from the old stools.

Corolla refers to the petals as a whole, usually when they are joined.

Crests: see p. 272.

Crucifer: member of Cabbage Family, Cruciferae.

Cluster

Composite

Deciduous: with leaves falling in autumn.

Deflexed: bent downwards.

Disc Florets: see p. 230.

Corolla

Dunes are areas of wind-blown, usually lime-rich shell sand near the sea, with intervening damp hollows, termed *slacks*. See also Ecology (p.309).

Epicalyx is a ring of sepal-like organs just below the true sepals (calyx). Common in Rose Family.

Deflexed

Falls: see p.272.
Female flowers contain styles only, no stamens.
Fen: a habitat on wet, lime-rich peat, not acid as in a bog (q.v.).
Florets are small flowers, part of a compound head.
Flower: the reproductive structure of a plant.
Flower parts comprise petals, sepals, stamens, styles, and sometimes other organs.
Fruits are composed of the seeds and structures surrounding them.

Disc florets

Head: used when flowers are grouped in more or less compact terminal groups.
Heath: area, often dominated by heathers or related shrubs, on acid soils.
Hips are usually brightly coloured false fruits, characteristic of roses.
Hoary: greyish with short hairs.
Honey-leaf; see Nectary.

Epicalyx

Introduced plants are not native to the area, but brought in by man.

Female flower

Keel: see p.118.

Labiate: belonging to the Labiate Family, Labiatae.
Lanceolate: spear-shaped, narrowly oval and pointed.
Lax: of a flower-head with the flowers well spaced; not dense.
Lime: strictly the product (Calcium oxide) of the burning of limestone rock (Calcium carbonate), but here used loosely to include limestone and chalk, and also soils formed on them; the opposite of acid. Lime-rich soils are those formed on limestone.
Linear: almost parallel-sided.
Lobed: of lvs deeply toothed, but not formed of separate leaflets. Cf *pinnatifid*.

Lobed

Male flowers contain stamens only, no styles.
Marsh: a community on wet ground, but not on peat.
Microspecies: species produced by complex reproductive processes which result in a large number of biologically distinct units, only distinguished with difficulty, and often on microscopic characters. They are here grouped into Aggregates (q.v.).
Midrib: central vein of a leaf, usually thick and raised.
Moor: usually heather covered, upland areas.
Morphology: the study of the shape, form or appearance of plants.

Male flower

Midrib

Nectar: sugary substance attractive to insects and secreted by many flowers.

Nectary

Nectary: organ in the flower producing nectar.
Net-veined: of a leaf with the veins not all parallel.
Node: point of origin of leaves on the stem.

Net-veined

Ochrea: see p.40.
Opposite: of leaves arising opposite each other on the stem.

Nodes

Palmate: with finger-like lobes or leaflets.
Parasites are plants, usually without green colouring, that obtain nutriment from other plants.
Peat is a soil composed of undecayed plant matter, often acid. See also Ecology (p.309).
Perennial: plant surviving more than 2 years; often stouter than annuals and sometimes woody.
Petals: usually conspicuous organs above the sepals, and surrounding the reproductive parts of the flower. See Figure.
Pinnate, Pinnatifid: see Figure.
Pod: fruit, usually long and cylindrical and never fleshy; as in peas.

Pinnate

Ray: see p.154.
Ray-floret: see p.230.
Rhizome: horizontal underground stem, and therefore bearing leaf-scars.
Rosette: flattened, rose-like group of leaves at the base of the stem.
Runners are horizontal above-ground stems, often rooting at the nodes.

Rhizome

Rosette

Runners

Samara: a winged key-shaped fruit.
Scale: small appendage not resembling a leaf. Normally small, brown, and papery.
Sepals form a ring immediately below the petals and are usually green or brown and less conspicuous.
Shrub: much-branched woody plant, shorter than a tree.
Shy flowerer: plant often passing whole years without flowering and for which other characters are therefore important for identification.
Silicula and Siliqua: see p.84.
Spadix: see p.270.
Spathe: see p.270.
Species: the basic unit of classification; see Introduction, p.10.
Spike: flower-head with flowers arranged up a central axis in a cylinder. Stalked spike has individual fls stalked.
Spine: straight, sharp-tipped appendage; cf. thorn.
Spreading: standing out horizontally or at a wide angle from the stem.
Stamens: the male organs in a flower, comprising a filament and a pollen sac, the anther (see Figure). Distinguished from styles (q.v.) by lying outside the centre of the flower usually in a ring, and by the usually coloured anthers.
Standard: see pp.118 and 272.
Stigma: the surface receptive to pollen at the tip of the style.
Stipules are scale- or leaf-like organs at the base of the leaf-stalk.

Samara

Scales

Sepals

Spike

Styles: the columns of filaments leading from the female organs to the stigma (q.v.); see Figure. Distinguished from stamens (q.v.) by lying in the centre of the flower, within the ring of stamens.

Stamens

Subspecies: the division in the classification of organisms immediately below the species. Subspecies are morphologically distinct from each other, but interbreed freely and are therefore included in the same species.

Stigma

Tendrils are twisted filaments forming part of a leaf or stem and used for climbing.
Thallus: plant body not differentiated into stems and lvs.
Thorn: sharp-tipped, woody appendage, straight or curved.
Tree: tall, woody plant, with a single woody stem at the base.
Trefoil, Trifoliate: with three leaflets.

Stipules

Umbel: a flowerhead with the flowers in a spike but the lower branches longer so that all are level. **Umbellifers** are members of the Umbelliferae (pp.154-66), characterised by having flowers in umbels.
Undershrub: low, often creeping woody perennial, often quite unlike taller shrubs.

Style

Valve: see Docks, p.40.
Variety: a distinct form of a plant, of even lower rank than a subspecies.

Tendrils

Waste places: areas much disturbed by man, but not cultivated.
Whorl: where several organs arise at the same point on a stem. Cf. opposite.
Winged: with a flange running down the stem or stalk.
Wings: see p.118.

Thorn

Winged fruit

Winged stem

Winged flower

Whorl

ERRATA.

On page 109, illustration number 2 should read 3, and vice versa.
On page 126, the right-hand group of illustrations at the bottom of the page should be captioned "**9*** Hybrids of Lucerne".
On page 164, plant 5 should read *Laserpitium latifolium,* plant 5b should read *L. siler,* and plant 5c should read *Laser trilobum.*

Plant Photography

It was once the custom for botanists to pick and press plants for their own herbaria, and this practice caused certain rare plants to become extinct or nearly so. Nowadays the botanist can keep a permanent photographic record which is more responsible, more informative, and more attractive. Here are a few hints.

Ideally use a single-lens reflex camera, enabling you to see your proposed picture through the lens. This is especially important for close-up work, for which you can obtain extension tubes.

Use slow speeds to give narrow apertures (high f values) and good depth of focus. A fast film helps here.

Correct exposure setting requires a light meter, because plants at ground level and among vegetation are often in very unusual light conditions.

For brightly-coloured or white flowers move up half-a-stop or a stop on the f-scale to avoid over-exposure.

If you need to move surrounding vegetation to take your picture, be sure to replace it afterwards to protect the plant.

And look where you tread!

Societies to Join

Botanical Society of the British Isles, c/o Natural History Museum, Cromwell Road, London S.W.7. The leading national society for serious students of the British flora, both amateur and professional.

Wild Flower Society, Rams Hill House, Horsmonden, Tonbridge, Kent, is for amateur lovers of wild flowers, who enjoy keeping annual diaries, and enter for various flower-hunting and recording competitions.

Council for Nature, c/o Zoological Gardens, Regent's Park, London N.W.1, will supply addresses of local natural history societies, many of which have botanical sections, and almost all of which run field meetings where beginners are helped to learn the local wild flowers. The Council will also supply the address of your local County Naturalists' Trust, which all lovers of British wild flowers who are seriously interested in their conservation ought to join.

Further Reading

Britain

Flora of the British Isles, by A. R. Clapham, T. G. Tutin, and E. F. Warburg (Cambridge) is the standard British flora for serious workers.

The Pocket Guide to Wild Flowers, by David McClintock and R. S. R. Fitter (Collins) is a complete field guide, fully illustrated.

Finding Wild Flowers, by Richard Fitter (Collins). A comprehensive key to the British flora, with details of good sites for flower hunting.

The Concise British Flora in Colour, by W. Keble Martin (Ebury Press/ Michael Joseph). A tour-de-force, all British wild flowers painted by one man. Text rather brief.

The New Naturalist Series (Collins) includes *Wild Flowers*, by John Gilmour and Max Walters; *Wild Orchids of Britain*, by V. S. Summerhayes; *Wild Flowers of Chalk and Limestone*, by J. E. Lousley; *Flowers of the Coast*, by Ian Hepburn; *Mountain Flowers*, by John Raven and Max Walters; and *British Plant Life*, by W. B. Turrill.

Europe

Flora Europaea, edited by T. G. Tutin and others (Cambridge) is the standard European flora, eventually in five volumes, with three so far published.

Flowers of Europe: a field guide, by Oleg Polunin (Oxford). Covers a wider area less thoroughly than the present volume. Illustrated by colour photographs.

Index of English Names

Abraham, Isaac and Jacob, 192
Acacia, False, 120
Aconite, Winter, 66
Adderstongue, 270
Agrimony, 106
  Bastard, 106
  Fragrant, 106
  Hemp, 230
Alder, 30
  Green, 298
  Grey, 30
Alexanders, 166
  Biennial, 166
Alison, Hoary, 96
  Mountain, 84
  Small, 84
  Sweet, 96
Alkanet, 192
  Green, 192
  Large Blue, 192
  Yellow, 188
Allseed, 132
  Four-leaved, 56
Alyssum, Golden, 84
Amaranth, Common, 48
  Green, 48
  White, 48
Amelanchier, 114
Amsinckia, 188
Androsace, Northern, 176
Anemone, Blue, 76
  Wood, 76
  Yellow, 76
Angelica, 160
  Garden, 160
Apple, Crab, 114
  Cultivated, 114
Apple of Peru, 206
Archangel, Yellow, 200
Arnica, 236
Aronia, 307
Arrow-grass, Marsh, 220
  Sea, 220
Arrowhead, 258
  Canadian, 258
Artichoke, Jerusalem, 236
Arum, Bog, 270
Asarabacca, 38
Ash, 36
  Flowering, 36
Asparagus, Wild, 268
Aspen, 30
Asphodel, Bog, 262
  German, 260
  Scottish, 260
Aster, Sea, 232
  Stink, 230
Astrantia, 154
Auricula, 174
Avens, Alpine, 110
  Mountain, 110
  Water, 110
Awlwort, 290
Azalea, Wild, 172

Balm, 202
  Bastard, 198
Balsam, Himalayan, 138
  Orange, 138
  Small, 138
  Touch-me-not, 138

Baneberry, 72
Barberry, 78
Bartsia, Alpine, 212
  Red, 212
  Yellow, 212
Basil, Cow, 62
  Wild, 204
Bayberry, 298
Bearberry, Alpine, 172
  Black, 172
Bedstraw, Fen, 186
  Heath, 186
  Hedge, 186
  Lady's, 186
  Limestone, 186
  Marsh, 186
  Northern, 186
  Slender, 186
  Slender Marsh, 186
  Wall, 186
Beech, 32
Beet, Sea, 46
Beggar Ticks, 238
Bellflower, Bearded, 226
  Clustered, 226
  Creeping, 226
  Giant, 226
  Ivy-leaved, 226
  Nettle-leaved, 226
  Peach-leaved, 226
  Rampion, 226
  Spreading, 226
Bermuda Buttercup, 132
Betony, 202
Bilberry, 172
  Northern, 172
Bindweed, Black, 42
  Copse, 42
  Field, 184
  Great, 184
  Hairy, 184
  Hedge, 184
  Sea, 184
Birch, Downy, 30
  Dwarf, 30
  Silver, 30
Birdsfoot, 128
  Orange, 128
Birdsfoot Trefoil, 128
  Greater, 128
  Hairy, 128
  Narrow-leaved, 128
  Slender, 128
Birdsnest, Yellow, 168
Birthwort, 38
Bistort, 40
  Alpine, 40
  Amphibious, 40
Bittercress, Coralroot, 90
  Hairy, 94
  Large, 90
  Narrow-leaved, 94
  Wavy, 94
Bittersweet, 206
Blackberry, 108
Black-eyed Susan, 236
Blackthorn, 114
Bladder-nut, 140
Bladderseed, 158
Bladderwort, Greater, 290
  Intermediate, 290
  Lesser, 290

Blinks, 42
Blood-drop Emlets, 212
Bluebell, 266
  Spanish, 266
Blue-eyed Grass, 272
Blue-eyed Mary, 190
Bogbean, 178
Bog Myrtle, 30
Bog Rosemary, 170
Borage, 192
Box, 140
Bramble, 108
  Arctic, 108
  Stone, 108
Briar, Sweet, 108
Brooklime, 214
Brookweed, 176
Broom, 120
  Black, 120
  Butcher's, 268
  Clustered, 120
  Winged, 120
Broomrape, Branched, 304
  Carrot, 304
  Clove-scented, 218
  Common, 218
  Greater, 218
  Ivy, 218
  Knapweed, 218
  Ox-tongue, 218
  Purple, 218
  Thistle, 218
  Thyme, 218
Bryony, Black, 266
  White, 148
Buckthorn, 140
  Alder, 140
  Sea, 140
Buckwheat, 40
Bugle, 196
  Blue, 196
  Pyramidal, 196
Bugloss, 192
  Purple Viper's, 192
  Viper's, 192
Bulrush, 296
  Lesser, 296
Burdock, Greater, 244
  Lesser, 244
Bur Marigold, Nodding, 238
  Trifid, 238
Burnet, Great, 106
  Salad, 106
Burnet Saxifrage, 158
  Greater, 158
Burning Bush, 138
Bur Parsley, Greater, 156
  Knotted, 156
  Small, 156
  Spreading, 156
Bur-reed, Branched, 296
  Floating, 306
  Least, 306
  Unbranched, 290
Butcher's Broom, 268
Butterbur, 238
  White, 238
Buttercup, Bermuda, 132
  Bulbous, 68
  Celery-leaved, 70
  Corn, 70
  Creeping, 68

Glacier, 70
Goldilocks, 70
Hairy, 68
Jersey, 68
Large White, 70
Meadow, 68
Small-flowered, 70
Woolly, 68
Butterfly Bush, 308
Butterwort, Alpine, 304
Common, 218
Large-flowered, 218
Pale, 218
Buttonweed, 238

Cabbage, Bastard, 88
Hare's-ear, 299
Isle of Man, 88
Lundy, 88
St. Patrick's, 104
Wallflower, 88
Warty, 86
Wild, 88
Calamint, Common, 204
Lesser, 204
Calypso, 274
Campion, Bladder, 58
Red, 60
Sea, 58
White, 60
Candytuft, Wild, 96
Canterbury Bell, 226
Caraway, 156
Whorled, 156
Carnation, 64
Carrot, Moon, 158
Wild, 156
Cassiope, 172
Catchfly, Alpine, 60
Berry, 62
Flaxfield, 58
Forked, 58
Italian, 58
Moss, 58
Night-flowering, 60
Northern, 58
Nottingham, 58
Rock, 58
Sand, 58
Small-flowered, 62
Spanish, 58
Sticky, 60
Sweet William, 58
White Sticky, 58
Catmint, 198
Hairless, 198
Catsear, Common, 254
Smooth, 254
Spotted, 254
Cattail, 296
Cedar, Western Red, 298
Celandine, Greater, 80
Lesser, 68
Celery, Wild, 162
Centaury, Common, 180
Guernsey, 180
Lesser, 180
Perennial, 180
Seaside, 180
Slender, 180
Yellow, 180
Chaffweed, 176
Chamomile, Corn, 232
Lawn, 232
Stinking, 232
Yellow, 236
Charlock, 88
Cherry, Bird, 114

Choke, 307
Cornelian, 154
Dwarf, 301
Rum, 301
St. Lucie's, 301
Wild, 114
Cherry Laurel, 307
Cherry-plum, 114
Chervil, Bur, 156
Garden, 156
Golden, 156
Rough, 156
Chestnut, Horse, 36
Sweet, 32
Water, 258
Chickweed, Common, 54
Greater, 54
Lesser, 54
Umbellate, 54
Upright, 54
Water, 54
Chickweed Wintergreen, 176
Chicory, 250
Chives, 264
Chondrilla, 252
Cinquefoil, Alpine, 112
Creeping, 112
Grey, 112
Hoary, 112
Marsh, 110
Norwegian, 112
Rock, 110
Shrubby, 112
Spring, 112
Sulphur, 112
White, 110
Clary, Meadow, 202
Whorled, 202
Wild, 202
Cleavers, Common, 186
Corn, 186
False, 186
Clematis, Alpine, 72
Cloudberry, 108
Clover, Alsike, 130
Burrowing, 130
Clustered, 301
Crimson, 130
Haresfoot, 130
Knotted, 130
Long-headed, 130
Mountain, 130
Mountain Zigzag, 301
Red, 130
Reversed, 301
Rough, 301
Sea, 130
Strawberry, 130
Suffocated, 130
Sulphur, 130
Twin-flowered, 301
Upright, 130
Western, 130
White, 130
Zigzag, 130
Club-rush, Common, 296
Cnidium, 164
Cocklebur, Rough, 304
Spring, 230
Stinking, 304
Coltsfoot, 240
Purple, 240
Columbine, 74
Comfrey, Common, 188
Dwarf, 303
Rough, 188
Russian, 188
Soft, 188

Tuberous, 188
Cone Flower, 236
Coral Necklace, 56
Coriander, 156
Corn Cockle, 60
Cornel, Dwarf, 154
Cornelian Cherry, 154
Cornflower, 248
Perennial, 248
Cornsalad, 222
Corydalis, Bulbous, 78
Climbing, 78
Yellow, 78
Cotoneaster, Himalayan, 307
Small-leaved, 307
Wall, 307
Wild, 114
Cottonweed, 238
Cowbane, 160
Cowberry, 172
Cow Basil, 62
Cowberry, 172
Cow Parsley, 156
Cowslip, 174
Cow-wheat, Common, 216
Crested, 216
Field, 216
Small, 216
Crab Apple, 114
Cranberry, 172
Small, 172
Cranesbill, Bloody, 134
Cut-leaved, 134
Dovesfoot, 134
Dusky, 134
French, 134
Hedgerow, 134
Knotted, 134
Long-stalked, 134
Marsh, 134
Meadow, 134
Pencilled, 134
Round-leaved, 134
Shining, 134
Small-flowered, 134
Wood, 134
Creeping Jenny, 174
Cress, Garden, 98
Hoary, 98
Mitre, 84
Shepherd's, 96
Thale, 94
Crocus, Autumn, 272
Sand, 272
Spring, 272
Crosswort, 186
Crowberry, 172
Crowfoot, Ivy-leaved, 70
Round-leaved, 70
Three-lobed, 70
Water, 70
Cuckoo Flower, 90
Cuckoo Pint, Large, 270
Cucumber, Prickly, 148
Cudweed, Broad-leaved, 234
Cape, 234
Common, 234
Dwarf, 234
Heath, 234
Highland, 234
Jersey, 234
Marsh, 234
Narrow-leaved, 234
Red-tipped, 234
Small, 234
Currant, Black, 140
Flowering, 307
Mountain, 140
Red, 140

Cyclamen, 176
Cyphel, 52
Cypress, Lawson's, 24

Daffodil, Tenby, 270
  Wild, 270
Daisy, 232
  Michaelmas, 232
  Ox-eye, 240
  Shasta, 240
Dame's Violet, 90
Dandelion, 254
  Lesser, 254
  Marsh, 254
  Red-veined, 254
Dead-nettle, Cut-leaved, 200
  Henbit, 200
  Red, 200
  Spotted, 200
  White, 200
Dewberry, 108
Diapensia, 168
Dill, 166
Dittander, 92
Dock, Argentine, 298
  Broad-leaved, 44
  Clustered, 44
  Curled, 44
  Fiddle, 44
  Golden, 44
  Marsh, 44
  Northern, 44
  Patience, 298
  Scottish, 44
  Shore, 44
  Water, 44
  Willow-leaved, 298
  Wood, 44
Dodder, Common, 184
  Flax, 303
  Greater, 184
Dogwood, 154
Dragon's Teeth, 128
Dropwort, 106
  Water, 162
Duckweed, Common, 292
  Fat, 306
  Greater, 306
  Ivy-leaved, 292
  Rootless, 292
Dusty Miller, 54

Eel-grass, 294
  Dwarf, 306
  Narrow-leaved, 306
Elder, 222
  Box, 307
  Dwarf, 222
  Ground, 158
  Red-berried, 222
Elecampane, 236
Elm, English, 34
  Fluttering, 34
  Small-leaved, 34
  Wych, 34
Enchanter's Nightshade, 148
  Alpine, 148
  Upland, 148
Eryngo, Field, 154
Escallonia, 307
Evening Primrose,
  Common, 148
  Fragrant, 148
  Large-flowered, 148
  Small-flowered, 148
Everlasting, Mountain, 234
  Pearly, 234
Eyebright, 216
  Irish, 216

Fat Hen, 46
Felwort, Marsh, 182
Fennel, 166
  Hog's, 164
Fenugreek, 130
  Classical, 126
Feverfew, 240
Fig, 307
  Hottentot, 42
Figwort, Balm-leaved, 208
  Common, 208
  French, 208
  Green, 208
  Water, 208
  Yellow, 208
Fir, Douglas, 24
  Silver, 24
Firethorn, 307
Flag, Purple, 272
  Sweet, 296
Flax, Common, 132
  Pale, 132
  Perennial, 132
  Purging, 132
  White, 302
  Yellow, 132
Fleabane, Alpine, 232
  Blue, 232
  Canadian, 230
  Common, 236
  Irish, 236
  Mexican, 232
  Small, 234
Fleawort, Field, 242
  Marsh, 242
Flixweed, 86
Fluellen, Round-leaved, 210
  Sharp-leaved, 210
Fool's Parsley, 158
Forgetmenot, Alpine, 303
  Bur, 190
  Changing, 190
  Creeping, 190
  Early, 190
  Field, 190
  Jersey, 190
  Pale, 190
  Tufted, 190
  Water, 190
  Wood, 190
Foxglove, 212
  Fairy, 212
  Large Yellow, 212
  Small Yellow, 212
Fringe Cups, 300
Fritillary, 262
Frogbit, 258
Fuchsia, 307
Fumana, Common, 148
Fumitory, Common, 78
  Ramping, 78
  Small, 78
  Wall, 78

Gagea, Belgian, 262
  Least, 262
  Meadow, 262
Gallant Soldier, 232
Garlic, Crow, 264
  Field, 264
  German, 264
  Keeled, 264
  Rosy, 264
Gentian, Alpine, 182
  Autumn, 182
  Bladder, 182
  Chiltern, 182

Cross, 182
  Dune, 182
  Early, 182
  Field, 182
  Fringed, 182
  Great Yellow, 180
  Marsh, 182
  Northern, 180
  Purple, 182
  Slender, 182
  Spotted, 182
  Spring, 182
  Willow, 182
Germander, Cut-leaved, 196
  Mountain, 196
  Wall, 196
  Water, 196
Gipsywort, 204
Gladiolus, 272
Glasswort, 48
  Perennial, 48
Globeflower, 68
Globularia, 218
Goatsbeard, 250
Goatsbeard Spiraea, 106
Goat's Rue, 118
Golden-rod, 230
  Canadian, 230
Gold of Pleasure, 84
Goldilocks, 230
Good King Henry, 46
Gooseberry, 140
Goosefoot, Fig-leaved, 46
  Green, 299
  Grey, 299
  Many-seeded, 46
  Maple-leaved, 46
  Nettle-leaved, 46
  Oak-leaved, 46
  Red, 46
  Small Red, 46
  Stinking, 46
  Upright, 46
Gorse, 120
  Dwarf, 120
  Western, 120
Grape-hyacinth, 266
  Small, 266
Grass of Parnassus, 100
Grass Poly, 150
Gratiola, 212
Greenweed, Dyer's, 120
  German, 120
  Hairy, 120
Gromwell, Common, 188
  Corn, 188
  Purple, 188
Ground Elder, 158
Ground Ivy, 198
Ground-pine, 196
Groundsel, 242
  Heath, 242
  Sticky, 242
Guelder Rose, 222
Gypsophila, Annual, 62
  Fastigiate, 62

Harebell, 226
Hare's-ear, Sickle, 166
  Slender, 164
  Small, 166
Hartwort, 164
Hawksbeard, Beaked, 256
  Bristly, 256
  French, 256
  Marsh, 256
  Northern, 256
  Rough, 256

Smooth, 254
Stinking, 256
Hawkbit, Autumn, 254
  Lesser, 254
  Rough, 254
Hawkweed, Alpine, 256
  Few-leaved, 256
  Leafy, 256
  Mouse-ear, 254
  Orange, 256
  Spotted, 256
Hawthorn, 114
  Midland, 114
Hazel, 32
Heath, Cornish, 170
  Cross-leaved, 170
  Dorset, 170
  Irish, 170
  Mackay's, 170
  Mountain, 170
  Prickly, 307
  St. Dabeoc's, 170
Heather, 170
  Bell, 170
Hedge Parsley, Upright, 156
Helichrysum, 234
Heliotrope, 188
  Winter, 238
Hellebore, Green, 66
  Stinking, 66
Helleborine, Broad-leaved, 284
  Dark Red, 284
  Dune, 284
  False, 260
  Green-flowered, 284
  Marsh, 284
  Narrow-leaved, 282
  Narrow-lipped, 284
  Red, 282
  Violet, 284
  White, 282
Hemlock, 160
  Western, 24
Hemp, 38
Hemp-nettle, Common, 200
  Downy, 200
  Large-flowered, 200
  Red, 200
Hen and Chickens Houseleek, 102
Henbane, 206
Hepatica, 76
Herb Bennet, 110
Herb Paris, 268
Herb Robert, 134
Hog's Fennel, 164
Hogweed, 160
  Giant, 160
Holly, 36
  Sea, 154
Hollyhock, 142
Honesty, 90
  Perennial, 90
Honewort, 158
Honeysuckle, 222
  Black-berried, 304
  Blue, 304
  Fly, 222
  Perfoliate, 222
Honeywort, Lesser, 188
Hop, 38
Horehound, Black, 198
  White, 198
Hornbeam, 32
Horned-poppy, Red, 80
  Yellow, 80
Hornwort, Rigid, 292

Soft, 306
Horse-radish, 92
Horsetail, Marsh, 296
  Water, 296
Hottentot Fig, 42
Houndstongue, 188
  Green, 189
Houseleek, Hen and Chickens, 102
Hutchinsia, 96
Hyacinth, Tassel, 266
Hyssop, 198

Iris, Butterfly, 272
  Garden, 272
  Siberian, 272
  Stinking, 272
  Yellow, 272
Ivy, 154
  Ground, 198

Jack-go-to-bed-at-noon, 250
Jacob's Ladder, 184
Joint Pine, 24
Juneberry, 307
Juniper, 24
Jupiter's Distaff, 202

Kale, Sea, 92
Knapweed, Black, 248
  Brown, 248
  Greater, 248
  Jersey, 248
Knawel, Annual, 56
  Perennial, 56
Knotgrass, 40
  Ray's, 40
  Sea, 40
Knotweed, Giant, 42
  Himalayan, 42
  Japanese, 42

Labrador Tea, 172
Laburnum, 120
Lady's Mantle, 106
  Alpine, 106
Lady's Slipper, 274
Lady's Tresses, Autumn, 282
  Creeping, 282
  Irish, 282
  Summer, 282
Larch, European, 24
  Japanese, 298
Larkspur, 74
  Eastern, 74
  Forking, 74
Laurel, Cherry, 307
  Spurge, 138
Leatherleaf, 172
Leek, Babington's, 264
  Cultivated, 305
  Few-flowered, 264
  Round-headed, 264
  Sand, 264
  Three-cornered, 264
  Wild, 264
Lentil, Wild, 118
Leopardsbane, 236
  Green, 236
Lettuce, Blue, 250
  Great, 252
  Least, 252
  Pliant, 305
  Prickly, 252
  Purple, 250
  Wall, 252
Lilac, 308
Lily, Kerry, 260
  Martagon, 262
  May, 260

Pyrenean, 262
  St. Bernard's, 260
  Snowdon, 260
Lily of the Valley, 260
Lime, Common, 36
  Large-leaved, 36
  Small-leaved, 36
Liquorice, Wild, 118
Little Robin, 134
Lobelia, Heath, 226
  Water, 290
London Pride, 300
Longleaf, 160
Loosestrife, Dotted, 174
  Purple, 150
  Tufted, 174
  Yellow, 174
Lords and Ladies, 270
Lousewort, 216
  Leafy, 216
  Marsh, 216
Lovage, 166
  Scots, 164
Love in a Mist, 66
Love Lies Bleeding, 48
Lucerne, 126
Lungwort, 190
  Narrow-leaved, 190
Lupin, Annual, 120
  Bitter Blue, 120
  Garden, 120
  Sweet, 120
  Tree, 120
  White, 120
  Wild, 120

Madder, Field, 184
  Wild, 186
Madwort, 190
Mallow, Chinese, 142
  Common, 142
  Dwarf, 142
  Least, 142
  Marsh, 142
  Musk, 142
  Rough, 142
  Small, 142
  Smaller Tree, 142
  Tree, 142
Maple, Field, 34
  Italian, 34
  Montpelier, 34
  Norway, 34
Marestail, 292
Marigold, 240
  Corn, 240
  Garden, 240
  Marsh, 68
  Nodding Bur, 238
  Trifid Bur, 238
Marjoram, 204
Marshwort, Creeping, 162
  Lesser, 162
Masterwort, 164
Mayweed, Pineapple, 232
  Scented, 232
  Scentless, 232
  Sea, 232
Meadow-rue, Alpine, 72
  Common, 72
  Greater, 72
  Lesser, 72
  Small, 72
Meadow Saffron, 266
Meadowsweet, 106
Medick, Black, 128
  Bur, 128
  Spotted, 128

Toothed, 128
Medlar, 307
Melilot, Ribbed, 126
  Small, 126
  Tall, 126
  White, 126
Mercury, Annual, 136
  Dog's, 136
Mezereon, 138
Michaelmas Daisy, 232
Micropus, 234
Mignonette, Corn, 100
  White, 100
  Wild, 100
Milfoil, Alternate Water, 292
  Spiked Water, 292
  Whorled Water, 292
Milk-parsley, Cambridge, 160
Milk-vetch, Alpine, 118
  Hairy, 118
  Mountain, 118
  Northern, 118
  Purple, 118
  Yellow, 118
  Yellow Alpine,118
Milkwort, Chalk, 302
  Common, 138
  Dwarf, 302
  Heath, 302
  Sea, 176
  Shrubby, 138
  Tufted, 302
Mind Your Own Business, 38
Mint, Corn, 204
  Round-leaved, 303
  Spear, 204
  Water, 204
Mistletoe, 38
Moneywort, Cornish, 216
Monkey Flower, 212
Monkshood, 74
Monk's Rhubarb, 298
Montbretia, 272
Moon Carrot, 158
Moonwort, 270
Moor-king, 216
Moschatel, 220
Motherwort, 200
  False, 200
Mouse-ear, Alpine, 54
  Arctic, 54
  Common, 54
  Dwarf, 54
  Field, 54
  Grey, 54
  Little, 54
  Sea, 54
  Starwort, 54
  Sticky, 54
Mousetail, 74
Mudwort, 290
  Welsh, 306
Mugwort, 238
  Chinese, 238
Mullein, Dark, 208
  Great, 208
  Hoary, 208
  Moth, 208
  Orange, 208
  Purple, 208
  Twiggy, 208
  White, 208
Musk, 212
Mustard, Ball, 84
  Black, 88
  Buckler, 84
  Garlic, 90
  Hedge, 86

Hoary, 88
Tower, 94
Treacle, 86
White, 88
White Ball, 92
Myrtle, Bog, 30

Naiad, Holly-leaved, 294
  Slender, 294
Narcissus, Poet's, 270
Navelwort, 100
Nettle, Stinging, 38
  Annual, 38
Nightshade, Black, 206
  Deadly, 206
  Enchanter's, 148
  Green, 206
  Hairy, 206
Nipplewort, 252
Nonea, 188
Nutmeg, Flowering, 308

Oak, Durmast, 32
  Evergreen, 32
  Holm, 32
  Pedunculate, 32
  Red, 32
  Sessile, 32
  Turkey, 32
  White, 32
Odontites, Yellow, 212
Old Man's Beard, 72
Onion, 305
  Welsh, 264
Orache, Common, 46
  Frosted, 46
  Grass-leaved, 46
  Shrubby, 307
  Spear-leaved, 46
  Stalked, 48
Orchid, Bee, 274
  Birdsnest, 286
  Black Vanilla, 274
  Bog, 286
  Broad-leaved Marsh, 278
  Bug, 276
  Burnt, 276
  Butterfly, 282
  Common Spotted, 278
  Coral-root, 286
  Dense-flowered, 280
  Early Marsh, 278
  Early Purple, 276
  Early Spider, 274
  Elder-flowered, 278
  False Musk, 280
  Fen, 286
  Flecked Marsh, 278
  Fly, 274
  Fragrant, 278
  Frog, 280
  Ghost, 286
  Greater Butterfly, 282
  Green-winged, 276
  Heath Spotted, 278
  Lady, 276
  Late Spider, 274
  Lesser Butterfly, 282
  Loose-flowered, 276
  Lizard, 280
  Man, 280
  Marsh, 278
  Military, 276
  Monkey, 276
  Musk, 280
  Northern Marsh, 278
  One-leaved Bog, 286
  Pale-flowered, 276

Pugsley's Marsh, 278
Pyramidal, 278
Southern Marsh, 278
Spotted, 278
Toothed, 276
Violet Birdsnest, 286
Wasp, 274
White Frog, 282
Oregon Grape, 78
Orpine, 102
Osier, 26
  Golden, 26
  Red, 154
Oxalis, Pink, 132
  Upright Yellow, 132
  Yellow, 132
Ox-eye, Large Yellow, 236
  Yellow, 236
Ox-eye Daisy, 240
Oxlip, 174
  False, 174
Ox-tongue, Bristly, 256
  Hawkweed, 256
Oyster Plant, 192

Pansy, Dwarf, 146
  Field, 146
  Mountain, 146
  Wild, 146
Parsley, Bur, 156
  Cambridge Milk, 160
  Corn, 162
  Cow, 156
  Fool's, 158
  Greater Bur, 156
  Knotted Bur, 156
  Milk, 164
  Small Bur, 156
  Spreading Bur, 156
  Stone, 162
  Upright Hedge, 156
  Wild, 162
Parsley Piert, 106
Parsnip, Wild, 166
Pasque Flower, 76
  Eastern, 76
  Pale, 76
  Small, 76
Pea, Black, 124
  Broad-leaved Everlasting, 124
  Marsh, 124
  Narrow-leaved Everlasting, 124
  Sea, 124
  Spring, 124
  Tuberous, 124
Pear, Cultivated, 114
  Plymouth, 301
  Wild, 114
Pearlwort, Alpine, 56
  Annual, 56
  Heath, 56
  Knotted, 56
  Procumbent, 56
  Sea, 56
  Snow, 56
Pellitory of the Wall, 38
Pennycress, Alpine, 98
  Field, 98
  Garlic, 98
  Mountain, 98
  Perfoliate, 98
Pennyroyal, 204
Pennywort, Marsh, 154
Peppermint, 204
Pepperwort, Field, 98
  Least, 98

323

Narrow-leaved, 98
Poor Man's, 98
Smith's, 98
Tall, 300
Periwinkle, Greater, 178
Lesser, 178
Persicaria, Pale, 40
Petty Whin, 120
Pheasant's Eye, 74
Large, 74
Summer, 74
Physocarpus, 307
Pick-a-back Plant, 300
Pignut, 158
Great, 158
Pimpernel, Blue, 176
Bog, 176
Scarlet, 176
Yellow, 174
Pine, Austrian, 24
Joint, 298
Maritime, 24
Scots, 24
Pink, Carthusian, 64
Cheddar, 64
Childling, 62
Clove, 64
Deptford, 64
Jersey, 64
Large, 64
Maiden, 64
Proliferous, 62
Wild, 64
Pipewort, 260, 290
Pirri-pirri Bur, 106
Pittosporum, 307
Plane, London, 34
Plantain, Branched, 220
Buckshorn, 220
Greater, 220
Hoary, 220
Ribwort, 220
Sea, 220
Pleurospermum, 160
Ploughman's Spikenard, 230
Plum, Wild, 114
Poet's Narcissus, 270
Polycnemum, 46
Pondweed, American, 306
Blunt-leaved, 306
Bog, 294
Broad-leaved, 294
Curled, 294
Fen, 294
Fennel, 294
Flat-stalked, 306
Grasswrack, 306
Hair-like, 306
Horned, 294
Lesser, 306
Loddon, 306
Long-stalked, 294
Opposite-leaved, 294
Perfoliate, 294
Reddish, 306
Sharp-leaved, 306
Shetland, 306
Shining, 294
Slender-leaved, 306
Small, 294
Various-leaved, 306
Poplar, Balsam, 30
Black, 30
Grey, 30
Italian, 30
White, 30
Poppy, Arctic, 80
Babington's, 80

Common, 80
Long-headed, 80
Opium, 80
Prickly, 80
Red Horned, 80
Rough, 80
Welsh, 80
Yellow Horned, 80
Primrose, 174
Birdseye, 174
Evening, 148
Scottish, 174
Primrose Peerless, 270
Privet, Garden, 178
Wild, 178
Purslane, Hampshire, 290
Iceland, 42
Pink, 42
Sea, 48
Water, 290

Quillwort, 290
Spring, 290
Quince, 307

Radish, Garden, 88
Sea, 88
Wild, 88
Ragged Robin, 60
Ragweed, 230
Ragwort, 242
Alpine, 242
Broad-leaved, 242
Fen, 242
Hoary, 242
Marsh, 242
Oxford, 242
Silver, 242
Rampion, Round-headed, 224
Spiked, 224
Ramsons, 264
Rannoch Rush, 220
Rape, 88
Raspberry, 108
Purple-flowering, 307
Rattle, Yellow, 216
Redshank, 40
Reed, 296
Reed Sweetgrass, 296
Reed-mace, 296
Lesser, 296
Rest-harrow, 126
Large Yellow, 126
Small, 126
Spiny, 126
Rhododendron, 170, 307
Arctic, 170
Yellow, 170, 307
River Beauty, 150
Rockcress, Alpine, 94
Annual, 94
Bristol, 94
Garden, 94
Hairy, 94
Northern, 94
Tall, 94
Rocket, Eastern, 86
False London, 86
Hairy, 88
London, 86
Sea, 92
Tall, 86
Rock-rose, Common, 148
Hoary, 148
Spotted, 148
White, 148
Rose, Burnet, 108
Dog, 108

Downy, 108
Field, 108
Guelder, 222
Provence, 108
Rock, 148
Rosemary, Bog, 170
Rose of Sharon, 144
Roseroot, 100
Rowan, 114
Rubus, 307
Rue, Goat's, 118
Rupturewort, Fringed, 56
Hairy, 299
Smooth, 56
Rush, Flowering, 258
Jointed, 296
Rannoch, 220
Soft, 296
Wood, 296

Safflower, 248
Sage, Wild, 202
Wood, 196
Sainfoin, 126
St. John's Wort, Flax-leaved, 144
Hairy, 144
Imperforate, 144
Irish, 144
Marsh, 144
Pale, 144
Perforate, 144
Slender, 144
Square-stalked, 144
Trailing, 144
Wavy, 144
St. Patrick's Cabbage, 104
Sallow, Eared, 26
Great, 26
Grey, 26
Salsify, 250
Saltwort, Prickly, 48
Samphire, Golden, 230
Rock, 166
Sandwort, Arctic, 52
Balearic, 52
Curved, 52
Fine-leaved, 52
Fringed, 52
Mossy, 52
Mountain, 52
Sea, 52
Spring, 52
Teesdale, 52
Three-veined, 52
Thyme-leaved, 52
Sanicle, 154
Savory, Winter, 198
Sawwort, 248
Alpine, 244
Saxifrage, Alternate
leaved Golden, 100
Arctic, 104
Burnet, 158
Drooping, 104
Greater Burnet, 158
Hawkweed, 300
Highland, 104
Irish, 104
Kidney, 300
Livelong, 104
Marsh, 104
Meadow, 104
Mossy, 104
Opposite-leaved Golden, 100
Pepper, 166
Purple, 104

324

Pyrenean, 300
Rue-leaved, 104
Starry, 104
Tufted, 104
Yellow, 104
Scabious, Devilsbit, 224
Field, 224
Sheepsbit, 224
Small, 224
Wood, 224
Yellow, 224
Scorpion Senna, 120
Scorpion Vetch, 120
Scurvy-grass,
Common, 92
Early, 92
English, 92
Greenland, 92
Sea-beet, 46
Seablite, Annual, 48
Hairy, 48
Shrubby, 48
Sea-heath, 148
Sea Holly, 154
Sea Kale, 92
Sea-lavender, Common,
178
Lax-flowered, 178
Matted, 178
Rock, 178
Sea Spurrey, Greater, 56
Green, 56
Lesser, 56
Rock, 56
Sedge, Bottle, 296
Greater Tussock, 296
Lesser Pond, 296
Self-heal, 198
Cut-leaved, 198
Large, 198
Senna, Bladder, 120
Scorpion, 120
Sermountain, 164
Service Tree, Wild, 301
Shaggy Soldier, 232
Shallon, 307
Shepherd's Cress, 96
Shepherd's Needle, 158
Shepherd's Purse, 96
Shoreweed, 290
Sibbaldia, 112
Silverweed, 112
Skullcap, 196
Lesser, 196
Spear-leaved, 196
Sloe, 114
Snapdragon, 210
Lesser, 210
Sneezewort, 238
Snowbell, Alpine, 174
Snowberry, 308
Snowdrop, 270
Snowdrop Windflower, 76
Snowflake, Spring, 270
Summer, 270
Soapwort, 60
Soldier, Gallant, 232
Shaggy, 232
Water, 258
Solomon's Seal, Angular, 268
Common, 268
Whorled, 268
Sorbaria, 307
Sorrel, Common, 44
French, 298
Mountain, 44
Sheep's, 44

Wood, 132
Sowbread, 176
Sow-thistle, Alpine, 250
Blue, 250
Marsh, 252
Perennial, 252
Prickly, 252
Smooth, 252
Spearwort, Adderstongue, 68
Creeping, 68
Greater, 68
Lesser, 68
Speedwell, Alpine, 214
American, 304
Blue Ivy-leaved, 214
Breckland, 304
Common Field, 214
Fingered, 304
French, 304
Germander, 214
Green Field, 214
Grey Field, 214
Heath, 214
Large, 214
Lilac Ivy-leaved, 214
Marsh, 214
Pink Water, 214
Rock, 214
Slender, 214
Spiked, 214
Spring, 304
Thyme-leaved, 214
Wall, 214
Water, 214
Wood, 214
Spikenard, Ploughman's, 230
Spignel, 158
Spindle-tree, 140
Spiraea, 307
Goatsbeard, 106
Spring Beauty, 42
Spruce, Norway, 24
Sitka, 24
Spurge, Broad-leaved, 136
Caper, 136
Cypress, 136
Dwarf, 136
Hairy, 302
Irish, 136
Leafy, 136
Marsh, 302
Petty, 136
Portland, 136
Purple, 136
Sea, 136
Sun, 136
Sweet, 302
Upright, 136
Wood, 136
Spurge Laurel, 138
Spurrey, Corn, 56
Greater Sea, 56
Greek Sea, 299
Lesser Sea, 56
Pearlwort, 299
Rock Sea, 56
Sand, 56
Squill, Alpine, 266
Autumn, 266
Spring, 266
Squinancywort, 184
Star-fruit, 258
Star of Bethlehem, Common,
268
Drooping, 268
Spiked, 268
Yellow, 262
Star-thistle, Cockspur, 248

Red, 248
Rough, 248
Yellow, 248
Starwort, Water, 292
Stink Aster, 230
Stitchwort, Bog, 54
Greater, 54
Lesser, 54
Marsh, 54
Wood, 54
Stock, Hoary, 92
Sea, 92
Stonecrop, Alpine, 102
Annual, 102
Biting, 102
Caucasian, 102
English, 102
Hairy, 102
Mossy, 102
Reflexed, 102
Rock, 102
Tasteless, 102
Thick-leaved, 102
White, 102
Storksbill, Common, 132
Musk, 132
Sea, 132
Strapwort, 56
Strawberry, Barren, 110
Garden, 110
Hautbois, 110
Pink Barren, 110
Wild, 110
Strawberry Tree, 172
Succory, Lamb's, 252
Sundew, Common, 100
Great, 100
Oblong-leaved, 100
Sunflower, 236
Perennial, 236
Swede, 88
Sweet Briar, 108
Sweet Cicely, 156
Sweet Gale, 30
Sweetgrass, Reed, 296
Sweet William, 64
Swine-cress, 98
Lesser, 98
Sycamore, 34

Tamarisk, 140
Tansy, 240
Tape-grass, 292
Tare, Hairy, 122
Slender, 122
Smooth, 122
Tarragon, 305
Tasselweed, Beaked, 306
Spiral, 294
Teasel, 224
Small, 224
Thistle, Cabbage, 244
Carline, 244
Cotton, 244
Creeping, 246
Dwarf, 246
Globe, 244
Great Marsh, 246
Marsh, 246
Meadow, 246
Melancholy, 246
Milk, 244
Musk, 246
Slender, 246
Spear, 246
Star, 248
Stemless Carline, 244
Tuberous, 246

Welted, 246
  Woolly, 246
  Yellow Melancholy, 244
Thorn-apple, 206
Thorow-wax, 166
  Narrow, 166
Thrift, 178
  Jersey, 303
Thyme, Basil, 204
  Large, 204
  Wild, 204
Thymelaea, Annual, 138
Tillaea, Water, 102
Toadflax, Alpine, 210
  Bastard, 38
  Common, 210
  Daisy-leaved, 210
  Ivy-leaved, 210
  Jersey, 210
  Pale, 210
  Prostrate, 210
  Purple, 210
  Sand, 210
  Small, 210
Tobacco Plant, Small, 206
Tomato, 206
Toothwort, 218
  Purple, 218
Tormentil, 112
  Trailing, 112
Towercress, 94
Traveller's Joy, 72
Tree of Heaven, 36
Trefoil, Birdsfoot, 128
  Hop, 128
  Large Hop, 128
  Lesser, 128
  Slender, 128
Tulip, Wild, 262
Tunic Flower, 62
Turnip, Wild, 88
Tutsan, 144
  Stinking, 144
  Tall, 144
Twayblade, Common, 280
  Lesser, 280
Twinflower, 222

Valerian, Common, 222
  Marsh, 222
  Red, 222
Venus's Looking Glass, 226
  Large, 226
Vervain, 192
Vetch, Bithynian, 122
  Bush, 122
  Common, 122
  Crown, 118
  Danzig, 122
  Fine-leaved, 122
  Fodder, 122
  Horseshoe, 128
  Kidney, 126
  Scorpion, 120
  Small Scorpion, 120
  Spring, 122
  Tufted, 122
  Upright, 122
  Wood, 122
  Yellow, 122
Vetchling, Bitter, 124
  Grass, 124
  Hairy, 124
  Meadow, 124
  Yellow, 124
Vincetoxicum, 178
Vine, 307
Violet, Common Dog, 146

Dame's, 90
Early Dog, 146
Fen, 146
Hairy, 146
Heath Dog, 146
Marsh, 146
Meadow, 146
Northern, 146
Pale Dog, 146
Sweet, 146
Teesdale, 146
Water, 176
White, 302
Yellow Wood, 146
Viper's Grass, 250
  Purple, 250
Virginia Creeper, 307

Wallflower, 86
Wall-rocket, Annual, 84
  Perennial, 84
Walnut, 30
Water Chestnut, 258
Watercress, 90
  Fool's, 162
  One-rowed, 90
Water Crowfoot, Brackish, 70
  Chalk-stream, 70
  Common, 70
  Fan-leaved, 70
  Pond, 70
  River, 70
  Thread-leaved, 70
Water Dropwort,
  Corky-fruited, 162
  Fine-leaved, 162
  Hemlock, 160
  Narrow-leaved, 162
  Parsley, 162
  River, 162
  Tubular, 162
Water-lily, Fringed, 178
  Least, 66
  White, 66
  Yellow, 66
Water Milfoil, Alternate, 306
  Spiked, 292
  Whorled, 292
Water Parsnip, Greater, 160
  Lesser, 162
Water-pepper, 40
  Small, 40
  Tasteless, 40
Water-plantain, Common, 258
  Floating, 258
  Lesser, 258
  Narrow-leaved, 258
  Parnassus-leaved, 258
  Ribbon-leaved, 258
Water Soldier, 258
Water-thyme, Curly, 292
  Greater, 306
  Large-flowered, 292
Water Tillaea, 102
Water Violet, 176
Waterweed, Canadian, 292
  Esthwaite, 292
Waterwort, 290
Wayfaring Tree, 222
Weld, 100
Whitebeam, 114
  Broad-leaved, 301
  Swedish, 301
Whitlow-grass, Common, 96
  Rock, 96
  Twisted, 96
  Wall, 96
  Yellow, 84

Willow, Almond, 26
  Bay, 26
  Bluish, 28
  Crack, 26
  Creeping, 28
  Cricket-bat, 26
  Dark-leaved, 28
  Downy, 28
  Dwarf, 28
  Eared, 26
  Goat, 26
  Grey, 26
  Mountain, 28
  Net-leaved, 28
  Purple, 26
  Tea-leaved, 28
  Violet, 26
  Weeping, 26
  White, 26
  Whortle-leaved, 28
  Woolly, 28
Willowherb, Alpine, 150
  American, 150
  Broad-leaved, 150
  Chickweed, 150
  Great, 150
  Hoary, 150
  Marsh, 150
  New Zealand, 150
  Pale, 150
  Rosebay, 150
  Short-fruited, 150
  Spear-leaved, 150
  Square-stemmed, 150
Windflower, Snowdrop, 76
Wineberry, 307
Wintercress, American, 86
  Common, 86
  Medium-flowered, 86
  Small-flowered, 86
Wintergreen, Chickweed, 176
  Common, 168
  Intermediate, 168
  One-flowered, 168
  Round-leaved, 168
  Toothed, 168
  Umbellate, 168
  Yellow, 168
Woad, 86
Wolfsbane, 74
  Northern, 74
Woodruff, 184
  Blue, 184
  Dyer's, 303
Wood-rush, 296
Wood-sorrel, 132
Wormwood, 238
  Field, 238
  Scottish, 238
  Sea, 238
Woundwort, Annual, 202
  Downy, 202
  Field, 202
  Hedge, 202
  Limestone, 202
  Marsh, 202
  Yellow, 202

Yarrow, 238
Yellowcress, Austrian, 299
  Creeping, 86
  Great, 86
  Iceland, 86
  Marsh, 86
Yellow-wort, 180
Yew, 24

Index of Scientific Names

Abies alba, 24
  grandis, 24
  procera, 24
Acaena anserinifolia, 106
Acer campestre, 34
  monspessulanus, 34
  negundo, 34
  opalus, 34
  platanoides, 34
  pseudoplatanus, 34
Aceras anthropophorum, 280
Achillea collina, 305
  millefolium, 238
  nobilis, 305
  pannonica, 305
  ptarmica, 238
  setacea, 305
Acinos arvensis, 204
Aconitum firmum, 74
  napellus, 74
  septentrionale, 74
  variegatum, 74
  vulparia, 74
Actaea erythrocarpa, 72
  spicata, 72
Adonis aestivalis, 74
  annua, 74
  flammea, 74
  vernalis, 74
Adoxa moschatellina, 220
Aegopodium podagraria, 158
Aesculus hippocastanum, 36
Aethusa cynapium, 158
Agrimonia eupatoria, 106
  procera, 106
Agrostemma githago, 60
Ailanthus altissima, 36
Ajuga chamaepitys, 196
  genevensis, 196
  pyramidalis, 196
  reptans, 196
Alcea rosea, 142
Alchemilla alpina, 106
  conjuncta, 106
  vulgaris, 106
Alisma gramineum, 258
  lanceolatum, 258
  plantago-aquatica, 258
Alliaria petiolata, 90
Allium ampeloprasum, 264
  angulosum, 305
  babingtonii, 264
  carinatum, 264
  cepa, 305
  fistulosum, 264
  oleraceum, 264
  paradoxum, 264
  porrum, 305
  roseum, 264
  rotundum, 264
  schoenoprasum, 264
  scorodoprasum, 264
  senescens, 264
  sphaerocephalon, 264
  strictum, 305
  suaveolens, 305
  triquetrum, 264
  ursinum, 264
  victorialis, 305
  vineale, 264

Alnus glutinosa, 30
  incana, 30
  viridis, 298
Althaea hirsuta, 142
  officinalis, 142
Alyssum alyssoides, 84
  montanum, 84
  saxatile, 84
Amaranthus albus, 48
  blitoides, 48
  bouchonii, 48
  caudatus, 48
  cruentus, 48
  graecizans, 48
  hybridus, 48
  lividus, 48
  retroflexus, 48
Ambrosia artemisiifolia, 230
Amelanchier grandiflora, 307
  ovalis, 114
  spicata, 307
Ammi majus, 156
  visnaga, 156
Amsinckia intermedia, 188
Anacamptis pyramidalis, 278
Anagallis arvensis, 176
  foemina, 176
  minima, 176
  tenella, 176
Anaphalis margaritacea, 234
Anarrhinum bellidifolium, 210
Anchusa azurea, 192
  ochroleuca, 188
  officinalis, 192
Andromeda polifolia, 170
Androsace elongata, 176
  lactea, 176
  maxima, 176
  septentrionalis, 176
Anemone apennina, 76
  narcissiflora, 76
  nemorosa, 76
  ranunculoides, 76
  sylvestris, 76
Anethum graveolens, 166
Angelica archangelica, 160
  palustris, 160
  sylvestris, 160
Antennaria alpina, 234
  carpatica, 234
  dioica, 234
  porsildii, 234
Anthemis arvensis, 232
  austriaca, 236
  cotula, 232
  tinctoria, 236
Anthericum liliago, 260
  ramosum, 260
Anthriscus caucalis, 156
  cerefolium, 156
  nitida, 308
  sylvestris, 156
Anthyllis vulneraria, 126
Antirrhinum majus, 210
Aphanes arvensis, 106
Apium graveolens, 162
  inundatum, 162
  nodiflorum, 162
  repens, 162
Aposeris foetida, 305
Aquilegia vulgaris, 74

Arabidopsis thaliana, 94
  suecica, 94
Arabis alpina, 94
  caucasica, 94
  glabra, 94
  hirsuta, 94
  pauciflora, 94
  recta, 94
  stricta, 94
  turrita, 94
Arbutus unedo, 172
Arctium lappa, 244
  minus, 244
  tomentosum, 244
Arctostaphylos uva-ursi, 172
Arctous alpina, 172
Aremonia agrimonioides, 106
Arenaria balearica, 52
  ciliata, 52
  gothica, 52
  humifusa, 52
  norvegica, 52
  serpyllifolia, 52
Aristolochia clematitis, 38
Armeria arenaria, 303
  maritima, 178
  maritima ssp. elongata, 178
Armoracia rusticana, 92
Arnica alpina, 236
  montana, 236
Arnoseris minima, 252
Aronia melanocarpa, 307
Artemisia absinthium, 238
  austriaca, 305
  borealis, 309
  campestris, 238
  dracunculus, 305
  laciniata, 305
  maritima, 238
  norvegica, 238
  pontica, 305
  rupestris, 305
  verlotorum, 238
  vulgaris, 238
Arthrocnemum perenne, 48
Arum italicum, 270
  maculatum, 270
Aruncus dioicus, 106
Asarum europaeum, 38
Asparagus officinalis, 268
Asperugo procumbens, 190
Asperula arvensis, 184
  cynanchica, 184
  tinctoria, 303
Aster alpinus, 232
  amellus, 305
  novae-angliae, 232
  novi-belgii, 232
  salignus, 232
  tripolium, 232
Astragalus alpinus, 118
  arenarius, 118
  cicer, 118
  danicus, 118
  frigidus, 118
  glycyphyllos, 118
  norvegicus, 118
  penduliflorus, 118
Astrantia major, 154
Athamanta cretensis, 158
Atriplex calotheca, 299

glabriuscula, 308
halimus, 307
hastata, 46
laciniata, 46
littoralis, 46
longipes, 308
oblongifolia, 299
patula, 46
rosea, 46
Atropa bella-donna, 206
Azolla filiculoides, 292

Baldellia ranunculoides, 258
repens, 258
Ballota alba, 198
nigra, 198
Barbarea intermedia, 86
stricta, 86
verna, 86
vulgaris, 86
Bartsia alpina, 212
Bassia hirsuta, 48
Bellidiastrum michelii, 305
Bellis perennis, 232
Berberis vulgaris, 78
Berteroa incana, 96
Berula erecta, 162
Beta vulgaris maritima, 46
Betonica officinalis, 202
Betula humilis, 298
nana, 30
pendula, 30
pubescens, 30
Bidens cernua, 238
connata, 238
frondosa, 238
radiata, 238
tripartita, 238
Bifora radians, 156
Bilderdykia convolvulus, 42
dumetorum, 42
Biscutella apricorum, 308
divoniensis, 308
guilloni, 308
laevigata, 84
neustriaca, 308
Blackstonia perfoliata, 180
serotina, 180
Borago officinalis, 192
Botrychium lunaria, 270
Brassica juncea, 88
napobrassica, 88
napus, 88
nigra, 88
oleracea, 88
rapa, 88
Braya linearis, 94
Bryonia alba, 148
cretica, 148
Buddleja davidii, 308
Bunias erucago, 86
orientalis, 86
Bunium bulbocastanum, 158
Bupleurum baldense, 166
falcatum, 166
gerardi, 164
lancifolium, 166
longifolium, 166
rotundifolium, 166
tenuissimum, 164
Buphthalmum salicifolium, 236
Butomus umbellatus, 258
Buxus sempervirens, 140

Cakile maritima, 92
Calamintha nepeta, 204
sylvatica, 204
Caldesia parnassifolia, 258

Calendula arvensis, 240
officinalis, 240
Calepina irregularis, 92
Calla palustris, 270
Callitriche hermaphroditica, 309
intermedia, 309
obtusangula, 309
palustris, 309
platycarpa, 309
stagnalis, 292
truncata, 309
Calluna vulgaris, 170
Caltha palustris, 68
Calypso bulbosa, 274
Calystegia pulchra, 184
sepium, 184
soldanella, 184
sylvatica, 184
Camelina alyssum, 84
macrocarpa, 84
microcarpa, 84
sativa, 84
Campanula barbata, 226
baumgartenii, 304
bononiensis, 226
cervicaria, 226
cochleariifolia, 304
glomerata, 226
latifolia, 226
medium, 226
patula, 226
persicifolia, 226
rapunculoides, 226
rapunculus, 226
rhomboidalis, 304
rotundifolia, 226
scheuchzeri, 304
trachelium, 226
uniflora, 304
Cannabis sativa, 38
Capsella bursa-pastoris, 96
rubella, 96
Cardamine amara, 90
bellidifolia, 90
bulbifera, 90
enneaphyllos, 90
flexuosa, 94
heptaphylla, 90
hirsuta, 94
impatiens, 90
nymanii, 308
palustris, 308
parviflora, 90
pentaphyllos, 90
pratensis, 90
raphanifolia, 90
Cardaminopsis arenosa, 94
halleri, 94
petraea, 94
Cardaria draba, 98
Carduus acanthoides, 246
crispus, 246
defloratus, 246
nutans, 246
palustre, 246
personata, 246
tenuiflorus, 246
Carex acutiformis, 296
paniculata, 296
rostrata, 296
Carlina acaulis, 244
vulgaris, 244
Carpinus betulus, 32
Carpobrotus edulis, 42
Carthamus lanatus, 248
tinctorius, 248
Carum carvi, 156
verticillatum, 156

Cassiope hypnoides, 172
tetragona, 172
Castanea sativa, 32
Caucalis platycarpos, 156
Centaurea aspera, 248
calcitrapa, 248
cyanus, 248
jacea, 248
melitensis, 248
montana, 248
nigra, 248
paniculata, 248
scabiosa, 248
solstitialis, 248
stoebe, 248
Centaurium erythraea, 180
glomeratum, 180
littorale, 180
pulchellum, 180
scilloides, 180
tenuifolium, 180
Centranthus ruber, 222
Cephalanthera damasonium, 282
longifolia, 282
rubra, 282
Cerastium alpinum, 54
arcticum, 54
arvense, 54
brachypetalum, 54
cerastoides, 54
diffusum, 54
dubium, 299
fontanum, 54
glomeratum, 54
pumilum, 54
semidecandrum, 54
tomentosum, 54
Ceratophyllum demersum, 292
submersum, 306
Cerinthe minor, 188
Chaenorhinum minus, 210
Chaerophyllum aromaticum, 156
aureum, 156
bulbosum, 156
hirsutum, 156
temulentum, 156
Chamaecyparis lawsoniana, 24
Chamaecytisus ratisbonensis, 300
supinus, 120
Chamaedaphne calyculata, 172
Chamaemelum nobile, 232
Chamaespartium sagittale, 120
Chamorchis alpina, 280
Cheiranthus cheiri, 86
Chelidonium majus, 80
Chenopodium album, 46
bonus-henricus, 46
botryodes, 46
ficifolium, 46
glaucum, 46
hybridum, 46
murale, 46
opulifolium, 299
polyspermum, 46
rubrum, 46
suecicum, 299
urbicum, 46
vulvaria, 46
Chimaphila umbellata, 168
Chondrilla juncea, 252
Chrysanthemum segetum, 240

328

Chrysosplenium
  alternifolium, 100
  oppositifolium, 100
  tetrandrum, 100
Cicendia filiformis, 180
Cicerbita alpina, 250
  macrophylla, 250
  plumieri, 250
  sibirica, 250
  tatarica, 250
Cichorium intybus, 250
Cicuta virosa, 160
Circaea alpina, 148
  × *intermedia*, 148
  lutetiana, 148
Cirsium acaule, 246
  arvense, 246
  dissectum, 246
  eriophorum, 246
  eristithales, 244
  heterophyllum, 246
  oleraceum, 244
  rivulare, 246
  spinosissimum, 244
  tuberosum, 246
  vulgare, 246
Clematis alpina, 72
  recta, 72
  vitalba, 72
Clinopodium vulgare, 204
Cnidium dubium, 164
Cochlearia aestuaria, 92
  anglica, 92
  danica, 92
  fenestrata, 92
  groenlandica, 92
  officinalis, 92
Coeloglossum viride, 280
Colchicum autumnale, 266
Colutea arborescens, 120
Conioselinum tataricum, 160
Conium maculatum, 160
Conopodium majus, 158
Conringia orientalis, 299
Consolida ambigua, 74
  orientalis, 74
  regalis, 74
Convallaria majalis, 260
Convolvulus arvensis, 184
Conyza canadensis, 230
Corallorhiza trifida, 286
Coriandrum sativum, 156
Corispermum leptopterum, 48
  marschalii, 48
Cornus alba, 154
  mas, 154
  sanguinea, 154
  sericea, 154
  suecica, 154
Coronilla coronata, 120
  emerus, 120
  minima, 120
  vaginalis, 120
  varia, 118
Coronopus didymus, 98
  squamatus, 98
Corrigiola litoralis, 56
Corydalis bulbosa, 78
  claviculata, 78
  intermedia, 78
  lutea, 78
  ochroleuca, 78
  pumila, 78
  solida, 78
Corylus avellana, 32
Cotoneaster horizontalis, 307
  integerrimus, 114
  microphyllus, 114
  nebrodensis, 301

  niger, 114
  simonsii, 307
Cotula coronopifolia, 238
Crambe maritima, 92
Crassula aquatica, 102
  tillaea, 102
  vaillantii, 102
Crataegus calycina, 301
  laevigata, 114
  monogyna, 114
Crepis biennis, 256
  capillaris, 254
  foetida, 256
  mollis, 256
  nicaeensis, 256
  paludosa, 256
  praemorsa, 256
  pulchra, 256
  setosa, 256
  tectorum, 256
  vesicaria, 256
Crinitaria linosyris, 230
Crithmum maritimum, 166
Crocosmia × *crocosmiiflora*, 272
Crocus albiflorus, 272
  nudiflorus, 272
Cruciata glabra, 186
  laevipes, 186
Cucubalus baccifer, 62
Cuscuta epilinum, 303
  epithymum, 184
  europaea, 184
Cyclamen europaeum, 176
  hederifolium, 176
  purpurascens, 176
Cydonia oblonga, 307
Cymbalaria muralis, 210
Cynoglossum germanicum, 188
  officinale, 188
Cypripedium calceolus, 274
Cytisus scoparius, 120

Daboecia cantabrica, 170
Dactylorhiza cruenta, 278
  fuchsii, 278
  incarnata, 278
  maculata, 278
  majalis, 278
  praetermissa, 278
  purpurella, 278
  sambucina, 278
  traunsteineri, 278
Daphne laureola, 138
  mezereum, 138
Damasonium alisma, 258
Datura stramonium, 206
Daucus carota, 156
Descurainia sophia, 86
Dianthus arenarius, 64
  armeria, 64
  barbatus, 64
  carthusianorum, 64
  caryophyllus, 64
  deltoides, 64
  gallicus, 64
  gratianopolitanus, 64
  plumarius, 64
  seguieri, 64
  superbus, 64
Diapensia lapponica, 168
Dictamnus albus, 138
Digitalis grandiflora, 212
  lutea, 212
  purpurea, 212
Diplotaxis muralis, 84
  tenuifolia, 84
  viminea, 84

Dipsacus fullonum, 224
  laciniatus, 224
  pilosus, 224
Doronicum pardalianches, 236
  plantagineum, 236
Dorycnium pentaphyllum, 301
Draba aizoides, 84
  alpina, 84
  cacuminum, 308
  cinerea, 308
  crassifolia, 84
  daurica, 96
  fladnizensis, 96
  glacialis, 308
  incana, 96
  muralis, 96
  nemorosa, 96
  nivalis, 96
  norvegica, 96
Dracocephalum ruyschiana, 198
  thymifolium, 198
Drosera anglica, 100
  intermedia, 100
  rotundifolia, 100
Dryas octopetala, 110

Echinocystis lobata, 148
Echinops commutatus, 244
  sphaerocephalus, 244
Echium lycopsis, 192
  vulgare, 192
Egeria densa, 292
Elatine alsinastrum, 290
  hexandra, 309
  hydropiper, 309
  triandra, 309
Elodea callitrichoides, 306
  canadensis, 292
  nuttallii, 292
Empetrum hermaphroditum, 308
  nigrum, 172
Endymion hispanicus, 266
  nonscriptus, 266
Ephedra distachya, 298
Epilobium adenocaulon, 150
  alpestre, 150
  alsinifolium, 150
  anagallidifolium, 150
  angustifolium, 150
  brunnescens, 150
  collinum, 150
  davuricum, 150
  dodonaei, 150
  duriaei, 308
  glandulosum, 150
  hirsutum, 150
  hornemanii, 150
  komarovianum, 150
  lactiflorum, 150
  lanceolatum, 150
  latifolium, 150
  montanum, 150
  nutans, 150
  obscurum, 150
  palustre, 150
  parviflorum, 150
  pedunculare, 150
  roseum, 150
  tetragonum, 150
Epipactis atrorubens, 284
  confusa, 284
  dunensis, 284
  helleborine, 284
  leptochila, 284
  microphylla, 284
  muelleri, 284

palustris, 284
phyllanthes, 284
purpurata, 284
Epipogium aphyllum, 286
Equisetum fluviatile, 296
  palustre, 296
Eranthis hyemalis, 66
Erica carnea, 170
  ciliaris, 170
  cinerea, 170
  erigena, 170
  mackaiana, 170
  tetralix, 170
  vagans, 170
Erigeron acer, 232
  alpinus, 232
  mucronatus, 232
  unalaschenensis, 232
  uniflorus, 232
Erinus alpinus, 212
Eriocaulon aquaticum, 260
Erodium cicutarium, 132
  maritimum, 132
  moschatum, 132
Erophila verna, 96
Erucastrum gallicum, 88
  nasturtiifolium, 88
Eryngium campestre, 154
  maritimum, 154
  planum, 154
Erysimum cheiranthoides, 86
  crepidifolium, 299
  hieracifolium, 299
  odoratum, 299
  repandum, 299
Escallonia macrantha, 307
Euonymus europaeus, 140
  latifolius, 140
Eupatorium cannabinum, 230
Euphorbia amygdaloides, 136
  brittingeri, 302
  cyparissias, 136
  dulcis, 302
  esula, 136
  exigua, 136
  helioscopia, 136
  hyberna, 136
  lathyrus, 136
  lucida, 136
  palustris, 302
  paralias, 136
  peplis, 136
  peplus, 136
  platyphyllos, 136
  portlandica, 136
  seguieriana, 136
  serrulata, 136
  villosa, 302
Euphrasia officinalis, 216
  salisburgensis, 216
Exaculum pusillum, 180

Fagopyrum esculentum, 40
  tataricum, 40
Fagus sylvatica, 32
Falcaria vulgaris, 160
Ficus carica, 307
Filago arvensis, 234
  gallica, 234
  lutescens, 234
  minima, 234
  neglecta, 234
  pyramidata, 234
  vulgaris, 234
Filipendula ulmaria, 106
  vulgaris, 106
Foeniculum vulgare, 166
Fragaria × ananassa, 110
  moschata, 110

vesca, 110
  viridis, 110
Frangula alnus, 140
Frankenia laevis, 148
Fraxinus excelsior, 36
  ornus, 36
Fritillaria meleagris, 262
Fuchsia magellanica, 307
Fumana procumbens, 148
Fumaria bastardii, 78
  capreolata, 78
  densiflora, 78
  martinii, 78
  muralis, 78
  occidentalis, 78
  officinalis, 78
  parviflora, 78
  purpurea, 78
  rostellata, 308
  schleicheri, 78
  schrammii, 308
  vaillantii, 78

Gagea arvensis, 262
  bohemica, 262
  lutea, 262
  minima, 262
  pratensis, 262
  spathacea, 262
Galanthus nivalis, 270
Galega officinalis, 118
Galeopsis angustifolia, 200
  bifida, 309
  dubium, 309
  ladanum, 309
  pubescens, 200
  segetum, 200
  speciosa, 200
  tetrahit, 200
Galinsoga ciliata, 232
  parviflora, 232
Galium aparine, 186
  boreale, 186
  debile, 186
  glaucum, 303
  mollugo, 186
  odoratum, 184
  palustre, 186
  parisiense, 186
  pumilum, 186
  rotundifolium, 186
  saxatile, 186
  spurium, 186
  sterneri, 186
  sylvaticum, 186
  timeroyi, 186
  tricornutum, 186
  trifidum, 186
  triflorum, 186
  uliginosum, 186
  verum, 186
Gaultheria shallon, 307
Genista anglica, 120
  germanica, 120
  pilosa, 120
  tinctoria, 120
Gentiana asclepiadea, 182
  clusii, 182
  cruciata, 182
  detonsa, 182
  lutea, 180
  nivalis, 182
  pneumonanthe, 182
  punctata, 182
  purpurea, 182
  utriculosa, 182
  verna, 182
Gentianella amarella, 182
  anglica, 182

aspera, 182
  aurea, 180
  baltica, 309
  campestris, 182
  ciliata, 182
  germanica, 182
  tenella, 182
  uliginosa, 182
Geranium bohemicum, 134
  columbinum, 134
  dissectum, 134
  divaricatum, 302
  endressii, 134
  lanuginosum, 308
  lucidum, 134
  molle, 134
  nodosum, 134
  palustre, 134
  phaeum, 134
  pratense, 134
  purpureum, 134
  pusillum, 134
  pyrenaicum, 134
  robertianum, 134
  rotundifolium, 134
  sanguineum, 134
  sylvaticum, 134
  versicolor, 134
Geum hispidum, 110
  × intermedium, 110
  montanum, 110
  rivale, 110
  urbanum, 110
Gladiolus illyricus, 272
  palustris, 272
Glaucium corniculatum, 80
  flavum, 80
Glaux maritima, 176
Glechoma hederacea, 198
Globularia vulgaris, 218
Glyceria maxima, 296
Gnaphalium luteoalbum, 234
  norvegicum, 234
  supinum, 234
  sylvaticum, 234
  uliginosum, 234
  undulatum, 234
Goodyera repens, 282
Gratiola officinalis, 212
Groenlandia densa, 294
Guizotia abyssinica, 236
Gymnadenia conopsea, 278
  odoratissima, 278
Gypsophila fastigiata, 62
  muralis, 62
  repens, 62

Hackelia deflexa, 303
Halimione pedunculata, 48
  portulacoides, 48
Hedera helix, 154
Hedysarum hedysaroides, 126
Helianthemum apenninum, 148
  canum, 148
  nummularium, 148
  oelandicum, 148
Helianthus annuus, 236
  rigidus, 236
  tuberosus, 236
Helichrysum arenarium, 234
Heliotropium europaeum, 188
Helleborus foetidus, 66
  viridis, 66
Hepatica nobilis, 76
Heracleum mantegazzianum, 160
  sphondylium, 160
Herminium monorchis, 280

Herniaria ciliolata, 56
  glabra, 56
  hirsuta, 299
Hesperis matronalis, 90
Hieracium alpinum, 256
  maculatum, 256
  murorum, 256
  umbellatum, 256
Himantoglossum hircinum,
  280
Hippophae rhamnoides, 140
Hippocrepis comosa, 128
Hippuris vulgaris, 292
Hirschfeldia incana, 88
Holosteum umbellatum, 54
Homogyne alpina, 240
Honkenya peploides, 52
Hornungia petraea, 96
Hottonia palustris, 176
Humulus lupulus, 38
Hydrocharis morsus-ranae,
  258
Hydrocotyle vulgaris, 154
Hymenolobus procumbens,
  96
Hyoscyamus niger, 206
Hypericum androsaemum,
  144
  calycinum, 144
  canadense, 144
  elegans, 308
  elodes, 144
  hircinum, 144
  hirsutum, 144
  humifusum, 144
  inodorum, 144
  linarifolium, 144
  maculatum, 144
  majus, 144
  montanum, 144
  mutilum, 144
  perforatum, 144
  pulchrum, 144
  tetrapterum, 144
  undulatum, 144
Hypochaeris glabra, 254
  maculata, 254
  radicata, 254
Hyssopus officinalis, 198

Iberis amara, 96
  intermedia, 96
  umbellata, 96
Ilex aquifolium, 36
Illecebrum verticillatum, 56
Impatiens capensis, 138
  glandulifera, 138
  noli-tangere, 138
  parviflora, 138
Inula britannica, 236
  conyza, 230
  crithmoides, 230
  ensifolia, 236
  germanica, 236
  graveolens, 230
  helenium, 236
  helvetica, 236
  hirta, 236
  salicina, 236
Iris aphylla, 272
  foetidissima, 272
  germanica, 272
  pseudacorus, 272
  sambucina, 272
  sibirica, 272
  spuria, 272
  versicolor, 272
Isatis tinctoria, 86
Isoetes lacustris, 290

  setacea, 290

Jasione montana, 224
  perennis, 224
Jovibarba sobolifera, 102
Juglans regia, 30
Juncus articulatus, 296
  effusus, 296
Juniperus communis, 24
Jurinea cyanoides, 248

Kernera saxatilis, 92
Kickxia elatine, 210
  spuria, 210
Knautia arvensis, 224
  sylvatica, 224
Kochia laniflora, 48
  scoparia, 48
Koenigia islandica, 42

Laburnum anagyroides, 120
Lactuca perennis, 250
  quercina, 252
  saligna, 252
  serriola, 252
  viminea, 252, 305
  virosa, 252
Lagarosiphon major, 292
Lagoseris sancta, 256
Lamiastrum galeobdolon, 200
Lamium album, 200
  amplexicaule, 200
  hybridum, 200
  maculatum, 200
  purpureum, 200
Lappula myosotis, 190
Lapsana communis, 252
Larix decidua, 24
  kaempferi, 298
Laser trilobum, 164
Laserpitium latifolium, 164
  prutenicum, 164
  siler, 164
Lathraea clandestina, 218
  squamaria, 218
Lathyrus aphaca, 124
  heterophyllus, 124
  hirsutus, 124
  japonicus, 124
  latifolius, 124
  montanus, 124
  montanus var. tenuifolius,
  124
  niger, 124
  nissolia, 124
  palustris, 124
  pannonicus, 124
  pratensis, 124
  sphaericus, 124
  sylvestris, 124
  tuberosus, 124
  vernus, 124
Lavatera arborea, 142
  cretica, 142
  thuringiaca, 142
Ledum palustre, 172
Legousia hybrida, 226
  speculum-veneris, 226
Lembotropis nigricans, 120
Lemna gibba, 306
  minor, 292
  polyrrhiza, 306
  trisulca, 292
Leontodon autumnalis, 254
  helveticus, 305
  hispidus, 254
  hyoseroides, 305
  incanus, 305
  taraxacoides, 254

Leonurus cardiaca, 200
  marrubiastrum, 200
Lepidium bonariense, 98
  campestre, 98
  densiflorum, 98
  divaricatum, 98
  graminifolium, 300
  heterophyllum, 98
  latifolium, 92
  neglectum, 98
  perfoliatum, 300
  ruderale, 98
  sativum, 98
  virginicum, 98
Leucanthemum maximum,
  240
  vulgare, 240
Leucojum aestivum, 270
  vernum, 270
Levisticum officinale, 166
Leycesteria formosa, 308
Ligusticum mutellinum, 164
  scoticum, 164
Ligustrum ovalifolium, 178
  vulgare, 178
Lilium bulbiferum, 262
  martagon, 262
  pyrenaicum, 262
Limodorum abortivum, 286
Limonium auriculae-
  ursifolium, 303
  bellidifolium, 178
  binervosum, 178
  humile, 178
  paradoxum, 309
  recurvum, 309
  transwallianum, 309
  vulgare, 178
Limosella aquatica, 290
  australis, 306
Linaria alpina, 210
  arenaria, 210
  arvensis, 210
  bipartita, 210
  genistifolia, 210
  pelisseriana, 210
  purpurea, 210
  repens, 210
  supina, 210
  vulgaris, 210
Lindernia pyxidaria, 303
Linnaea borealis, 222
Linum austriacum, 308
  bienne, 132
  catharticum, 132
  flavum, 132
  leonii, 308
  perenne, 132
  suffruticosum, 308
  tenuifolium, 302
  usitatissimum, 132
  viscosum, 302
Liparis loeselii, 286
Listera cordata, 280
  ovata, 280
Lithospermum arvense, 188
  officinale, 188
  purpuro-caeruleum, 188
Littorella uniflora, 290
Lloydia serotina, 260
Lobelia dortmanna, 290
  urens, 226
Lobularia maritima, 96
Loiseleuria procumbens, 172
Lomatogonium rotatum, 182
Lonicera alpigena, 304
  caerulea, 304
  caprifolium, 222
  nigra, 304

periclymenum, 222
xylosteum, 222
Lotus angustissimus, 128
corniculatus, 128
subbiflorus, 128
tenuis, 128
uliginosus, 128
Ludwigia palustris, 290
Lunaria annua, 90
rediviva, 90
Lupinus albus, 120
angustifolia, 120
arboreus, 120
luteus, 120
micranthus, 120
nootkatensis, 120
polyphyllus, 120
Luronium natans, 258
Lychnis alpina, 60
flos-cuculi, 60
viscaria, 60
Lycopersicum esculentum, 206
Lycopsis arvensis, 192
Lycopus europaeus, 204
Lysimachia nemorum, 174
nummularia, 174
punctata, 174
thyrsiflora, 174
vulgaris, 174
Lythrum borysthenicum, 306
hyssopifolia, 150
portula, 290
salicaria, 150

Maianthemum bifolium, 260
Mahonia aquifolium, 78
Malaxis monophyllos, 286
paludosa, 286
Malus domestica, 114
sylvestris, 114
Malva alcea, 142
moschata, 142
neglecta, 142
parviflora, 142
pusilla, 142
sylvestris, 142
verticillata, 142
Marrubium vulgare, 198
Matricaria matricarioides, 232
recutita, 232
Matthiola incana, 92
sinuata, 92
Meconopsis cambrica, 80
Medicago arabica, 128
falcata, 126
lupulina, 128
minima, 128
polymorpha, 128
sativa, 126
Melampyrum arvense, 216
cristatum, 216
nemorosum, 216
paludosum, 216
pratense, 216
sylvaticum, 216
Melilotus alba, 126
altissima, 126
dentata, 126
indica, 126
officinalis, 126
Melissa officinalis, 202
Melittis melissophyllum, 198
Mentha aquatica, 204
arvensis, 204
× piperita, 204
pulegium, 204
rotundifolia, 303
spicata, 204

× verticillata, 204
× villosa var.
alopecuroides, 303
Menyanthes trifoliata, 178
Mercurialis annua, 136
ovata, 136
perennis, 136
Mertensia maritima, 192
Mespilus germanica, 307
Meum athamanticum, 158
Micropus erectus, 234
Mimulus guttatus, 212
luteus, 212
moschatus, 212
Minuartia biflora, 52
hybrida, 52
mediterranea, 52
recurva, 52
rubella, 52
rubra, 52
sedoides, 52
setacea, 52
stricta, 52
verna, 52
viscosa, 52
Misopates orontium, 210
Moehringia muscosa, 52
trinervia, 52
Moenchia erecta, 54
Moneses uniflora, 168
Monotropa hypophegea, 308
hypopitys, 168
Montia fontana, 42
perfoliata, 42
sibirica, 42
Muscari atlanticum, 266
botryodes, 266
comosum, 266
neglectum, 266
Myagrum perfoliatum, 84
Mycelis muralis, 252
Myosotis alpestris, 303
arvensis, 190
baltica, 190
caespitosa, 190
discolor, 190
laxa, 309
ramosissima, 190
secunda, 190
scorpioides, 190
sicula, 190
sparsiflora, 303
stolonifera, 190
stricta, 190
sylvatica, 190
Myosoton aquaticum, 54
Myosurus minimus, 74
Myrica carolinensis, 298
gale, 30
Myricaria germanica, 302
Myriophyllum alterniflorum, 306
spicatum, 292
verticillatum, 292
Myrrhis odorata, 156

Najas flexilis, 294
marina, 294
minor, 306
Narcissus × biflorus, 270
obvallaris, 270
poeticus, 270
pseudonarcissus, 270
stellaris, 270
Narthecium ossifragum, 262
Nasturtium microphyllum, 90
officinale, 90
Neotinea intacta, 280
Neottia nidus-avis, 286

Nepeta cataria, 198
grandiflora, 198
nuda, 198
Neslia paniculata, 84
Nicandra physaloides, 206
Nicotiana rustica, 206
Nigella arvensis, 66
damascena, 66
Nigritella nigra, 274
Nonea pulla, 188
rosea, 188
Nuphar lutea, 66
pumila, 66
Nymphaea alba, 66
candida, 66
Nymphoides peltata, 178

Odontites jaubertiana, 212
lutea, 212
verna, 212
Oenanthe aquatica, 162
crocata, 160
fistulosa, 162
fluviatilis, 162
lachenalii, 162
peucedanifolia, 162
pimpinelloides, 162
silaifolia, 162
Oenothera ammophila, 308
biennis, 148
chicagoensis, 308
erythrosepala, 148
parviflora, 148
rubricaulis, 308
stricta, 148
strigosa, 308
suaveolens, 308
Omphalodes scorpioides, 190
verna, 190
Onobrychis arenaria, 126
viciifolia, 126
Ononis arvensis, 126
natrix, 126
pusilla, 126
reclinata, 126
repens, 126
spinosa, 126
Onopordum acanthium, 244
Onosma arenarium, 192
Ophioglossum vulgatum, 270
Ophrys apifera, 274
apifera var. trollii, 274
fuciflora, 274
insectifera, 274
sphegodes, 274
Orchis coriophora, 276
laxiflora, 276
mascula, 276
militaris, 276
morio, 276
pallens, 276
palustris, 276
purpurea, 276
simia, 276
tridentata, 276
ustulata, 276
Origanum vulgare, 204
Orlaya grandiflora, 156
Ornithogalum nutans, 268
pyrenaicum, 268
umbellatum, 268
Ornithopus perpusillus, 128
pinnatus, 128
Orobanche alba, 218
alsatica, 304
amethystea, 304
arenaria, 304
caryophyllacea, 218
elatior, 218

flava, 304
gracilis, 304
hederae, 218
maritima, 304
minor, 218
picridis, 218
purpurea, 218
ramosa, 304
rapum-genistae, 218
reticulata, 218
Orthilia secunda, 168
Otanthus maritimus, 238
Oxalis acetosella, 132
articulata, 132
corniculata, 132
corymbosa, 302
europaea, 132
latifolia, 302
pes-caprae, 132
stricta, 302
Oxyria digyna, 44
Oxytropis campestris, 118
deflexa, 118
halleri, 118
lapponica, 118
pilosa, 118

Papaver argemone, 80
dahlianum, 308
dubium, 80
hybridum, 80
laestadianum, 308
lapponicum, 308
lecoqii, 80
radicatum, 80
rhoeas, 80
somniferum, 80
Parentucellia viscosa, 212
Parietaria judaica, 38
officinalis, 38
Paris quadrifolia, 268
Parnassia palustris, 100
Parthenocissus inserta, 307
quinquefolia, 307
Pastinaca sativa, 166
Pedicularis flammea, 216
foliosa, 216
hirsuta, 216
lapponica, 216
oederi, 216
palustris, 216
sceptrum-carolinum, 216
sylvatica, 216
Pentaglottis sempervirens,
192
Pernettya mucronata, 307
Petasites albus, 238
fragrans, 238
frigidus, 238
hybridus, 238
japonicus, 305
spurius, 238
Petrorhagia nanteulii, 62
prolifera, 62
saxifraga, 62
Petroselinum crispum, 162
segetum, 162
Peucedanum alsaticum, 164
carvifolia, 164
cervaria, 164
gallicum, 164
lancifolium, 164
officinale, 164
oreoselinum, 164
ostruthium, 164
palustre, 164
Phragmites australis, 296
Phyllodoce caerulea, 170

Physocarpus opulifolius, 307
Physospermum cornubiense,
158
Phyteuma nigrum, 224
orbiculare, 224
spicatum, 224
tenerum, 224
Picea abies, 24
sitchensis, 24
Picris echioides, 256
hieracioides, 256
Pilosella aurantiacum, 256
lactucella, 305
officinarum, 254
Pimpinella major, 158
saxifraga, 158
Pinguicula alpina, 304
grandiflora, 218
lusitanica, 218
villosa, 304
vulgaris, 218
Pinus contorta, 24
mugo, 24
nigra, 24
pinaster, 24
radiata, 24
strobus, 24
sylvestris, 24
Pittosporum crassifolium, 307
Plantago coronopus, 220
indica, 220
lanceolata, 220
major, 220
maritima, 220
media, 220
Platanthera bifolia, 282
chlorantha, 282
Platanus hybrida, 34
Pleurospermum austriacum,
160
Podospermum laciniatum,
250
Polemonium acutifolium, 303
caeruleum, 184
Polycarpon diphyllum, 56
tetraphyllum, 56
Polycnemum arvense, 299
majus, 46
Polygala amarella, 302
calcarea, 302
chamaebuxus, 138
comosa, 302
serpyllifolia, 302
vulgaris, 138
Polygonatum × hybridum,
268
multiflorum, 268
odoratum, 268
verticillatum, 268
Polygonum amphibium, 40
arenastrum, 40
aviculare, 40
bistorta, 40
boreale, 40
foliosum, 308
hydropiper, 40
lapathifolium, 40
maritimum, 40
minus, 40
mite, 40
oxyspermum, 40
patulum, 40
persicaria, 40
polystachyum, 42
rurivagum, 40
viviparum, 40
Populus alba, 30
× canadensis, 30

canescens, 30
gileadensis, 30
nigra, 30
nigra var. italica, 30
tremula, 30
Potamogeton acutifolius, 306
alpinus, 306
berchtoldii, 294
coloratus, 294
compressus, 306
crispus, 294
epihydrus, 306
filiformis, 306
friesii, 306
gramineus, 306
lucens, 294
natans, 294
nodosus, 306
obtusifolius, 306
pectinatus, 294
perfoliatus, 294
polygonifolius, 294
praelongus, 294
pusillus, 306
rutilus, 306
trichoides, 306
Potentilla alba, 110
anglica, 112
anserina, 112
argentea, 112
aurea, 112
chamissonis, 308
cinerea, 112
collina, 112
crantzii, 112
erecta, 112
fruticosa, 112
heptaphylla, 112
hyparctica, 112
inclinata, 112
intermedia, 112
micrantha, 110
montana, 110
multifida, 300
neglecta, 308
nivea, 112
norvegica, 112
palustris, 110
pusilla, 112
recta, 112
reptans, 112
rupestris, 110
sterilis, 110
supina, 300
tabernaemontani, 112
thuringiaca, 112
Prenanthes purpurea, 250
Primula auricula, 174
elatior, 174
farinosa, 174
scandinavica, 174
scotica, 174
sibirica, 174
stricta, 174
veris, 174
veris × vulgaris, 174
vulgaris, 174
Prunella grandiflora, 198
laciniata, 198
vulgaris, 198
Prunus avium, 114
cerasifera, 114
cerasus, 301
domestica, 114
fruticosa, 301
laurocerasus, 307
mahaleb, 301
padus, 114

serotina, 301
spinosa, 114
virginiana, 307
Pseudorchis albida, 282
Pseudotsuga menziesii, 24
Ptychotis saxifraga, 158
Pulicaria dysenterica, 236
vulgaris, 234
Pulmonaria angustifolia, 303
longifolia, 190
mollis, 303
officinalis, 190
Pulsatilla patens, 76
pratensis, 76
vernalis, 76
vulgaris, 76
Pyracantha coccinea, 307
Pyrola chlorantha, 168
media, 168
minor, 168
norvegica, 168
rotundifolia, 168
Pyrus communis, 114
cordata, 301
pyraster, 114
salvifolia, 301

Quercus cerris, 32
ilex, 32
petraea, 32
pubescens, 32
robur, 32
rubra, 32

Radiola linoides, 132
Ranunculus aconitifolius, 299
acris, 68
aquatilis, 70
arvensis, 70
auricomus, 70
baudotii, 70
bulbosus, 68
circinatus, 70
cymbalaria, 68
ficaria, 68
flammula, 68
fluitans, 70
glacialis, 70
hederaceus, 70
hyperboreus, 68
illyricus, 68
lanuginosus, 68
lapponicus, 68
lingua, 68
montanus, 68
nemorosus, 68
nivalis, 299
ololeucos, 70
omiophyllus, 70
ophioglossifolius, 68
oreophilus, 308
paludosus, 68
parviflorus, 70
peltatus, 70
penicillatus, 70
platanifolius, 70
polyanthemos, 68
pygmaeus, 299
repens, 68
reptans, 68
sardous, 68
sceleratus, 70
sulphureus, 299
trichophyllus, 70
tripartitus, 70
Raphanus maritimus, 88
raphanistrum, 88
sativus, 88
Rapistrum perenne, 88
rugosum, 88

Reseda alba, 100
lutea, 100
luteola, 100
phyteuma, 100
Reynoutria japonica, 42
sachalinensis, 42
Rhamnus catharticus, 140
saxatilis, 302
Rhinanthus minor, 216
Rhodiola rosea, 100
Rhododendron lapponicum, 170
luteum, 170, 307
ponticum, 170, 307
Rhynchosinapis
cheiranthos, 88
monensis, 88
wrightii, 88
Ribes alpinum, 140
nigrum, 140
petraeum, 302
rubrum, 140
sanguineum, 307
spicatum, 308
uva-crispa, 140
Robinia pseudacacia, 120
Romulea columnae, 272
Rorippa amphibia, 86
austriaca, 299
islandica, 86
palustris, 86
pyrenaica, 86
sylvestris, 86
Rosa arvensis, 108
canina, 108
foetida, 307
gallica, 108
mollis, 308
pimpinellifolia, 108
rubiginosa, 108
rugosa, 307
scabriuscula, 308
sempervirens, 307
sherardii, 308
stylosa, 108
tomentosa, 108
villosa, 308
virginiana, 307
Rubia peregrina, 186
Rubus arcticus, 108
caesius, 108
chamaemorus, 108
fruticosus, 108
idaeus, 108
odoratus, 307
phoenicolasius, 307
saxatilis, 108
spectabilis, 307
Rudbeckia laciniata, 236
serotina, 236
Rumex acetosa, 44
acetosella, 44
alpinus, 298
aquaticus, 44
arifolius, 298
conglomeratus, 44
crispus, 44
frutescens, 298
hydrolapathum, 44
longifolius, 44
maritimus, 44
obtusifolius, 44
palustris, 44
patientia, 298
pseudonatronatus, 298
pulcher, 44
rupestris, 44
sanguineus, 44
scutatus, 298

thyrsiflorus, 298
triangulivalvis, 298
Ruppia cirrhosa, 294
maritima, 306
Ruscus aculeatus, 268

Sagina apetala, 56
caespitosa, 308
intermedia, 56
maritima, 56
nodosa, 56
procumbens, 56
saginoides, 56
subulata, 56
Sagittaria natans, 258
rigida, 258
sagittifolia, 258
Salicornia europaea, 48
Salix alba, 26
alba ssp. coerulea, 26
alba ssp. vitellina, 26
appendiculata, 26
arbuscula, 28
arenaria, 308
atrocinerea, 308
aurita, 26
babylonica, 26
bicolor, 308
borealis, 308
caprea, 26
cinerea, 26
coaetanea, 308
daphnoides, 26
elaeagnos, 26
fragilis, 26
glandulifera, 308
glauca, 28
hastata, 28
herbacea, 28
hibernica, 308
lanata, 28
lapponum, 28
myrsinites, 28
myrtilloides, 28
nigricans, 28
pentandra, 26
phylicifolia, 28
polaris, 28
purpurea, 26
repens, 28
reticulata, 28
rosmarinifolia, 308
starkeana, 28
stipulifera, 308
triandra, 26
viminalis, 26
xerophila, 308
Salsola kali, 48
Salvia glutinosa, 202
horminoides, 202
pratensis, 202
sylvestris, 202
verbenaca, 202
verticillata, 202
Sambucus ebulus, 222
nigra, 222
racemosa, 222
Samolus valerandi, 176
Sanguisorba minor, 106
officinalis, 106
Sanicula europaea, 154
Saponaria officinalis, 60
Satureja montana, 198
Saussurea alpina, 244
Saxifraga adscendens, 300
aizoides, 104
cernua, 104
cespitosa, 104
cotyledon, 104

334

foliosa, 104
× geum, 300
granulata, 104
hartii, 308
hieracifolia, 300
hirculus, 104
hirsuta, 300
hypnoides, 104
moschata, 300
nivalis, 104
osloensis, 300
oppositifolia, 104
paniculata, 104
× polita, 300
rivularis, 104
rosacea, 104
rotundifolia, 300
spathularis, 104
stellaris, 104
tenuis, 300
tridactylites, 104
umbrosa, 300
× urbium, 300
Scabiosa canescens, 224
columbaria, 224
ochroleuca, 224
Scandix pecten-veneris, 158
Scheuchzeria palustris, 220
Scilla autumnalis, 266
bifolia, 266
verna, 266
Scirpus lacustris, 296
Scleranthus annuus, 56
perennis, 56
Scorzonera austriaca, 250
hispanica, 250
humilis, 250
purpurea, 250
Scrophularia auriculata, 208
canina, 208
nodosa, 208
scorodonia, 208
umbrosa, 208
vernalis, 208
Scutellaria columnae, 196
galericulata, 196
hastifolia, 196
minor, 196
Sedum acre, 102
album, 102
alpestre, 102
andegavense, 102
anglicum, 102
annuum, 102
dasyphyllum, 102
forsteranum, 102
hirsutum, 102
hispanicum, 102
reflexum, 102
rubens, 102
sexangulare, 102
spurium, 102
telephium, 102
villosum, 102
Selinum carvifolia, 160
pyrenaeum, 160
Sempervivum tectorum, 300
Senecio aquaticus, 242
cineraria, 242
crispatus, 242
erraticus, 242
erucifolius, 242
fluviatilis, 242
fuchsii, 242
integrifolius, 242
jacobaea, 242
paludosus, 242
palustris, 242
spathulifolius, 242

squalidus, 242
sylvaticus, 242
vernalis, 242
viscosus, 242
vulgaris, 242
Serratula tinctoria, 248
Sesamoides canescens, 100
Seseli annuum, 158
hippomarathrum, 158
libanotis, 158
montanum, 158
Sherardia arvensis, 184
Sibbaldia procumbens, 112
Sibthorpia europaea, 216
Sideritis montana, 202
Silaum silaus, 166
Silene acaulis, 58
alba, 60
armeria, 58
conica, 62
dichotoma, 58
dioica, 60
furcata, 58
gallica, 62
gallica var. quinquevulnera,
62
italica, 58
linicola, 58
maritima, 58
noctiflora, 60
nutans, 58
otites, 58
rupestris, 58
tatarica, 58
viscosa, 58
vulgaris, 58
wahlbergella, 58
Silybum marianum, 244
Simethis planifolia, 260
Sinapis, alba, 88
arvensis, 88
Sison amomum, 162
Sisymbrium altissimum, 86
austriacum, 86
irio, 86
loeselii, 86
officinale, 86
orientale, 86
strictissimum, 86
supinum, 94
volgense, 86
Sisyrinchium bermudiana,
272
Sium latifolium, 160
Smyrnium olusatrum, 166
perfoliatum, 166
Solanum dulcamara, 206
nigrum, 206
sarrachoides, 206
villosum, 206
Soldanella alpina, 174
Soleirolia soleirolii, 38
Solidago canadensis, 230
gigantea, 230
graminifolia, 230
virgaurea, 230
Sonchus arvensis, 252
asper, 252
oleraceus, 252
palustris, 252
Sorbaria sorbifolia, 307
Sorbus alba, 114
aucuparia, 114
intermedia, 301
latifolia, 301
torminalis, 301
Sparganium angustifolium,
306
erectum, 296

minimum, 306
simplex, 290
Spergula arvensis, 56
morisonii, 299
pentandra, 299
Spergularia bocconii, 299
echinosperma, 299
marina, 56
media, 56
rubra, 56
rupicola, 56
segetalis, 299
Spiraea alba, 309
douglasii, 309
salicifolia, 307
tomentosa, 309
Spiranthes aestivalis, 282
romanzoffiana, 282
spiralis, 282
Stachys alpina, 202
annua, 202
arvensis, 202
germanica, 202
palustris, 202
recta, 202
sylvatica, 202
Staphylea pinnata, 140
Stellaria alsine, 54
calycantha, 299
crassifolia, 299
crassipes, 299
graminea, 54
holostea, 54
humifusa, 299
longifolia, 299
media, 54
neglecta, 54
nemorum, 54
pallida, 54
palustris, 54
Stenactis annua, 305
strigosa, 305
Stratiotes aloides, 258
Streptopus amplexifolius, 268
Suaeda maritima, 48
vera, 48
Subularia aquatica, 290
Succisa pratensis, 224
Swertia perennis, 182
Symphoricarpos rivularis, 308
Symphytum asperum, 188
grandiflorum, 303
officinale, 188
orientale, 188
tuberosum, 188
uplandicum, 188
Syringa vulgaris, 307

Tamarix gallica, 140
Tamus communis, 266
Tanacetum corymbosum, 240
parthenium, 240
vulgare, 240
Taraxacum
Sect. Erythrosperma, 254
Sect. Palustris, 254
Sect. Spectabilia, 254
Sect. Vulgaria, 254
Taxus baccata, 24
Teesdalia nudicaulis, 96
Telekia speciosa, 236
Tellima grandiflora, 300
Tetragonolobus maritimus,
128
Teucrium botrys, 196
chamaedrys, 196
montanum, 196
scordium, 196
scorodonia, 196

Thalictrum alpinum, 72
  aquilegifolium, 72
  flavum, 72
  lucidum, 72
  minus, 72
  morisonii, 72
  simplex, 72
Thesium alpinum, 38
  bavarum, 38
  ebracteatum, 38
  humifusum, 38
  linophyllon, 38
  pyrenaicum, 38
  rostratum, 38
Thlaspi alliaceum, 98
  alpestre, 98
  arvense, 98
  montanum, 98
  perfoliatum, 98
Thuja plicata, 298
Thymelaea passerina, 138
Thymus pulegioides, 204
  serpyllum, 204
Tilia cordata, 36
  platyphyllos, 36
  × vulgaris, 36
Tofieldia calyculata, 260
  pusilla, 260
Tolmiea menziesii, 300
Tordylium maximum, 164
Torilis arvensis, 156
  japonica, 156
  nodosa, 156
Trachystemon orientalis, 192
Tragopogon dubium, 250
  porrifolius, 250
  pratensis, 250
Trapa natans, 258
Tribulus terrestris, 302
Trientalis europaea, 176
Trifolium alpestre, 301
  arvense, 130
  aureum, 128
  badium, 128
  bocconei, 301
  campestre, 128
  dubium, 128
  fragiferum, 130
  glomeratum, 301
  hybridum, 130
  incarnatum, 130
    ssp. molinerii, 130
  medium, 130
  michelianum, 130
  micranthum, 128
  montanum, 130
  occidentale, 130
  ochroleucon, 130
  ornithopodioides, 130
  patens, 128
  pratense, 130
  repens, 130
  resupinatum, 301
  rubens, 301
  scabrum, 301
  spadiceum, 128
  squamosum, 130
  striatum, 130
  strictum, 130
  subterraneum, 130
  suffocatum, 130
Triglochin maritima, 220
  palustris, 220
Trigonella foenum-graecum,
  126
  monspeliaca, 126
Trinia glauca, 158

Tripleurospermum inodorum,
  232
  maritimum, 232
Trollius europaeus, 68
Tsuga heterophylla, 24
Tuberaria guttata, 148
Tulipa sylvestris, 262
Turgenia latifolia, 156
Tussilago farfara, 240
Typha angustifolia, 296
  latifolia, 296

Ulex europaeus, 120
  gallii, 120
  minor, 120
Ulmus glabra, 34
  laevis, 34
  minor, 34
  procera, 34
Umbilicus rupestris, 100
Urtica dioica, 38
  urens, 38
Utricularia bremii, 309
  intermedia, 290
  minor, 290
  neglecta, 309
  ochroleuca, 309
  vulgaris, 290

Vaccaria pyramidata, 62
Vaccinium microcarpum, 172
  myrtillus, 172
  oxycoccos, 172
  uliginosum, 172
  vitis-idaea, 172
Valeriana dioica, 222
  montana, 222
  officinalis, 222
  sambucifolia, 309
  tripteris, 304
Valerianella carinata, 309
  coronata, 304
  dentata, 309
  eriocarpa, 309
  locusta, 222
  rimosa, 309
Vallisneria spiralis, 292
Veratrum album, 260
Verbascum blattaria, 208
  lychnitis, 208
  nigrum, 208
  phlomoides, 208
  phoeniceum, 208
  pulverulentum, 208
  thapsiforme, 208
  thapsus, 208
  virgatum, 208
Verbena officinalis, 192
Veronica acinifolia, 304
  agrestis, 214
  alpina, 214
  anagallis-aquatica, 214
  anagalloides, 304
  arvensis, 214
  austriaca, 303
  beccabunga, 214
  catenata, 214
  chamaedrys, 214
  dillenii, 304
  filiformis, 214
  fruticans, 214
  hederifolia, 214
  humifusa, 214
  longifolia, 303
  montana, 214
  officinalis, 214
  opaca, 214

  peregrina, 304
  persica, 214
  polita, 214
  praecox, 304
  prostrata, 303
  scutellata, 214
  serpyllifolia, 214
  spicata, 214
  sublobata, 214
  teucrium, 214
  triphyllos, 304
  urticaefolia, 304
  verna, 304
Viburnum lantana, 222
  opulus, 222
Vicia bithynica, 122
  cassubica, 122
  cracca, 122
  dumetorum, 122
  hirsuta, 122
  lathyroides, 122
  lutea, 122
  narbonensis, 122
  orobus, 122
  pannonica, 122
  pisiformis, 122
  sativa, 122
  sepium, 122
  sylvatica, 122
  tenuifolia, 122
  tenuissima, 122
  tetrasperma, 122
  villosa, 122
Vinca major, 178
  minor, 178
Vincetoxicum hirundinaria,
  178
Viola alba, 302
  arvensis, 146
  biflora, 146
  canina, 146
  collina, 302
  elatior, 302
  epipsila, 302
  hirta, 146
  hispida, 146
  kitaibeliana, 146
  lactea, 146
  lutea, 146
  mirabilis, 302
  odorata, 146
  palustris, 146
  persicifolia, 146
  pumila, 146
  reichenbachiana, 146
  riviniana, 146
  rupestris, 146
  selkirkii, 146
  suavis, 302
  tricolor, 146
  uliginosa, 302
Viscum album, 38
Vitis vinifera, 307

Wahlenbergia hederacea, 226
Willemetia stipitata, 252
Wolffia arrhiza, 292

Xanthium echinatum, 304
  spinosum, 230
  strumarium, 304

Zannichellia palustris, 294
Zostera angustifolia, 306
  marina, 294
  noltii, 306